AF869841

Springer Proceedings in Mathematics

Volume 3

For other titles in this series go to
www.springer.com/series/8806

Springer Proceedings in Mathematics

The book series will feature volumes of selected contributions from workshops and conferences in all areas of current research activity in mathematics. Besides an overall evaluation, at the hands of the publisher, of the interest, scientific quality, and timeliness of each proposal, every individual contribution will be refereed to standards comparable to those of leading mathematics journals. It is hoped that this series will thus propose to the research community well-edited and authoritative reports on newest developments in the most interesting and promising areas of mathematical research today.

Emmanuil H. Georgoulis • Armin Iske
Jeremy Levesley
Editors

Approximation Algorithms for Complex Systems

Proceedings of the 6th International Conference on Algorithms for Approximation, Ambleside, UK, August 31st – September 4th, 2009

Springer

Editors
Emmanuil H. Georgoulis
University of Leicester
Department of Mathematics
Leicester LE1 7RH, UK
eg64@le.ac.uk

Jeremy Levesley
University of Leicester
Department of Mathematics
Leicester LE1 7RH, UK
jll@le.ac.uk

Armin Iske
University of Hamburg
Department of Mathematics
D-20146 Hamburg, Germany
iske@math.uni-hamburg.de

ISSN 2190-5614
ISBN 978-3-642-16875-8 e-ISBN 978-3-642-16876-5
DOI 10.1007/978-3-642-16876-5
Springer Heidelberg Dordrecht London New York

Mathematics Subject Classification (2010): 65Dxx, 65D15, 65D05, 65D07

© Springer-Verlag Berlin Heidelberg 2011
This work is subject to copyright. All rights are reserved, whether the whole or part of the material is concerned, specifically the rights of translation, reprinting, reuse of illustrations, recitation, broadcasting, reproduction on microfilm or in any other way, and storage in data banks. Duplication of this publication or parts thereof is permitted only under the provisions of the German Copyright Law of September 9, 1965, in its current version, and permission for use must always be obtained from Springer. Violations are liable to prosecution under the German Copyright Law.
The use of general descriptive names, registered names, trademarks, etc. in this publication does not imply, even in the absence of a specific statement, that such names are exempt from the relevant protective laws and regulations and therefore free for general use.

Cover design: deblik, Berlin

Printed on acid-free paper

Springer is part of Springer Science+Business Media (www.springer.com)

Preface

It appears that the role of the mathematician is changing, as the world requires sophisticated tailored solutions to specific problems. Approximation methods are of vital importance in many of these problems, as we usually do not have perfect information, but require means by which robust solutions can be garnered from complex and often evolutionary situations. This book collects papers from world experts in a broad variety of relevant applications of approximation theory, including dynamical systems, multiscale modelling of fluid flow, metrology, and geometric modelling to mention a few.

The 14 papers in this volume document modern trends in approximation through recent theoretical developments, important computational aspects and multidisciplinary applications. The book is arranged in seven invited surveys, followed by seven contributed research papers. The surveys of the first seven chapters are addressing the following relevant topics.

Emergent behaviour in large electrical networks
by Darryl P. Almond, Chris J. Budd, and Nick J. McCullen;
Algorithms for multivariate piecewise constant approximation
by Oleg Davydov;
Anisotropic triangulation methods in adaptive image approximation
by Laurent Demaret and Armin Iske;
Form assessment in coordinate metrology
by Alistair B. Forbes and Hoang D. Minh;
Discontinuous Galerkin methods for linear problems
by Emmanuil H. Georgoulis;
A numerical analyst's view of the lattice Boltzmann method
by Jeremy Levesley, Alexander N. Gorban, and David Packwood;
Approximation of probability measures on manifolds
by Jeremy Levesley and Xingping Sun.

Moreover, the diverse contributed papers of the remaining seven chapters reflect recent developments in approximation theory, approximation practice and their applications. Graduate students who wish to discover the state of the art in a number of important directions of approximation algorithms will find this a valuable volume. Established researchers from statisticians through to fluid modellers will find interesting new approaches to solving familiar but challenging problems.

This book grew out of the sixth in the conference series on *Algorithms for Approximation*, which took place from 31st August to September 4th 2009 in Ambleside in the Lake District of the United Kingdom. The conference was supported by the EPSRC (Grant EP/H018026) and the London Mathematical Society, and had around 70 delegates from 20 different countries.

The conference included also two workshops, the *Model Reduction Workshop* and *New Mathematics for the Computational Brain*. The papers of the first workshop are appearing in a Springer volume *Coping with Complexity: Model Reduction and Data Analysis* (Lecture Notes in Computational Science and Engineering, Vol. 75, Nov 29, 2010), edited by Alexander N. Gorban and Dirk Roose, and the second as a special issue of *International Journal of Neural Systems*, Vol. 20, No. 3 (2010). The interaction between approximation theory, model reduction, and neuroscience, was very fruitful, and resulted in a number of new projects which would never happened without collecting such a broad-based group of people.

Finally, we are grateful to all the authors who have submitted their fine papers for this volume, especially for their patience with the editors. The contributions to this volume have all been refereed, and thanks go out to all the referees for their timely and considered comments. Finally, we very much appreciate the cordial relationship we have had with Springer-Verlag, Heidelberg, through Martin Peters.

Leicester, August 2010

Emmanuil H. Georgoulis
Armin Iske
Jeremy Levesley

Contents

List of Contributors

Darryl P. Almond
Bath Institute for Complex Systems
University of Bath
Bath BA2 7AY, UK
D.P.Almond@bath.ac.uk

Chris J. Budd
Bath Institute for Complex Systems
University of Bath
Bath BA2 7AY, UK
cjb@maths.bath.ac.uk

Maurice G. Cox
National Physical Laboratory
Teddington TW11 0LW, UK
maurice.cox@npl.co.uk

Oleg Davydov
University of Strathclyde
Dept of Mathematics and Statistics
Glasgow G1 1XH, UK
oleg.davydov@strath.ac.uk

Laurent Demaret
HelmholtzZentrum münchen
Institute for Biomathematics
Neuherberg, Germany
laurent.demaret@helmholtz-zentrum.de

Alistair B. Forbes
National Physical Laboratory
Teddington TW11 0LW, UK
alistair.forbes@npl.co.uk

Emmanuil H. Georgoulis
University of Leicester
Department of Mathematics
Leicester LE1 7RH, UK
eg64@le.ac.uk

Peter Giesl
University of Sussex
Department of Mathematics
Falmer BN1 9QH, UK
p.a.giesl@sussex.ac.uk

Alexander N. Gorban
University of Leicester
Department of Mathematics
Leicester LE1 7RH, UK
ag153@le.ac.uk

Armin Iske
University of Hamburg
Department of Mathematics
D-20146 Hamburg, Germany
iske@math.uni-hamburg.de

Tahir S. Khan
University of Leicester
Department of Mathematics
Leicester LE1 7RH, UK
tsk5@le.ac.uk

Jeremy Levesley
University of Leicester
Department of Mathematics
Leicester LE1 7RH, UK
jl1@le.ac.uk

Brian Li
University of Leicester
Department of Mathematics
Leicester LE1 7RH, UK
ml111@le.ac.uk

Ivan Markovsky
University of Southampton
School Electronics & Comput. Science
Southampton SO17 1BJ, UK
im@ecs.soton.ac.uk

Nick J. McCullen
Bath Institute for Complex Systems
University of Bath
Bath BA2 7AY, UK
n.mccullen@bath.ac.uk

Hoang D. Minh
National Physical Laboratory
Teddington TW11 0LW, UK
minh.hoang@npl.co.uk

David Packwood
University of Leicester
Department of Mathematics
Leicester LE1 7RH, UK
dp123@le.ac.uk

A. Fabien Rabarison
University of Strathclyde
Dept of Mathematics and Statistics
Glasgow G1 1XQ, UK
fabi.rabarison@strath.ac.uk

Xingping Sun
Missouri State University
Department of Mathematics
Springfield, MO 65897, U.S.A.
xsun@missouristate.edu

Johan de Villiers
Stellenbosch University
Department of Mathematical Sciences
Private Bag X1, Matieland 7602, SA
jmdv@sun.ac.za

Xin-She Yang
National Physical Laboratory
Teddington TW11 0LW, UK
xin-she.yang@npl.co.uk

Part I

Invited Surveys

Emergent Behaviour in Large Electrical Networks

Darryl P. Almond, Chris J. Budd, and Nick J. McCullen

Bath Institute for Complex Systems, University of Bath, BA2 7AY, UK

Summary. Many complex systems have emergent behaviour which results from the way in that the components of the system interact, rather than their individual properties. However, it is often unclear as to what this emergent behaviour can be, or inded how large the system should be for such behaviour to arise. In this paper we will address these problems for the specific case of an electrical network comprising a mixture of resistive and reactive elements. Using this model we will show, using some spectral theory, the types of emergent behaviour that we expect and also how large a system we need for this to be observed.

1 Introduction and Overview

The theory of complex systems offers great potential as a way of describing and understanding many phenomena involving large numbers of interacting agents, varying from physical systems (such as the weather) to biological and social systems [1]. A system is *complex* rather than just *complicated* if the individual components interact strongly and the resulting system behaviour is a product more of these interactions than of the individual components. Such behaviour is generally termed *emergent behaviour* and we can colloquially that the complex system is demonstrating behaviour which is *more than the sum of its parts.* However, such descriptions of complexity are really rather vague and leave open many scientific questions. These include: how large does a system need to be before it is complex, what sort of interactions lead to emergent behaviour, and can the types of emergent behaviour be classified. More generally, how can we analyse a complex system? We do not believe that these questions can be answered *in general*, however, we can find answers to them in the context of specific complex problems. This is the purpose of this paper, which will study a complex system comprising a large binary electrical network, which can be used to model certain material behaviours.

Such large binary networks comprise disordered mixtures of two different but interacting components. These, arise both directly, in electrical circuits

E.H. Georgoulis, A. Iske, J. Levesley (eds.), *Approximation Algorithms for Complex Systems*, Springer Proceedings in Mathematics 3, DOI: 10.1007/978-3-642-16876-5_1,
© Springer-Verlag Berlin Heidelberg 2011

[2, 5, 16], or mechanical structures [13], and as models of other systems such as disordered materials with varying electrical [6], thermal or mechanical properties in the micro-scale which are then coupled at a meso-scale. Many systems of condensed matter have this form [14, 9, 7]. Significantly, such systems are often observed to have *macroscopic emergent properties* which can have *emergent power-law behaviour* over a wide range of parameter values which is **different from any power law behaviour of the individual elements of the network**, and is a product of the way in which the responses of the components *combine.* As an example of such a binary disordered network, we consider a (set of random realisations of a) binary network comprising a random mixture with a proportion of $(1 - p)$ resistors with frequency independent admittance $1/R$ and p capacitors with complex admittance $i\omega C$ directly proportional to frequency ω. This network when subjected to an applied alternating voltage of frequency ω has a the total admittance $Y(\omega)$. We observe that over a wide range of frequencies $0 < \omega_1 < \omega < \omega_2$, the admittance displays *power law emergent characteristics*, so that $|Y|$ is proportional to ω^α, for an appropriate exponent α. Significantly, α is not equal to zero or one (the power law of the response of the individual circuit elements) but depends upon the proportion of the capacitors in the network. For example when this proportion takes the *critical value* of $p = 1/2$, then $\alpha = 1/2$. **The effects of network size, and** component proportion, are important in that ω_1 and ω_2 depend upon both p and N. In the case of $p = 1/2$ this is a strong dependence and we will see that ω_1 is inversely proportional to N and ω_2 directly proportional to N, as N increases to infinity. It is in this frequency range that both the resistors and capacitors share the (many) current paths through the network, and they interact strongly. The emergent behaviour is a result of this interaction. For $0 < \omega < \omega_1$ and $\omega > \omega_2$ we see a transition in the behaviour. In these ranges either the resistors or the capacitors act as conductors, and there are infrequent current paths, best described by percolation theory. In these ranges the emergent power law behaviour changes and we see instead the individual component responses. Hence we see in this example of a complex system (i) an emergent region with a power law response depending on the proportion but not the arrangement or number of the components (ii) a more random percolation region with a response dominated by that of individual components and (iii) a transition between these two regions at frequency values which depends on the number and proportion of components in the system. The purpose of this paper is to partly explain this behaviour.

The layout of the remainder of this paper is as follows. In Section 2 we will give a series of numerical results which illustrate the various points made above on the nature of the network response. In Section 3 we will formulate the matrix equations describing the network and the associated representation of the admittance function in terms of poles and zeros. In Section 4 we will discuss, and derive, a series of statistical results concerning the distribution of the poles ad zeros. In Section 5 we will use these statistical results to derive the asymptotic form of the admittance $|Y(\omega)|$ in the critical case of $p = 1/2$. In Section 6 we compare the asymptotic predictions with the numerical computations. Finally in Section 7 we will draw some conclusions from this work.

2 Simple Network Models and Their Responses

In this section we show the basic models for composite materials and associated random binary electrical networks, and present the graphs of their responses. In particular we will look in detail at the existence of a power law emergent region, and will obtain empirical evidence for the effects of network size N and capacitor proportion p, on both this region and the 'percolation behaviour' when $CR\,\omega \ll 1$ and $CR\,\omega \gg 1$.

2.1 Modelling Composites as Complex Rectangular Networks

An initial motivation for studying binary networks comes from models of composite materials. Disordered two-phase composite materials are found to exhibit power-law scaling in their bulk responses over several orders of magnitude in the contrast ratio of the components [10, 5], and this effect has been observed [2, 4] in both physical and numerical experiments on conductor-dielectric composite materials. In the electrical experiments this was previously referred to as "Universal Dielectric Response" (UDR), and it has been observed [14, 9] that this is an emergent property arising out of the random nature of the mixture. A simple model of such a conductor-dielectric mixtures with fine structure is a large electrical circuit replacing the constituent conducting and dielectric parts with a linear C-R network of $N \gg 1$ resistors and capacitors, respectively as illustrated in Figure 1.

For a binary disordered mixture, the different components can be assigned randomly to bonds on a lattice [15]. with bonds assigned randomly as either C or R, with probability p, $1-p$ respectively. The components are distributed in a two-dimensional lattice between two bus-bars. On of which is grounded and the other is raised to a potential $V(t) = V\exp(i\omega t)$. This leads to a current $I(t) = I(\omega)\exp(i\omega t)$ between the bus-bars, and we measure the macroscopic (complex) admittance given by $Y(\omega) = I(\omega)/V$. A large review of this and binary disordered networks can be found in [8, 5].

We now present an overview of the results found for the admittance conduction of the C-R network, explaining the PLER and its bounds. In particular we need to understand the difference between percolation behaviour and power law emergent behaviour. To motivate these results we consider initially the cases of very low and very large frequency.

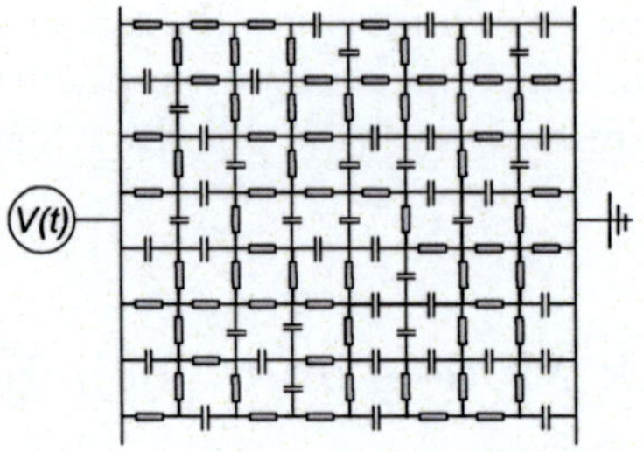

Fig. 1. The layout of the binary electrical circuit.

Percolation and Power-Law Emergent Behaviour

As described in the introduction, in the case of *very low frequency* $CR\,\omega \ll 1$, the capacitors act as open circuits and the resistors become the main conducting paths with far higher admittance than the capacitors. The circuit then becomes a *percolation network* [3, 11] in which the bonds are either conducting with probability $(1-p)$ or non-conducting with probability p. The network conducts only if there is a percolation path from one electrode to the other. It is well known [3] that if $p > 1/2$ then there is a very low probability that such a percolation path exists. In contrast, if $p < 1/2$ then such a path exists with probability approaching one as the network size increases. The case of $p = 1/2$ is critical with a 50% probability that such a path exists. This implies that if $p < 1/2$ then for low frequencies the conduction is almost certainly resistive and the overall admittance is independent of angular frequency ω. In contrast, if $p > 1/2$ then the conduction is almost certainly capacitative and the overall admittance is directly proportional to ω. If $p = 1/2$ (the critical percolation probability for a 2D square lattice) then half of the realisations will give an admittance response independent of ω and half an admittance response proportional to ω. In the case of *very high frequencies* $CR\,\omega \gg 1$, we see an opposite type of response. In this case the capacitors act as almost short circuits with far higher admittance than the resistors. Again we effectively see percolation behaviour with the resistors behaving as approximately open circuits in this case. Thus if $p > 1/2$ we again expect to see a response proportional to ω and if $p < 1/2$ a response independent of ω. The case of $p = 1/2$ again leads to both types of response with equal likelihood of occurrence depending upon the network configuration. Note that this implies that if $p = 1/2$ then there are ***four possible qualitatively different*** types of response for any random realisation of the system.

For *intermediate values* of ω the values of the admittance of the resistors and the capacitors are much closer to each other and there are many current paths through the network, In Figure 2 we see the current paths in the three cases of (a) percolation, (b) transition between percolation and emergence (c) emergence.

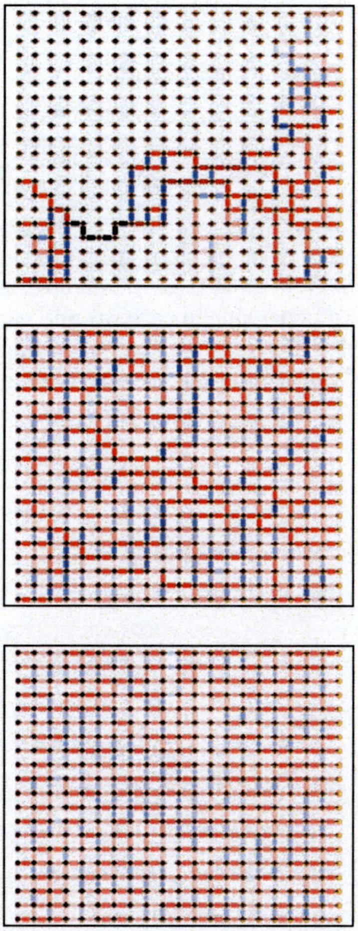

Fig. 2. An illustration of the three different types of current path observed in the percolation, transition and emergent regions.

The emergence region has power-law emergent behaviour. This is characterised by two features: *(i)* an admittance response that is proportional to ω^α for some $0 < \alpha < 1$ over a range $\omega \in (\omega_1, \omega_2)$. *(ii)* In the case of $p = 1/2$ a response that is not randomly dependent upon the network configuration. Figures 3 and 4 plot the admittance response as a function of ω in the cases of $p = 0.4, p = 0.6$ and $p = 1/2$. The figures clearly demonstrate the forms of behaviour described above. Observe that in all cases we see quite a sharp transition between the percolation type behaviour and the emergent power law behaviour as ω varies.

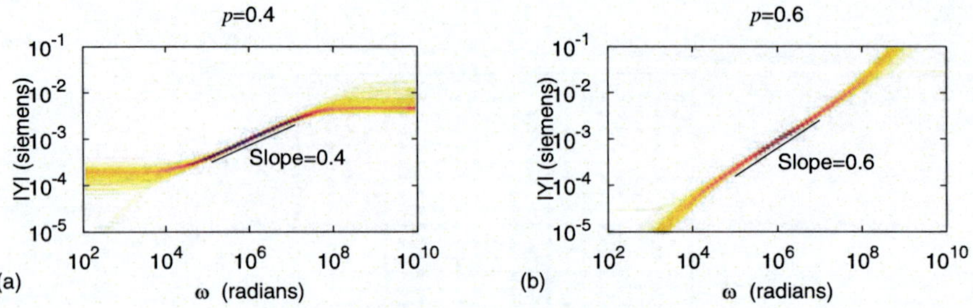

Fig. 3. Typical responses of network simulations for values of $p \neq 1/2$ which give qualitatively different behaviour so that in the percolation region with $CR\,\omega \ll 1$ or $CR\,\omega \gg 1$, we see resistive behaviour in case (a) and capacitative behaviour in case (b). The figures presented are density plots of 100 random realisations for a 20×20 network. Note that all of the realisations give very similar results.

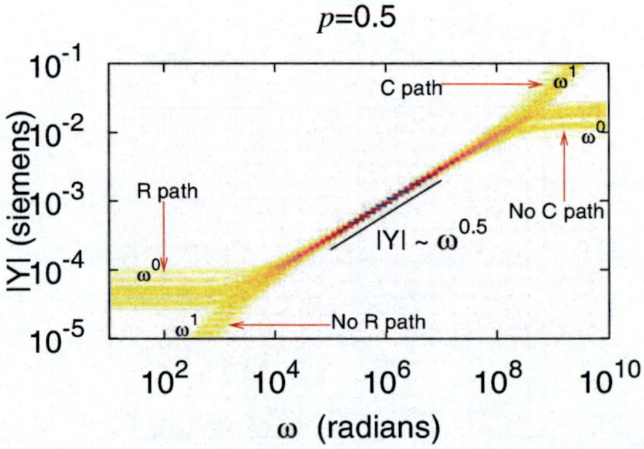

Fig. 4. Responses for 100 realisations at $p = 1/2$ showing four different qualitative types of response for different realisations. Here, about half of the responses have a resistive percolation path and half have a capacitive one at low frequencies with a similar behaviour at high frequencies. The responses at high and low values of $CR\,\omega$ indicate which of these cases exist for a particular realisation. The power-law emergent region can also be seen in which the admittance scales as $\sqrt{\omega}$ and all of the responses of the different network realisations coincide

The Effects of Network Size N and Capacitor Proportion p.

We have seen above how the response of the network depends strongly upon p. It also depends (more weakly) upon the network size N. Figure 5 shows the response for the critical value of $p = 1/2$ for different values of N. Observe that in this case the width of the power-law emergent region increases apparently without bound, as N increases, as do the magnitude of the responses for small and large frequencies. In contrast in Figure 6 we plot the response for $p = 0.4$ and again increase N In contrast to the former case, away from the critical percolation probability the size of the power-law emergent region appears to scale with N for small N before becoming asymptotic to a finite value for larger values of N.

These results are consistent with the predictions of the Effective-Medium-Approximation (EMA) calculation [12, 5] which uses a homogenisation argument to determine the response of a network with an infinite number ($N = \infty$) of components. In particular, the EMA calculation predicts that in this limiting case, the response for $p = 1/2$ is always proportional to $\sqrt{\omega}$ and that it $p < 1/2$ then the response is asymptotic to $\epsilon \equiv 1/2 - p$ as $\omega \to 0$ and to $1/\epsilon$ as $\omega \to \infty$. However the EMA calculation does not include the effects of the network size.

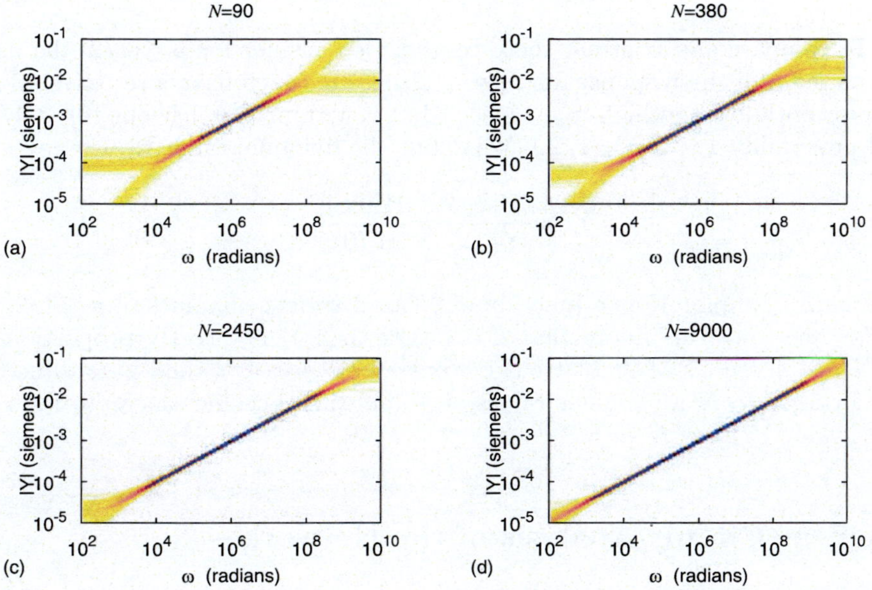

Fig. 5. The effect of network size N on the width of the power-law emergent region for $p = 1/2$, in which we see this region increasing without bound.

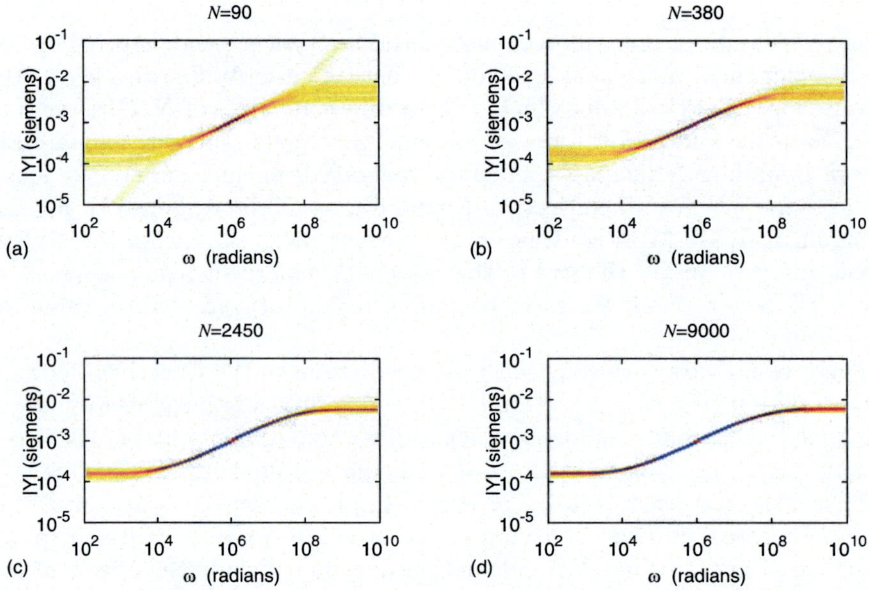

Fig. 6. The effect of the network size N on the power-law emergent region for $p = 0.4$, in which we see this region becoming asymptotic to a finite set as $N \to \infty$.

To compare and contrast these results, we consider for $p \leq 1/2$, the *dynamic range* of the response for those realisations which have a resistive solution for both low and high frequencies (that is with probability one if $p < 1/2$ and probability 1/4 if $p = 1/2$.). We define the dynamic range to be

$$\hat{Y} = \frac{|Y|_{max}}{|Y|_{min}} = \frac{|Y(\infty)|}{|Y(0)|}.$$

In Figure 7 we plot $\hat{Y}$ as a function of N for a variety of values of $p \leq 1/2$.

We see from this figure that if $p = 1/2$ then $\hat{Y}$ is directly proportional to N for all values of N plotted. In contrast, if $p < 1/2$ then $\hat{Y}$ is directly proportional to N for smaller values of N and then becomes asymptotic to a finite value $\hat{Y}(p)$ as $N \to \infty$.

3 Linear Circuit Analysis of the Network

We now describe in detail how the disordered material is modelled by by a general network model. In this we consider two components of admittance y_1 and y_2 so that the proportion of the first component is $(1-p)$ and the second is p. These will have admittance ratio $\mu = \frac{y_2}{y_1}$. For a capacitor-resistor (C-R)

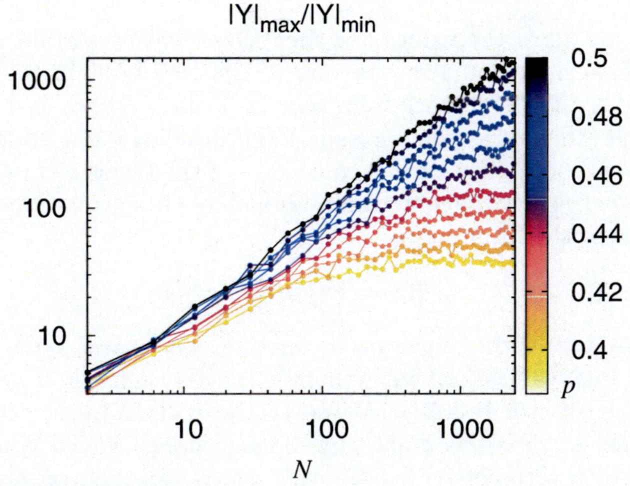

Fig. 7. The variation of the dynamic range $\hat{Y} = |Y|_{max}/|Y|_{min}$ as a function of N and p.

network with a proportion p of capacitors with admittance $y_2 = i\omega C$ and resistors admittance $y_1 = 1/R$ we have

$$\mu = i\omega CR \quad \text{is purely imaginary.}$$

3.1 Linear Circuit Formulation

Now consider a 2D N node square lattice network, with all of the nodes on the left-hand-side (LHS) connected via a bus-bar to a time varying voltage $V(t) = Ve^{i\omega t}$ and on the right-hand-side (RHS) via a bus-bar to earth ($0V$). We assign a (time-varying) voltage v_i with $i = 1 \ldots N$ to each (interior) node, and set $\mathbf{v} = (v_1, v_2, v_3 \ldots v_N)^T$ the be the vector of voltage unknowns. We will also assume that adjacent nodes are connected by a bond of admittance $y_{i,j}$. Here we assume further that $y_{i,j} = y_1$ with probability $1 - p$ and $y_{i,j} = y_2$ with probability p. ¿From Kirchhoff's current law, at any node all currents must sum to zero, so there are no sinks or sources of current other than at the boundaries. In particular, the current from the node i to an adjoining node at j is given by $I_{i,j}$ where

$$I_{i,j} = (v_i - v_j)y_{i,j}.$$

It then follows that if i is *fixed* and j is allowed to vary over the four nodes adjacent to the node at i then

$$\sum_j y_{i,j}(v_i - v_j) = 0. \tag{1}$$

If we consider all of the values of i then there will be certain values of i that correspond to nodes adjacent to one of the two boundaries. If i is a node adjacent to the *left* boundary then certain of the terms v_j in (1) will take the value of the (known) applied voltage $V(t)$. Similarly, if a node is adjacent to the right hand boundary then then certain of the terms v_j in (1) will take the value of the ground voltage 0. Combining all of these equations together leads to a system of the form

$$K\mathbf{v} = V(t)\mathbf{b} = Ve^{i\omega t}\mathbf{b},$$

where $K \equiv K(\omega)$ is the (constant in time) $N \times N$ *sparse symmetric Kirchhoff matrix* for the system formed by combining the individual systems (1), and the adjacency vector $\mathbf{b} \equiv \mathbf{b}(\omega)$ is the vector of the admittances of the bonds between the left hand boundary and those nodes which connected to this boundary, with zero entries for all other nodes. As this is a *linear system*, we can take

$$\mathbf{v} = \mathbf{V}e^{i\omega t}$$

so that the (constant in time) vector $\mathbf{V}$ satisfies the linear algebraic equation

$$K\mathbf{V} = V\mathbf{b}.$$

If we consider the total current flow I from the LHS boundary to the RHS boundary then we have

$$I = \mathbf{b}^T(V\mathbf{e} - \mathbf{V}) \equiv \mathbf{b}^T\mathbf{V} - Vc$$

where $\mathbf{e}$ is the vector comprising ones for those nodes adjacent to the left boundary and zeroes otherwise and $c = \mathbf{b}^T\mathbf{V}$. Combining these expressions, the equations describing the system are then given by

$$K\mathbf{V} - \mathbf{b}V \equiv \mathbf{0}, \quad cV - \mathbf{b}^T\mathbf{V} = I. \tag{2}$$

The *bulk admittance* $Y(\mu)$ of the whole system is then given by $Y = I/V$ so that

$$Y(\mu) = c - \mathbf{b}^T K^{-1}\mathbf{b}.$$

Significantly, the symmetric Kirchoff matrix K can be separated into the two sparse symmetric $N \times N$ component matrices $K = K_1 + K_2$ which correspond to the conductance paths along the bonds occupied by each of the two types of components. Furthermore

$$K_1 = y_1 L_1 \quad \text{and} \quad K_2 = y_2 L_2 = \mu y_1 L_2$$

where the terms of the sparse symmetric connectivity matrices L_1 and L_2 are constant and take the values $-1, 0, 1, 2, 3, 4$. Note that K is a *linear affine function* of μ. Furthermore,

$$\Delta = L_1 + L_2$$

is the discrete (negative definite symmetric) Laplacian for a 2D lattice. Similarly we can decompose the adjacency vector into two components $\mathbf{b}_1$ and $\mathbf{b}_2$ so that

$$\mathbf{b} = \mathbf{b}_1 + \mathbf{b}_2 = y_1\mathbf{e}_1 + y_2\mathbf{e}_2$$

where $\mathbf{e}_1$ and $\mathbf{e}_2$ are *orthogonal vectors* comprising ones and zeros only corresponding to the two bond types adjacent to the LHS boundary. Observe again that $\mathbf{b}$ is a linear affine function of μ. A similar decomposition can applied to the scalar $c = c_1 + \mu c_2$.

3.2 Poles and Zeroes of the Admittance Function

To derive formulæ for the expected admittances in terms of the admittance ratio $\mu = i\omega CR$ we now examine the structure of the admittance function $Y(\mu)$. As the matrix K, the adjacency vector $\mathbf{b}$ and the scalar c are all affine functions of the parameter μ it follows immediately from Cramer's rule applied to (2) that the admittance of the network $Y(\mu)$ is rational function of the parameter μ, taking the form of the ratio of two complex polynomials $P(\mu)$ and $Q(\mu)$ of respective degrees $r \leq N$ and $s \leq N$, so that

$$Y(\mu) = \frac{Q(\mu)}{P(\mu)} = \frac{q_0 + q_1\mu + q_2\mu^2 + \dots q_r\mu^r}{p_0 + p_1\mu + p_2\mu^2 + \dots p_s\mu^s}.$$

We require that $p_0 \neq 0$ so that the response is physically realisable, with $Y(\mu)$ bounded as ω and hence $\mu \to 0$. Several properties of the network can be immediately deduced from this formula. First consider the case of ω *small*, so that $\mu = i\omega CR$ is also small. From the discussions in Section 2, we predict that either there is (a) a resistive percolation path in which case $Y(\mu) \sim \mu^0$ as $\mu \to 0$ or (b) such a path does not exist, so that the conduction is capacitative with $Y(\mu) \sim \mu$ as $\mu \to 0$. The case (a) arises when $p_0 \neq 0$ and the case (b) when $q_0 = 0$. Observe that this implies that the absence of a resistive percolation path as $\mu \to 0$ is equivalent to the polynomial $Q(\mu)$ *having a zero when* $\mu = 0$.. Next consider the case of ω and hence μ large. In this case

$$Y(\mu) \sim \frac{q_r}{p_s}\mu^{r-s} \quad \text{as} \quad \mu \to \infty.$$

Again we may have (c) a resistive percolation path at high frequency with response $Y(\mu) \sim \mu^0$ as $\mu \to \infty$, or a capacitative path with $Y(\mu) \sim \mu$. In case (c) we have $s = r$ and $p_r \neq 0$ and in case (d) we have $s = r - 1$ so that we can think of taking $p_r = 0$. Accordingly, we identify four types of network defined in terms of there percolation paths for low and high frequencies, which correspond to the cases (a),(b),(c),(d) so that

(a)	$p_0 \neq 0$
(b)	$p_0 = 0$
(c)	$p_r \neq 0$
(d)	$p_r = 0$

The polynomials $P(\mu)$ and $Q(\mu)$ can be factorised by determining their respective roots $\mu_{p,k},, k = 1 \ldots s$ and $\mu_{z,k},, k = 1 \ldots r$ which are the *poles and zeroes* of the network. We will collectively call these zeroes and poles the *resonances* of the network. It will become apparent that the emergent response of the network is in fact a manifestation of certain regularities of the locations of these resonances. Note that in Case (b) we have $\mu_{z,1} = 0$. Accordingly the network admittance can be expressed as

$$Y(\mu, N) = D(N) \frac{\prod_{k=1}^{r} (\mu - \mu_{z,k})}{\prod_{k=1}^{s} (\mu - \mu_{p,k})}. \tag{3}$$

Here $D(N)$ is a function which does not depend on μ but does depend on the characteristics of the network.

It is a feature of the stability of the network (bounded response), and the affine structure of the symmetric linear equations which describe it [5], that the poles and zeros of $Y(\mu)$ are all *negative real numbers, and interlace* so that

$$0 \geq \mu_{z,1} > \mu_{p,1} > \mu_{z,2} > \mu_{p,2} \ldots > \mu_{z,s} > \mu_{p,s} (> \mu_{z,r}).$$

Because of this, we may recast the equation (3) in terms of ω so that

$$|Y(\mu, N)| = |D(N)| \frac{\prod_{k=1}^{r} |\omega - iW_{z,k}|}{\prod_{k=1}^{s} |\omega - iW_{p,k}|}$$

where $W_{z,k} \geq 0, W_{p,k} > 0$.

4 The Distribution of the Resonances

The previous section has described the network response in terms of the location of the poles and the zeros. We now consider the statistical distribution of these values and claim that it is this distribution which leads to the observed emergent behaviour. We note that if we consider the elements of the network to be assigned randomly (with the components taking each of the two possible values with probabilities p and $1-p$ then we can consider the resonances to be random variables. There are three interesting questions to ask, which become relevant for the calculations in the next section, namely

1. What is the statistical distribution of $W_{p,k}$ if N is large.
2. Given that the zeros interlace the poles, what is the statistical distribution of the location of a zero between its two adjacent poles.
3. What are the ranges of $W_{p,k}$, in particular, how do $W_{p,1}$ and $W_{p,N}$ vary with N and p.

In each case we will find good numerical evidence for strong statistical regularity of the poles, especially in the critical case of $p = 1/2$, leading to partial answers to each of the above questions. For the remainder of this paper we will now only consider this critical case.

4.1 Pole Location

In the critical case, the two matrices L_1 and L_2 representing the connectivity of the two components, have a statistical duality, so that any realisation which leads to a particular matrix L_1 is equally likely to lead to the same matrix L_2. Because of this, if μ is an observed eigenvalue of the pair (L_1, L_2) then it is equally likely for there to be an observed eigenvalue of the pair (L_2, L_1). The latter being precisely $1/\mu$. Thus in any set of realisations of the system we expect to see the eigenvalues μ and $1/\mu$ occuring with equal likelihood. It follows from this simple observation that the variable $\log|\mu|$ should be expected to have a symmetric probability distribution with mean zero. Applying the central limit theorem in this case leads to the expectation that $\log|\mu|$ should follow a normal distribution with mean zero (so that μ has a *log-normal* distribution centred on $\mu = -1$). Similarly, if μ_1 is the smallest value of μ and μ_N the largest value then $\mu_1 = -1/\mu_N$. In fact we will find that in ths case of $p = 1/2$ we have $\mu_1 \sim -1/N$ and $\mu_N \sim -N$. In terms of the frequency response, as $\mu_{p,k} = -CR\, W_{p,k}$, it follows that $\log(W_{p,k})$ is expected to have a mean value of $-\log(CR)$. Following this initial discussion, we now consider some actual numerical computations of the distribution of the poles in a C-R network for which $CR = 10^{-6}$ so that $-\log_{10}(CR) = 6$. We take a single realisation of a network with $N \approx 380$ nodes, and determine the location of $CR\, W_{p,k}$ for this case. We then plot the location of the logarithm of the poles as a function of k. The results are given in Fig 8 for the case of $p = 1/2$. Two features of this figure are immediately obvious. Firstly, the terms $W_{p,k}$ appear to be the point values of a regular function $f(k)$. Secondly, as predicted above, the logarithm of the pole location shows a strong degree of symmetry about zero. We compare the form of this graph with that of the inverse error function, that is we compare erf($\log(CR\, W_{pk})$) with k. The correspondence is very good, strongly indicating that $\log(f)$ takes the form of the inverse error function. .

4.2 Pole-Zero Spacing

The above has considered the distribution of the poles. As a next calculation we consider the statistical distribution of the location of the zeros with respect

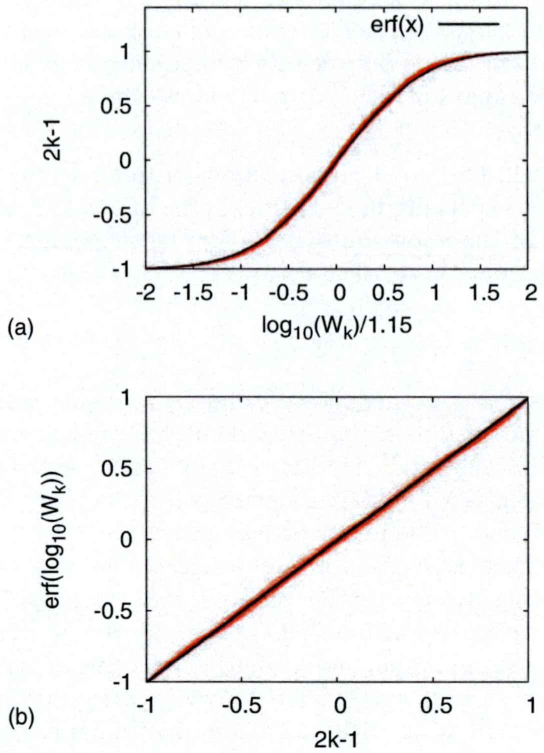

Fig. 8. The location of the logarithm of the poles as a function of k and a comparison with the inverse error function.

to the poles. In particular we consider the variable δ_k given by

$$\delta_k \equiv \frac{W_{p,k} - W_{z,k}}{W_{p,k} - W_{p,k-1}}$$

which expresses the *relative location* of each zero in terms of the poles which it interlaces. In Figure 9 we plot the distribution of the mean value $\bar{\delta}_k$ of δ^k over 100 *realisations* of the network. plotted as a function of the mean location of $\log(W_{p,k})$ for $p = 1/2$. This figure is remarkable as it indicates that when $p = 1/2$ the mean value of δ_k is equal to $1/2$ almost independently of the value of $\log(W_{p,k})$. There is some deviation from this value at the high and low ends of the range, presumably due to the existence of the degenerate poles at zero and at infinity, and there is some evidence for a small asymmetry in the results, but the constancy of the mean near to $1/2$ is very convincing. This shows a remarkable duality between the zeros and the poles in the case of $p = 1/2$, showing that not only do they interlace, but that on average the zeros are mid-way between the poles and the poles are mid-way between the zeros.

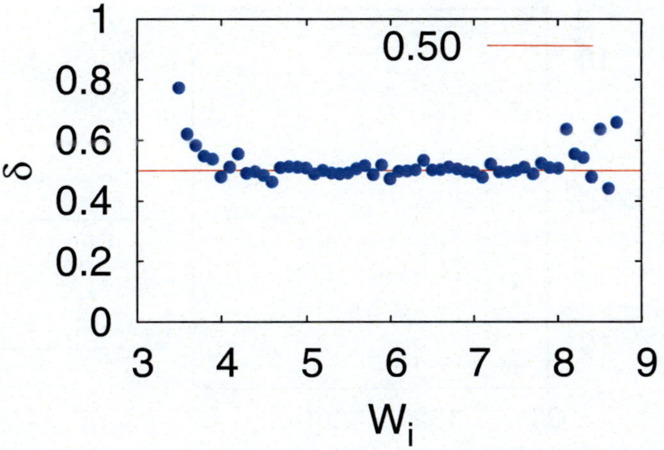

Fig. 9. Figure showing how the mean value $\bar{\delta}_k$, taken over many realisations of the critical network, varies with the mean value of $\log(W_{p,k})$.

4.3 Limiting Finite Resonances

Let N' be the number of *finite non-zero resonances*, the location of the first non-zero pole and zero be $W_{z,1}, W_{p,1}$, and the location of the last finite pole and zero be $W_{p,N'}, W_{z,N'}$. Observe that in the case of $p = 1/2$ we expect a symmetrical relation so that $W_{p,1}$ and $W_{p,N'}$ might be expected to take reciprocal values. The value of N' can be considered statistically, and represents probability of a node contributing to the current paths and not being part of an isolated structure. Statistical arguments presented in [5] indicate that when $p = 1/2$ this is given by

$$N' = 3\left(2 - \sqrt{3}\right) N = 0.804\ldots N.$$

We now consider the values of $W_{p,1}$ and of $W_{p,N'}$. These will become very important when we look at the *transition between emergent type behaviour and percolation type behaviour* which is one of the objectives of this research. A log-log plot of the values of $W_{z,1}, W_{p,1}$ and of $W_{z,N'}, W_{p,N'}$ as functions of N for the case of $p = 1/2$ is given in Figure 10 . There is very clear evidence from these plots that each of $W_{z,1}, W_{p,1}$ and $W_{z,N'}, W_{p,N'}$ both have a *strong power law dependence* upon N for *all values of* N. Indeed we have from a careful inspection of this figure that

$$W_{z,1}, W_{p,1} \sim N^{-1} \quad \text{and} \quad W_{z,N'}, W_{p,N'} \sim N.$$

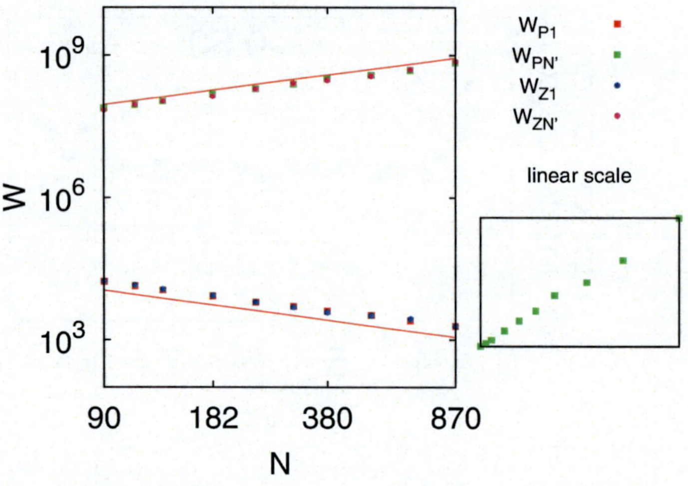

Fig. 10. Figure showing how the maximum pole and zero locations $W_{p,N'}, W_{z,N'}$ scale as N and the minimum pole and zero locations $W_{p,1}, W_{z,1}$ scale as $1/N$.

4.4 Summary

The main conclusions of this section are that there is a strong statistical regularity in the location of the poles and the zeros of the admittance function. In particular we may make the following conclusions based on the calculations reported in this section.

1. $W_{p,k} \sim f(k)$ for an appropriate continuous function $f(k)$ where f depends upon N very weakly.
2. If $p = 1/2$ and if $W_{z,1} \neq 0$, then

$$W_{1p}, W_{1z} \sim N^{-1}, \quad W_{N'p}, W_{N'z} \sim N.$$

3. If $p = 1/2$ then $\bar{\delta}_k \approx 1/2$.

All of these conclusions point towards a good degree of statistical regularity in the pole distribution, and each can be justified to a certain extent by statistical and other arguments. We will now show how these statistical regularities lead to power law emergent region, and how this evolves into a percolation response.

5 Asymptotic Analysis of the Power Law Emergent Response

We will use the results in the summary of the previous section to derive the form of the conductance in the case of a critical C-R network. The formulae that we derive will take one of four forms, depending upon the nature of the percolation paths for low and high frequencies.

5.1 Derivation of the Response

We now consider the formulae for the absolute value of the admittance of the C-R network at a frequency of ω given by

$$|Y(\omega)| = |D(N)| \frac{\prod_{k=1}^{r} |\omega - iW_{z,k}|}{\prod_{k=1}^{s} |\omega - iW_{p,k}|}$$

where the zero interlacing theorem implies that

$$0 \leq W_{z,1} < W_{p,2} < W_{z,2} < W_{2p} < \ldots < W_{p,s}(< W_{z(s+1)}).$$

Here we assume that we have $s = N'$ poles, but consider situations with different percolation responses at low and high frequencies depending upon whether the first zero $W_{z,1} = 0$ and on the existence or not of there is a final zero $W_{z,(N'+1)}$. These four cases lead to four functional forms for the conductance, all of which are realisable in the case of $p = 1/2$. In this section we derive each of these four forms from some simple asymptotic arguments. At this stage the constant $D(N)$ is undetermined, but we will be able to deduce its value from our subsequent analysis. Although simple, these arguments lead to remarkably accurate formulae when $p = 1/2$, when compared with the numerical calculations, that predict not only the PLER but also the limits of this region.

Simple trigonometric arguments imply that

$$|Y(\omega)| = |D(N)| \frac{\prod_{k=1}^{r} \sqrt{\omega^2 + W_{z,k}^2}}{\prod_{k=1}^{s} \sqrt{\omega^2 + W_{p,k}^2}} \tag{4}$$

To obtain an asymptotic formula from this identity, we will assume that $s = N'$ is large, and that there is a *high density* of poles and zeros along the imaginary axis. From the results in the previous section we know that asymptotically the poles $W_{p,k}$ follow a regular distribution and that the the zeroes have a regular spacing between the poles. The conclusions of the previous section on the distribution of the poles and the zeros leads to the following formulae

$$
\begin{aligned}
W_{p,k} &\sim f(k) \\
\frac{W_{p,(k+1)} - W_{z,k}}{W_{p,(k+1)} - W_{p,k}} &= \delta_k, \\
W_{p,(k+1)} - W_{pk)} &\sim f'(k) \\
W_{z,(k+1)} &\sim f(k) + (1-\delta_k)_k f'(k)
\end{aligned}
$$

Here, as we have seen, the function $\log(f(k))$ is given by the inverse of the error function, but its precise form does not matter too much for the next calculation. To do this we firstly consider the contributions to the product in (4) which arise from the terms involving the first pole to the final zero, so that we consider the following product

$$P \equiv |D(N)| \prod_{k=1}^{N'} \frac{\sqrt{\omega^2 + W_{z,(k+1)}^2}}{\sqrt{\omega^2 + W_{p,k}^2}}$$

Note that this product has implicitly assumed the existence of a final zero $W_{z,(N'+1)}$. This contribution will be corrected in cases for which such a final zero does not exist. Using the results in (5), in particular on the mean spacing of the zeros between the poles, we may express P as

$$
\begin{aligned}
P^2 &= |D(N)|^2 \prod_{k=1}^{N'} \frac{\omega^2 + (f(k) + (1-\bar{\delta}_k)f'(k))^2}{\omega^2 + f^2(k)} \\
&= |D(N)|^2 \prod_{k=1}^{N'} \frac{\omega^2 + f^2(k) + 2(1-\bar{\delta}_k)f(k)f'(k) + HOT}{\omega^2 + f^2(k)} \\
&= |D(N)|^2 \prod_{k=1}^{N'} 1 + \frac{2(1-\bar{\delta}_k)f(k)f'(k)}{\omega^2 + f^2(k)}.
\end{aligned}
$$

Now take the logarithm of both sides and using the approximation $\log(1+x) \approx x$ for small x, we have approximately

$$\log(P^2) \approx \log(|D(N)|^2) + \sum_{k=1}^{N'} \frac{2(1-\bar{\delta}_k)f(k)f'(k)}{\omega^2 + f^2(k)}. \tag{5}$$

We now approximate the sum in (5) by an integral, so that

$$\log(P^2) \approx \log(|D(N)|^2) + \int_{k=1}^{N'} (1-\bar{\delta}_k)\frac{2f(k)f'(k)}{\omega^2 + f^2(k)} dk.$$

Making a change of variable from k to f, gives

$$\log(P^2) \approx \log(|D(N)|^2) + \int_{W_{p,1}}^{W_{p,N'}} (1 - \bar{\delta}(f)) \frac{2f\,df}{\omega^2 + f^2} \tag{6}$$

We look at the special form that the above equation takes when $p = 1/2$. In this case, from the results of the last section, we know that $\bar{\delta}_k$ is very close to being constant at 1/2, so that in (6) we have $1 - \bar{\delta} = 1/2$. We can then integrate the expression for P exactly to give

$$\log(P^2) \approx \log(|D(N)|^2) + \frac{1}{2}\log\left(\frac{(W_{p,N'})^2 + \omega^2}{(W_{p,1})^2 + \omega^2}\right)$$

so that

$$P \approx |D(N)| \left(\frac{(W_{p,N'})^2 + \omega^2}{(W_{p,1})^2 + \omega^2}\right)^{\frac{1}{4}}.$$

In this *critical case* it is equally likely that we will/will not have percolation paths at both small and large values of ω. Accordingly, we must consider four equally likely cases of the distribution of the poles and zeros which could arise in any random realisation of the network. Thus to obtain the four possible responses of the network, we must now consider the contribution of the first zero and also of the last zero.

Case 1: First zero at the origin, last zero at $N' + 1$

In this case we multiply P by ω to give $|Y_1|(\omega)$ so that

$$|Y_1(\omega)| \approx |D(N)_1|\omega \left(\frac{(W_{p,N'})^2 + \omega^2}{(W_{p,1})^2 + \omega^2}\right)^{\frac{1}{4}}$$

Case 2: First zero not at the origin, last zero at $N' + 1$.

In this case we multiply P by $\sqrt{W_{z,1}^2 + \omega^2}$ to give $|Y(\omega)|$. We also use the result from the previous section that asymptotically $W_{z,1}$ and $W_{p,1}$ have the same form. This gives

$$|Y_2(\omega)| \approx |D(N)_2|(W_{p,N'}^2 + \omega^2)^{\frac{1}{4}} \left(W_{p,1}^2 + \omega^2\right)^{\frac{1}{4}}$$

Case 3: First zero at the origin, last zero at N'

In this case we multiply P by ω and divide by $\sqrt{W_{z,N'}^2 + \omega^2}$ to give $|Y|$. Exploiting the fact that asymptotically $W_{p,N'} \sim W_{z,N'}$ we have

$$|Y_3(\omega)| \approx |D(N)_3| \frac{\omega}{\left(W_{p,N'}^2 + \omega^2\right)^{1/4} (W_{p,1}^2 + \omega^2)^{\frac{1}{4}}}$$

Case 4: First zero not at the origin, last zero at N'

In this case we multiply P by $\sqrt{W_{z,1}^2 + \omega^2}$ and divide by $\sqrt{W_{z,N'}^2 + \omega^2}$ to give $|Y(\omega)|$. Again, exploiting the fact that asymptotically $W_{p,N'} \sim W_{z,N'}$ we have

$$|Y_4(\omega)| \approx |D(N)_4| \left(\frac{(W_{p,1})^2 + \omega^2}{(W_{p,N'})^2 + \omega^2} \right)^{\frac{1}{4}}$$

We know, further, from the calculations in the previous section that for all sufficiently large values of N

$$CR\, W_{p,1} \sim \frac{1}{N} \quad \text{and} \quad CR\, W_{p,N'} \sim N.$$

Substituting these values into the above formulae gives:

$$|Y_1(\omega)| \approx |D(N)|\, \omega \left(\frac{(N/CR)^2 + \omega^2}{(1/NCR)^2 + \omega^2} \right)^{\frac{1}{4}} \tag{7}$$

with similar formulae for Y_2, Y_3, Y_4. The values for the constants $|D(N)|$ can, in each case, be determined by considering the mid range behaviour of each of these expressions. In each case, the results of the classical Keller duality theory [12] predict that each of these expressions takes the same form in the range $1/N \ll CR\,\omega \ll N$ and has the scaling law given by

$$|Y_i(\omega)| \approx \sqrt{\frac{\omega C}{R}}, \quad i = 1, 2, 3, 4.$$

Note that this is a true emergent expression. It has a different form from the individual component power laws, and it is also independent of the percolation path types for low and high frequencies. It is precisely the expression expected from an infinite lattice with $p = 1/2$ due to the Keller duality theorem [12], in that $|Y|^2 = \omega C/R$ is equal to the product of the conductances of the two separate components. As we have seen, the origin of this expression lies in the effect of averaging the contributions of each of the poles and zeros (and hence the associated simple linear circuits) through the approximation of the sum by an integral. In the case of (say) Y_1 we see that the mid-range form of the expression (7) is given by

$$|Y_1| = |D(N)| \frac{\sqrt{N\omega}}{\sqrt{CR}}.$$

This then implies that $|D(N)| = C/\sqrt{N}$ so that

$$|Y_1(\omega)| \approx \frac{\omega C}{\sqrt{N}} \left(\frac{(N/CR)^2 + \omega^2}{(1/NCR)^2 + \omega^2} \right)^{\frac{1}{4}} \tag{8}$$

Very similar arguments lead to the following expressions in the other three cases:

$$|Y_2(\omega)| \approx \frac{C}{\sqrt{N}} \left((N/CR)^2 + \omega^2\right)^{\frac{1}{4}} \left((1/NCR)^2 + \omega^2\right)^{\frac{1}{4}} \tag{9}$$

$$|Y_3(\omega)| \approx \frac{\sqrt{N}}{R} \frac{\omega}{\left((N/CR)^2 + \omega^2\right)^{\frac{1}{4}} \left((1/NCR)^2 + \omega^2\right)^{\frac{1}{4}}} \tag{10}$$

$$|Y_4(\omega)| \approx \frac{\sqrt{N}}{R} \left(\frac{(1/NCR)^2 + \omega^2}{(N/CR)^2 + \omega^2}\right)^{\frac{1}{4}} \tag{11}$$

The four formulae above give a very complete asymptotic description of the response of the C-R network when $p = 1/2$. In particular they allow us to see the transition between the power-law emergent region and the percolation regions and they also describe the form of the expressions in the percolation regions. We see a *clear transition between the emergent and the percolation regions* at the two frequencies

$$\omega_1 = \frac{1}{NCR} \quad \text{and} \quad \omega_2 = \frac{N}{CR}.$$

Hence, the number of components in the system for $p = 1/2$ has a strong influence on the *boundaries* of the emergent region and also on the *percolation response*. However the emergent behaviour itself is *independent of* N. Observe that these frequencies *correspond directly to the limiting pole and zero values.* This gives a partial answer to the question, *how large does* N have to be to see an emergent response from the network. The answer is that N has to be sufficiently large so that $1/NCR$ and N/CR are widely separated frequencies. The behaviour in the percolation regions in then given by the following

$$|Y_1(CR\,\omega \ll 1)| \approx \omega C\sqrt{N}, \quad |Y_1(CR\,\omega \gg 1)| \approx \frac{\omega C}{\sqrt{N}}.$$

$$|Y_2(CR\,\omega \ll 1)| \approx \frac{1}{\sqrt{N}R}, \quad |Y_2(CR\,\omega \gg 1)| \approx \frac{\omega C}{\sqrt{N}}.$$

$$|Y_3(CR\,\omega \ll 1)| \approx \omega C\sqrt{N}, \quad |Y_3(CR\,\omega \gg 1)| \approx \frac{\sqrt{N}}{R}.$$

$$|Y_4(CR\,\omega \ll 1)| \approx \frac{1}{\sqrt{N}R}, \quad |Y_4(CR\,\omega \gg 1)| \approx \frac{\sqrt{N}}{R}.$$

We note that these percolation limits, with the strong dependence upon $\sqrt{N}$ are exactly as observed in Section 2.

6 Comparison of the Asymptotic and Numerical Results for the Critical Case

We can compare the four formulae (8,9,10.11) with the numerical calculations of the network conductance as a function of ω for four different configurations of the system, with different percolation paths for low and high frequencies. The results of this comparison are shown in Figure 11 in which we plot the numerical calculations together with the asymptotic formulae for a range of values of N given by $N = S(S-1)$ with $S = 10, 20, 50, 100$. We can see from this that the predictions of the asymptotic formulae (8,9,10.11) fit perfectly with the results of the numerical computations over all of the values of N considered. Indeed they agree both in the power law emergent region and in the four possible percolation regions. The results and the asymptotic formulae clearly demonstrate the effect of the network size in these cases.

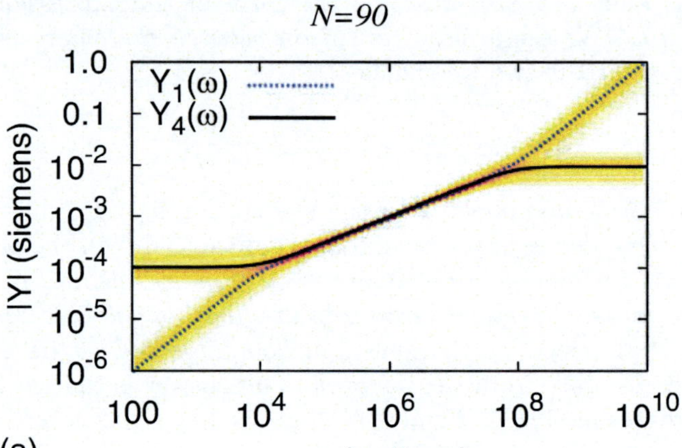

(a)

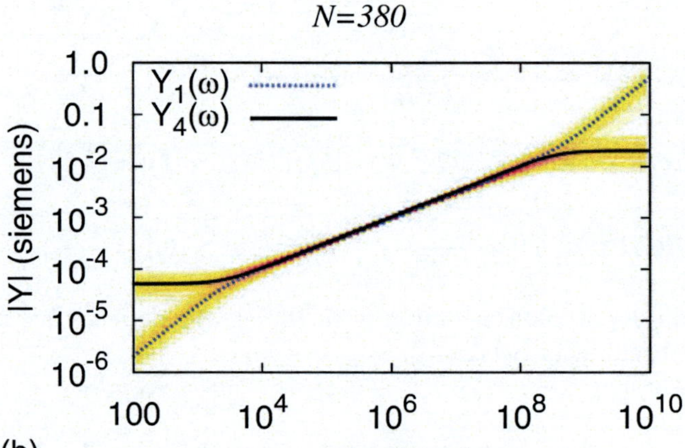

(b)

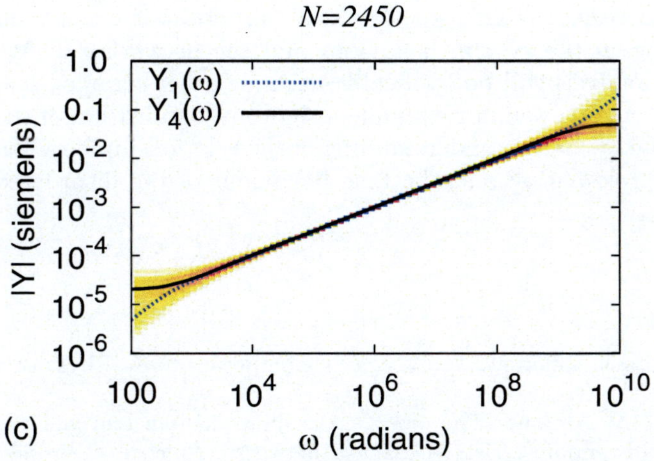

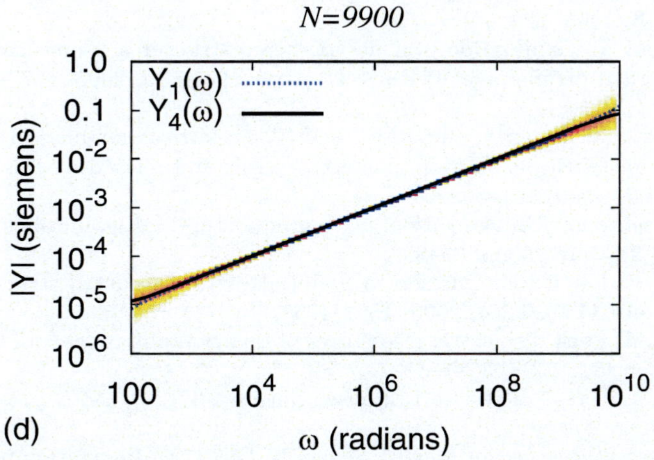

Fig. 11. Comparison of the asymptotic formulae with the numerical computations for the $C - R$ network over many runs, with $p = 1/2$ and network sizes sizes $S = 10, 20, 50, 100$, $N = S(S - 1)$.

7 Discussion

In this paper we have seen that the electrical network approximation to a complex disordered material has a remarkably rich behaviour. In this we see both percolation behaviour (which reflects that of the individual components) and emergent behaviour which follows a power-law quite different from that of the original components. The analysis of this system has involved studying the statistical properties of the resonances of the response. Indeed we could argue that both the percolation and emergent power-law responses are *simply manifestations of this spectral regularity.* The agreement between the asymptotic and the numerical results is very good, which supports our claim, and shows a clear link between the system behaviour and the network size. We hope that this form of analysis will be applicable to many other complex systems. Many questions remain open, for example a rigorous justification of the observed spectral regularity and an understanding of the relationship between the network model approximation and the true behaviour of the disordered material.

References

1. P.W. Anderson: More is different. Biology and Computation: A Physicist's Choice, 1994.
2. R. Bouamrane and D.P. Almond: The emergent scaling phenomenon and the dielectric properties of random resistor-capacitor networks. Journal of Physics, Condensed Matter **15**(24), 2003, 4089–4100.
3. S.R. Broadbent and J.M. Hammersley: Percolation processes I,II. Proc. Cambridge Philos. Soc. **53**, 1953, 629–641.
4. C. Brosseau: Modelling and simulation of dielectric heterostructures: a physical survey from an historical perspective. J. Phys. D: Appl. Phys. **39**, 2005, 1277–1294.
5. J.P. Clerc, G. Giraud, J.M. Laugier, and J.M. Luck: The electrical conductivity of binary disordered systems, percolation clusters, fractals and related models. Adv. Phys. **39**(3), June 1990, 191–309.
6. J.C. Dyre and T.B. Schrøder: Universality of ac conduction in disordered solids. Rev. Mod. Phys. **72**(3), July 2000, 873–892.
7. K. Funke and R.D. Banhatti: Ionic motion in materials with disordered structures. Solid State Ionics **177**(19-25), 2006, 1551–1557.
8. T. Jonckheere and J.M. Luck: Dielectric resonances of binary random networks. J. Phys. A: Math. Gen. **31**, 1998, 3687–3717.
9. A.K. Jonscher: The universal dielectric response. Nature **267**(23), 1977, 673–679.
10. A.K. Jonscher: *Universal Relaxation Law: A Sequel to Dielectric Relaxation in Solids.* Chelsea Dielectrics Press, 1996.
11. C.D. Lorenz and R.M. Ziff: Precise determination of the bond percolation thresholds and finite-size scaling corrections for the sc, fcc, and bcc lattices. Physical Review E **57**(1), 1998, 230–236.
12. G.W. Milton: Bounds on the complex dielectric constant of a composite material. Applied Physics Letters **37**, 1980, 300.
13. K.D. Murphy, G.W. Hunt, and D.P. Almond: Evidence of emergent scaling in mechanical systems. Philosophical Magazine **86**(21), 2006, 3325–3338.
14. K.L. Ngai, C.T. White, and A.K. Jonscher: On the origin of the universal dielectric response in condensed matter. Nature **277**(5693), 1979, 185–189.
15. V.-T. Truong and J.G. Ternan: Complex conductivity of a conducting polymer composite at microwave frequencies. Polymer **36**(5), 1995, 905–909.
16. B. Vainas, D.P. Almond, J. Luo, and R. Stevens: An evaluation of random RC networks for modelling the bulk ac electrical response of ionic conductors. Solid State Ionics **126**(1), 1999, 65–80.

Algorithms and Error Bounds for Multivariate Piecewise Constant Approximation

Oleg Davydov

Department of Mathematics and Statistics, University of Strathclyde, G1 1XH, UK

Summary. We review the surprisingly rich theory of approximation of functions of many variables by piecewise constants. This covers for example the Sobolev-Poincaré inequalities, parts of the theory of nonlinear approximation, Haar wavelets and tree approximation, as well as recent results about approximation orders achievable on anisotropic partitions.

1 Introduction

Let Ω be a bounded domain in $\mathbb{R}^d$, $d \geq 2$. Suppose that Δ is a *partition* of Ω into a finite number of subsets $\omega \subset \Omega$ called *cells*, where the default assumptions are just these: $|\omega| := \text{meas}(\omega) > 0$ for all $\omega \in \Delta$, $|\omega \cap \omega'| = 0$ if $\omega \neq \omega'$, and $\sum_{\omega \in \Delta} |\omega| = |\Omega|$. For a finite set D we denote its cardinality by $|D|$, so that $|\Delta|$ stands for the number of cells ω in Δ. Given a function $f : \Omega \to \mathbb{R}$, we are interested in the error bounds for its approximation by piecewise constants in the space

$$S(\Delta) = \Big\{ \sum_{\omega \in \Delta} c_\omega \chi_\omega : c_\omega \in \mathbb{R} \Big\}, \qquad \chi_\omega(x) := \begin{cases} 1, & \text{if } x \in \omega, \\ 0, & \text{otherwise.} \end{cases}$$

The best approximation error is measured in the L_p-norm $\|\cdot\|_p := \|\cdot\|_{L_p(\Omega)}$,

$$E(f, \Delta)_p := \inf_{s \in S(\Delta)} \|f - s\|_p, \qquad 1 \leq p \leq \infty,$$

and various methods are known for the generation of the sequences of partitions Δ_N such that $E(f, \Delta_N)_p \to 0$ as $N \to \infty$ under certain smoothness assumptions on f, such as $f \in W_q^r(\Omega)$, where $W_q^r(\Omega)$ is the Sobolev space.

Note that the *simple functions* (measurable functions that take only finitely many values) used in the definition of Lebesgue integral are piecewise constants in the above sense. Given a function $f \in L_\infty(\Omega)$, we can generate a partition Δ_N as follows. Let $m, M \in \mathbb{R}$ be the essential infimum and essential supremum of f in Ω, respectively. Note that $\|f\|_\infty =$

E.H. Georgoulis, A. Iske, J. Levesley (eds.), *Approximation Algorithms for Complex Systems*, Springer Proceedings in Mathematics 3, DOI: 10.1007/978-3-642-16876-5_2,
© Springer-Verlag Berlin Heidelberg 2011

$\max\{-m, M\} \geq (M-m)/2$. Split the interval $[m, M]$ into N subintervals $I_k = [m+(k-1)h, m+kh)$, $k = 1, \ldots, N-1$, $I_N = [m+(N-1)h, M]$, $h = (M-m)/N$, and set

$$s_N = \sum_{k=1}^{N} c_k \chi_{\omega_k}, \qquad \omega_k = f^{-1}(I_k), \quad c_k = m + (k - \tfrac{1}{2})h.$$

Then

$$\|f - s_N\|_\infty \leq \tfrac{M-m}{2N} \leq N^{-1}\|f\|_\infty.$$

If f is continuous on $\overline{\Omega}$ and $m = -M \neq 0$, then the above splitting of $[m, M]$ can be used to show that $E(f, \Delta)_\infty \geq N^{-1}\|f\|_\infty$ for any partition Δ with $|\Delta| \leq N$. Clearly, the above partition Δ_N is in general very complicated because the cells ω may be arbitrary measurable sets and so the above s_N cannot be stored using a finite number of real parameters.

Therefore piecewise constant approximation algorithms are practically useful only if the resulting approximation can be efficiently encoded. In the spirit of optimal recovery we will measure the complexity of an approximation algorithm by the maximum number of real parameters needed to store the piecewise constant function s it produces. If the algorithm produces an explicit partition Δ and defines s by $s = \sum_{\omega\in\Delta} c_\omega \chi_\omega$, then the constants c_ω give N such parameters, where $N = |\Delta|$. As in all 'partition based' algorithms discussed in this paper the partition Δ can be described using $\mathcal{O}(N)$ parameters, their overall complexity is $\mathcal{O}(N)$. The same is true for the 'dictionary based' algorithms such as Haar wavelet thresholding, with N being the number of basis functions that are active in an approximation.

In this paper we review a variety of algorithms for piecewise constant approximation for which the error bounds ('Jackson estimates') are available for functions in classical function spaces (Sobolev spaces or Besov spaces). We do not discuss 'Bernstein estimates' and the characterization of approximation spaces, and refer the interested reader to the original papers and the survey [14] that extensively covers this topic. However, we present a number of 'saturation' theorems that give a limit on the accuracy achievable by certain methods on general smooth functions. With only one exception (in the beginning of Section 3) we do not discuss the approximation of functions of one variable, where we again refer to [14].

The paper is organized as follows. Section 2 is devoted to a simple linear approximation algorithm based on a uniform subdivision of the domain and local approximation by constants. In addition, we show that the approximation order $N^{-1/d}$ cannot be improved on isotropic partitions and give a review of the results on the approximation by constants (Sobolev-Poincaré inequalities) on general domains. Section 3 is devoted to the methods of nonlinear approximation restricted to our topic of piecewise constants. We discuss adaptive partition based methods such as Birman-Solomyak's algorithm and tree approximation, as well as dictionary based methods such as Haar wavelet thresholding and best n-term approximation. Finally, in Section 4 we present a simple algorithm with the approximation order $N^{-2/(d+1)}$ of piecewise constants on anisotropic polyhedral partitions, which cannot be further improved if the cells of a partition are required to be convex.

2 Linear Approximation on Isotropic Partitions

Given $s = \sum_{\omega\in\Delta} c_\omega \chi_\omega$, we have

$$\|f - s\|_p = \begin{cases} \left(\sum_{\omega\in\Delta} \|f - c_\omega\|^p_{L_p(\omega)}\right)^{1/p} & \text{if } p < \infty, \\ \sup_{\omega\in\Delta} \|f - c_\omega\|_{L_\infty(\omega)} & \text{if } p = \infty. \end{cases} \tag{1}$$

Hence the best approximation on a fixed partition Δ is achieved when c_ω are the best approximating constants $c^*_\omega(f)$ such that

$$\|f - c^*_\omega(f)\|_{L_p(\omega)} = \inf_{c\in\mathbb{R}} \|f - c\|_{L_p(\omega)} =: E(f)_{L_p(\omega)}.$$

In the case $p = \infty$ obviously

$$c^*_\omega(f) = \frac{1}{2}(M_\omega f + m_\omega f), \quad E(f)_{L_\infty(\omega)} = \frac{1}{2}(M_\omega f - m_\omega f),$$

where

$$M_\omega f := \operatorname*{ess\,sup}_{x\in\omega} f(x), \quad m_\omega f := \operatorname*{ess\,inf}_{x\in\omega} f(x).$$

For any $1 \le p \le \infty$, it is easy to see that the average value of f on ω,

$$f_\omega := |\omega|^{-1} \int_\omega f(x)\,dx,$$

satisfies $\|f_\omega - c\|_{L_p(\omega)} \le \|f - c\|_{L_p(\omega)}$ for any constant c, in particular for $c = c^*_\omega$. Therefore

$$\|f - f_\omega\|_{L_p(\omega)} \le 2E(f)_{L_p(\omega)},$$

and we conclude that the approximation

$$s_\Delta(f) := \sum_{\omega\in\Delta} f_\omega \chi_\omega \in S(\Delta) \tag{2}$$

is near best in the sense that

$$\|f - s_\Delta(f)\|_p \le 2E(f, \Delta)_p, \qquad 1 \le p \le \infty.$$

If $f_{|\omega}$ belongs to the Sobolev space $W^1_p(\omega)$, and the domain ω is sufficiently smooth then the error $\|f - f_\omega\|_{L_p(\omega)}$ may be estimated with the help of the Poincaré inequality

$$\|f - f_\omega\|_{L_p(\omega)} \le C_\omega \operatorname{diam}(\omega)\, |f|_{W_p^1(\omega)}, \quad f \in W_p^1(\omega), \tag{3}$$

where C_ω may still depend on ω in a scale-invariant way. For example, if ω is a Lipschitz domain, then C_ω can be found depending only on d, p and the Lipschitz constant of the boundary. At the end of this section we provide some more detail about the Poincaré inequality as well as the more general Sobolev-Poincaré inequalities available for various types of domains.

If the partition Δ is such that $C_\omega \le C$, where C is independent of ω (but may depend for example on the Lipschitz constant of the boundary of Ω), then (1) and (3) imply

$$\|f - s_\Delta(f)\|_p \le C \operatorname{diam}(\Delta)\, |f|_{W_p^1(\Omega)}, \quad \operatorname{diam}(\Delta) := \max_{\omega\in\Delta} \operatorname{diam}(\omega).$$

This estimate suggests looking for partitions Δ that minimize $\operatorname{diam}(\Delta)$ provided the number of cells $N = |\Delta|$ is fixed. Clearly, $\operatorname{diam}(\Delta) \ge CN^{-1/d}$ for some constant C independent of N, and the order $N^{-1/d}$ is achieved if we for example choose a (hyper)cube Q containing Ω, split it uniformly into N^d equal subcubes Q_i, and define the cells of Δ by intersecting Ω with these subcubes, $\omega_i = \Omega \cup Q_i$, see Figure 1. This gives a simple algorithm for piecewise constant approximation with approximation order $N^{-1/d}$ for all $f \in W_p^1(\Omega)$.

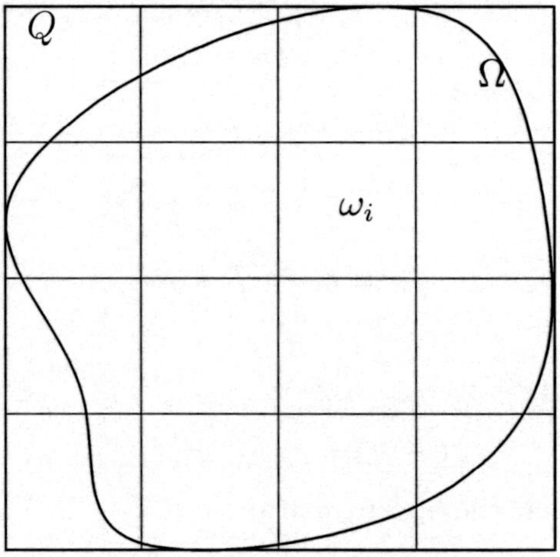

Fig. 1. Uniform partition.

For the sake of simplicity we formulate this and all other algorithms only for the case when Ω is a cube $(0, 1)^d$.

Algorithm 1

Define Δ by splitting $\Omega = (0,1)^d$ into $N = m^d$ cubes $\omega_1, \ldots, \omega_N$ of edge length $h = 1/m$. Let $s_\Delta(f)$ be given by (2).

Theorem 1. *The error of the piecewise constant approximation $s_\Delta(f)$ generated by Algorithm 1satisfies*

$$\|f - s_\Delta(f)\|_p \leq C(d,p)N^{-1/d}\,|f|_{W_p^1(\Omega)}, \quad f \in W_p^1(\Omega),\ 1 \leq p \leq \infty. \tag{4}$$

The order $N^{-1/d}$ in (4) means that an approximation with error $\|f - s_\Delta(f)\|_p = \mathcal{O}(\varepsilon)$ is only achieved using $(\frac{1}{\varepsilon})^d$ degrees of freedom, which grows exponentially fast with the number of dimensions d. This phenomenon is often referred to as *curse of dimensionality.*

The approximation order $N^{-1/d}$ in (4) cannot be improved in general. See [14, Section 6.2] for a discussion of *saturation and inverse theorems*, where certain smoothness properties of f are deduced from appropriate assumptions about the order of its approximation by multivariate piecewise polynomials. For example, assuming that $E(f,\Delta)_\infty = o(\mathrm{diam}(\Delta))$ as $\mathrm{diam}(\Delta) \to 0$ for all partitions Δ, we can easily show that f is a constant function. Indeed, for any $x, y \in \Omega$ we can find a partition Δ such that x and y belong to the same cell ω, and $\mathrm{diam}(\omega) = \mathrm{diam}(\Delta) \leq 2\|x-y\|_2$. Then $|f(x) - f(y)| \leq |f(x) - f_\omega| + |f_\omega - f(y)| \leq 4E(f,\Delta)_\infty = o(\mathrm{diam}(\omega))$. Hence $|f(x) - f(y)| = o(\|x-y\|_2)$ as $y \to x$, which implies that f has a zero differential at every $x \in \Omega$, that is f is a constant.

A saturation theorem in terms of the number of cells holds for any sequence of 'isotropic' partitions. We say that a sequence of partitions $\{\Delta_N\}$ is *isotropic* if there is a constant γ such that

$$\mathrm{diam}(\omega) \leq \gamma\rho(\omega) \quad \text{for all } \omega \in \bigcup_N \Delta_N,$$

where $\rho(\omega)$ is the maximum diameter of d-dimensional balls contained in ω. Note that an isotropic partition may contain cells of very different sizes, see for example Figure 2 below.

Theorem 2. *Assume that $f \in C^1(\Omega)$ and there is an isotropic sequence of partitions $\{\Delta_N\}$ with $\lim_{N\to\infty} \mathrm{diam}(\Delta_N) = 0$ such that*

$$E(f,\Delta_N)_\infty = o(|\Delta_N|^{-1/d}), \quad N \to \infty.$$

Then f is a constant.

Proof. If f is not constant, then the gradient $\nabla f := [\partial f/\partial x_i]_{i=1}^d$ is nonzero at a point $\hat{x} \in \Omega$. Since the gradient of f is continuous, there is $\delta > 0$, a unit vector σ and a cube $Q \subset \Omega$ with edge length h containing $\hat{x}$ such that $D_\sigma f(x) \geq \delta$

for all $x \in Q$, where $D_\sigma f = \nabla f^T \sigma$ denotes the directional derivative of f. The cube $\tilde{Q} := \{x \in Q : \text{dist}(x, \partial Q) > h/4\}$ has edge length $h/2$ and volume $(h/2)^d$. Assume that N is large enough to ensure that $\text{diam}(\Delta_N) < h/4$. Then any cell $\omega \in \Delta_N$ that has nonempty intersection with $\tilde{Q}$ is contained in Q. If $[x_1, x_2]$ is an interval in ω parallel to σ, then $|f(x_2) - f(x_1)| \geq \delta \|x_2 - x_1\|_2$, which implies $M_\omega f - m_\omega f \geq \delta \rho(\omega)$ and hence

$$\varepsilon_N := E(f, \Delta_N)_\infty \geq E(f)_{L_\infty(\omega)} \geq \frac{\delta}{2}\rho(\omega) \geq \frac{\delta}{2\gamma}\,\text{diam}(\omega).$$

Therefore

$$|\omega| \leq \mu_d \left(\frac{\gamma}{\delta}\right)^d \varepsilon_N^d,$$

where μ_d denotes the volume of the d-dimensional ball of radius 1. Since $\tilde{Q}$ is covered by such cells ω, we conclude that

$$\left(\frac{h}{2}\right)^d = |\tilde{Q}| \leq \sum_{\omega \cap \tilde{Q} \neq \emptyset} |\omega| \leq \mu_d \left(\frac{\gamma}{\delta}\right)^d \varepsilon_N^d |\Delta_N|,$$

which implies

$$E(f, \Delta_N)_\infty \geq \frac{h\delta}{2\gamma\mu_d^{1/d}} |\Delta_N|^{-1/d},$$

contrary to the assumption. □

Sobolev-Poincaré inequalities

Sobolev-Poincaré inequalities provide bounds for the error of $f - f_\omega$. They hold on domains satisfying certain geometric conditions, for example the interior cone condition or the Lipschitz boundary condition. In some cases even a necessary and sufficient condition for ω to admit such an inequality is known. A domain $\omega \subset \mathbb{R}^d$ is called a *John domain* if there is a fixed point $x_0 \in \omega$ and a constant $c_J > 0$ such that every point $x \in \omega$ can be connected to x_0 by a curve $\gamma \subset \omega$ such that

$$\text{dist}(y, \partial\omega) \geq c_J \ell(\gamma(x, y)), \quad \text{for all } y \in \gamma,$$

where $\ell(\gamma(x, y))$ denotes the length of the segment of γ between x and y. Every domain with the interior cone condition is a John domain, but not otherwise. In particular, there are John domains with fractal boundary of Hausdorff dimension greater than $d - 1$.

The following Sobolev inequality holds for all John domains $\omega \subset \mathbb{R}^d$, see [18] and references therein,

$$\|f - f_\omega\|_{L_{q^*}(\omega)} \leq C(d, q, \lambda) \|\nabla f\|_{L_q(\omega)}, \quad f \in W_q^1(\omega), \quad 1 \leq q < d, \qquad (5)$$

where $q^* = dq/(d-q)$ is the Sobolev conjugate of q, and λ is the John constant of ω. Note that $\|\nabla f\|_{L_q(\omega)}$ denotes the L_q-norm of the euclidean norm of ∇f, that is $\|\nabla f\|^q_{L_q(\omega)} = \int_\omega \left(\sum_{i=1}^d |\partial f/\partial x_i|^2\right)^{q/2} dx$, which is equivalent to the more standard seminorm of the Sobolev space $W^1_q(\omega)$ given by $|f|^q_{W^1_q(\omega)} = \sum_{i=1}^d \int_\omega |\partial f/\partial x_i|^q\, dx$. We prefer using $\|\nabla f\|_{L_q(\omega)}$ because of the explicit expressions for the constant C in (7) available in certain cases, see the end of this section.

According to [7], if the Sobolev inequality (5) holds for some $1 \le q < d$ and certain mild separation condition (valid for example for any simply connected domain in $\mathbb{R}^2$) is satisfied, then ω is a John domain.

Assuming $1 \le p < \infty$, let $\tau = \max\{\frac{d}{1+d/p}, 1\}$. Then $\tau^* \ge p$ and $1 \le \tau < d$. If $|\omega| < \infty$, then Hölder inequality and (5) imply for any $q \ge \tau$

$$\begin{aligned}\|f - f_\omega\|_{L_p(\omega)} &\le |\omega|^{\frac{1}{p}-\frac{1}{\tau^*}} \|f - f_\omega\|_{L_{\tau^*}(\omega)} \\ &\le C(d,\tau,\lambda)|\omega|^{\frac{1}{p}-\frac{1}{\tau^*}} \|\nabla f\|_{L_\tau(\omega)} \\ &\le C(d,\tau,\lambda)|\omega|^{\frac{1}{d}+\frac{1}{p}-\frac{1}{q}} \|\nabla f\|_{L_q(\omega)},\end{aligned}$$

and we arrive at the following Sobolev-Poincaré inequality for all p, q such that $1 \le p < \infty$ and $\tau \le q \le \infty$,

$$\|f - f_\omega\|_{L_p(\omega)} \le C(d,p,\lambda)|\omega|^{\frac{1}{d}+\frac{1}{p}-\frac{1}{q}} \|\nabla f\|_{L_q(\omega)}, \quad f \in W^1_q(\omega). \tag{6}$$

In particular, since $\tau \le p$, we can choose $q = p$, which leads to the Poincaré inequality for bounded John domains for all $1 \le p < \infty$ in the form

$$\|f - f_\omega\|_{L_p(\omega)} \le C\,\mathrm{diam}(\omega) \|\nabla f\|_{L_p(\omega)}, \quad f \in W^1_p(\omega), \tag{7}$$

where C depends only on d, p, λ.

Poincaré inequality in the case $p = \infty$ has been considered in [25]. If $\omega \subset \mathbb{R}^d$ is a bounded path-connected domain, then

$$E(f)_{L_\infty(\omega)} \le r(\omega)\|\nabla f\|_{L_\infty(\omega)}, \quad \text{with } r(\omega) := \inf_{x\in\omega} \sup_{y\in\omega} \rho_\omega(x,y),$$

where $\rho_\omega(x,y)$ is the geodesic distance, i.e. the infimum of the lengths of the paths in ω from x to y.

If ω is star-shaped with respect to a point, then $r(\omega) \le \mathrm{diam}(\omega)$, and so (7) holds with $C = 2$ for all such domains if $p = \infty$. Moreover, as observed in [25], the arguments of [2, 11] can be applied to show that for any bounded star-shaped domain

$$\|f - f_\omega\|_{L_p(\omega)} \le \tfrac{2}{1-d/p}\,\mathrm{diam}(\omega)\|\nabla f\|_{L_p(\omega)}, \quad f \in W^1_p(\omega),\ d < p \le \infty. \tag{8}$$

In particular, (8) applies to star-shaped domains with cusps that fail to be John domains.

If ω is a bounded convex domain in $\mathbb{R}^d$, then (7) holds for all $1 \le p \le \infty$ with a constant C depending only on d [12]. Moreover, optimal constants are known for $p = 1, 2$: $C = 1/\pi$ for $p = 2$ [3, 22] and $C = 1/2$ for $p = 1$ [1]. Since $r(\omega) = \frac{1}{2}\,\mathrm{diam}(\omega)$, it follows that (7) holds with $C = 1$ if $p = \infty$.

Note that similar estimates are available for the approximation by polynomials of any degree, where in the case $p = q$ the corresponding result is usually referred to as the Bramble-Hilbert lemma, see [6, Chapter 4]. Moreover, instead of Sobolev spaces, the smoothness of f can be measured in some other function spaces (e.g. Besov spaces), or with the help of a modulus of smoothness (Whitney estimates), see [13, 14].

3 Nonlinear Approximation

We have seen in Section 2 that the approximation order $N^{-1/d}$ is the best achievable on isotropic partitions. Nevertheless, by using more sophisticated algorithms the estimate (4) can be improved in the sense that the norm $|f|_{W_p^1(\Omega)}$ in its right hand side is replaced by a weaker norm, for example $|f|_{W_q^1(\Omega)}$ with $q < p$. This improvement is often quite significant because the norm $|f|_{W_q^1(\Omega)}$ is finite for functions with more substantial singularities than those allowed in the space $W_p^1(\Omega)$, see [14] for a discussion.

Recall that in Algorithm 1 the partition Δ is independent of the target function f, and so $s_\Delta(f)$ depends linearly on f. A simple example of a nonlinear algorithm is given by Kahane's approximation method for continuous functions of bounded total variation on an interval, see [14, Section 3.2]. To define a partition of the interval (a, b), the points $a = t_0 < t_1 < \cdots < t_N = b$ are chosen such that $\mathrm{var}_{(t_{i-1},t_i)}(f) = \frac{1}{N}\,\mathrm{var}_{(a,b)}(f)$, $i = 1, \ldots, N-1$. By setting $\omega_i = (t_{i-1}, t_i)$, $c_i = (M_{\omega_i} f + m_{\omega_i} f)/2$, we see that the piecewise constant function $s = \sum_{i=1}^{N-1} c_i \chi_{\omega_i}$ satisfies $\|f - s\|_\infty \le \frac{1}{2N}\,\mathrm{var}_{(a,b)}(f)$. Thus, for the partition $\Delta = \{\omega_i\}_{i=1}^{N-1}$,

$$E(f, \Delta)_\infty \le \frac{1}{2N}\,\mathrm{var}_{(a,b)}(f) = \frac{1}{2N}|f|_{BV(a,b)} \le \frac{1}{2N}|f|_{W_1^1(a,b)},$$

where the last inequality presumes that f belongs to $W_1^1(a, b)$, that is it is absolutely continuous and its derivative is absolutely integrable.

In the multivariate case the first algorithm of this type was given in [5]. It is based on *dyadic partitions* Δ of $\Omega = (0, 1)^d$ that consist of the *dyadic cubes* of the form

$$2^{-jd}(k_1, k_1 + 1) \times \cdots \times (k_d, k_d + 1), \quad j = 0, 1, 2, \ldots, \; 0 \le k_i < 2^{jd},$$

produced adaptively by successive *dyadic subdivisions* of a cube into 2^d equal subcubes with halved edge length, see Figure 2.

The following lemma plays a crucial role in [5].

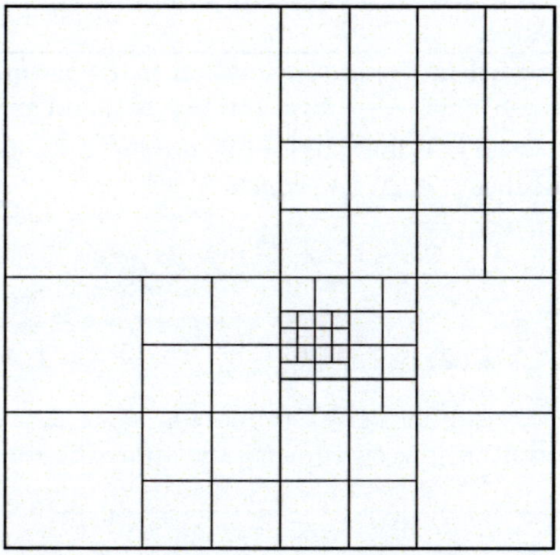

Fig. 2. Example of a dyadic partition.

Lemma 1. *Let $\Phi(\omega)$ be a nonnegative function of sets $\omega \subset \Omega$ which is subadditive in the sense that $\Phi(\omega') + \Phi(\omega'') \le \Phi(\omega' \cup \omega'')$ as soon as ω', ω'' are disjoint subdomains of Ω. Given $\alpha > 0$, we set*

$$g_\alpha(\omega) := |\omega|^\alpha \Phi(\omega), \quad \omega \subset \Omega,$$

and, for any partition Δ of Ω,

$$G_\alpha(\Delta) := \max_{\omega \in \Delta} g_\alpha(\omega).$$

Assume that a sequence of partitions $\{\Delta_k\}_{k=0}^\infty$ of $\Omega = (0,1)^d$ into dyadic cubes is obtained recursively as follows. Set $\Delta_0 = \{\Omega\}$. Obtain Δ_{k+1} from Δ_k by the dyadic subdivision of those cubes $\omega \in \Delta_k$ for which

$$g_\alpha(\omega) \ge 2^{-d\alpha} G_\alpha(\Delta_k).$$

Then

$$G_\alpha(\Delta_k) \le C(d,\alpha) |\Delta_k|^{-(\alpha+1)} \Phi(\Omega), \quad k = 0, 1, \ldots.$$

This lemma can be used with $\Phi(\omega) = |f|^q_{W^1_q(\omega)}$, $1 \le q < \infty$, which is obviously subadditive, giving rise to the following algorithm which we only formulate for piecewise constants even though the results in [5] also apply to the higher order piecewise polynomials.

Algorithm 2 ([5])

Suppose we are interested in the approximation in L_p norm, $1 < p \leq \infty$. Choose $1 \leq q < \infty$ such that $q > \tau := \frac{d}{1+d/p}$ ($\tau = d$ if $p = \infty$), and assume that $f \in W_q^1(\Omega)$, $\Omega = (0,1)^d$. Set $\Delta_0 = \{\Omega\}$. While $|\Delta_k| < N$, obtain Δ_{k+1} from Δ_k by the dyadic subdivision of those cubes $\omega \in \Delta_k$ for which

$$g_\alpha(\omega) \geq 2^{-d\alpha} \max_{\omega \in \Delta} g_\alpha(\omega),$$

where

$$g_\alpha(\omega) := |\omega|^\alpha |f|^q_{W_q^1(\omega)}, \quad \alpha = \tfrac{q}{\tau} - 1.$$

Since $|\Delta_k| < |\Delta_{k+1}|$, the subdivisions terminate at some $\Delta = \Delta_m$ with $|\Delta| \geq N$ and $|\Delta| = \mathcal{O}(N)$. The resulting piecewise constant approximation $s_\Delta(f)$ of f is given by (2).

Theorem 3 ([5]). *The error of the piecewise constant approximation $s_\Delta(f)$ generated by Algorithm 2 satisfies*

$$\|f - s_\Delta(f)\|_p \leq C(d,p,q) N^{-1/d} |f|_{W_q^1(\Omega)}, \quad f \in W_q^1(\Omega). \tag{9}$$

Proof. We only consider the case $p < \infty$. For any $\omega \in \Delta$ it follows by the Sobolev-Poincaré inequality (6) for cubes that

$$\|f - f_\omega\|^p_{L_p(\omega)} \leq C_1 |\omega|^{\frac{p}{d}+1-\frac{p}{q}} |f|^p_{W_q^1(\omega)} = C_1 g_\alpha^{p/q}(\omega) \leq C_1 G_\alpha^{p/q}(\Delta),$$

where C_1 depends only on d, p, q. Hence

$$\|f - s_\Delta(f)\|_p^p = \sum_{\omega \in \Delta} \|f - f_\omega\|^p_{L_p(\omega)} \leq C_1 |\Delta| \, G_\alpha^{p/q}(\Delta).$$

Now Lemma 1 implies

$$\|f - s_\Delta(f)\|_p \leq C_2 |\Delta|^{1/p} |\Delta|^{-(\alpha+1)/q} \Phi^{1/q}(\Omega) = C_2 |\Delta|^{-1/d} |f|_{W_q^1(\Omega)}. \quad \square$$

If $q \geq p$, then the estimate (9) is also valid for the much simpler Algorithm 1. Therefore the scope of Algorithm 2 is when $f \in W_q^1(\Omega)$ for some q satisfying $\tau < q < p$ but $f \notin W_p^1(\Omega)$ or if $|f|_{W_q^1(\Omega)}$ is significantly smaller than $|f|_{W_p^1(\Omega)}$. Note that the computation of $g_\alpha(\omega)$ in Algorithm 2 requires first order partial derivatives of f. Algorithms 2 is nonlinear (in contrast to Algorithm 1) because the partition Δ depends on the target function f.

An adaptive algorithm based on the local approximation errors rather than local Sobolev norm of f was studied in [17]. We again restrict to the piecewise constant case.

Algorithm 3

Assume $f \in L_p(\Omega)$, $\Omega = (0,1)^d$, for some $0 < p \leq \infty$ and choose $\varepsilon > 0$. Set $\Delta_0 = \{\Omega\}$. For $k = 0, 1, \ldots$, obtain Δ_{k+1} from Δ_k by the dyadic subdivision of those cubes $\omega \in \Delta_k$ for which

$$\|f - f_\omega\|_{L_p(\omega)} > \varepsilon.$$

Since $\|f - f_\omega\|_{L_p(\omega)} \to 0$ as $|\omega| \to 0$, the subdivisions terminate at some $\Delta = \Delta_m$. The resulting piecewise constant approximation $s_\Delta(f)$ of f is given by (2).

Now in contrast to [5], $0 < p < 1$ is also allowed. The error bounds are obtained for functions in Besov spaces rather than Sobolev spaces. Recall that f belongs to the Besov space $B^\alpha_{q,\sigma}(\Omega)$, $\alpha > 0$, $0 < q, \sigma \leq \infty$, if

$$|f|_{B^\alpha_{q,\sigma}(\Omega)} = \begin{cases} \left(\int_0^\infty (t^{-\alpha}\omega_r(f,t)_q)^\sigma \frac{dt}{t} \right)^{1/\sigma} & \text{if } 0 < \sigma < \infty, \\ \sup_{t>0} t^{-\alpha}\omega_r(f,t)_q & \text{if } \sigma = \infty. \end{cases}$$

is finite, where $r = [\alpha] + 1$ is the smallest integer greater than α, and $\omega_r(f,t)_q$ denotes the r-th modulus of smoothness of f in L_q. In particular, $B^\alpha_{q,\infty}(\Omega) = \mathrm{Lip}(\alpha, L_q(\Omega))$ for $0 < \alpha < 1$.

Theorem 4 ([17]). *Let $0 < \alpha < 1$, $q > \frac{d}{\alpha + d/p}$ and $0 < \sigma \leq \infty$. If $f \in B^\alpha_{q,\sigma}(\Omega)$, then for any N there is an $\varepsilon > 0$ such that the partition Δ produced by Algorithm 3 satisfies $|\Delta| \leq N$ and*

$$\|f - s_\Delta(f)\|_p \leq C(d,p,q) N^{-\alpha/d} |f|_{B^\alpha_{q,\sigma}(\Omega)}.$$

The set of all dyadic cubes is a tree $\mathcal{T}^{dc}$, where the children of a cube ω are the cubes $\omega_1, \ldots, \omega_{2^d}$ obtained by its dyadic subdivision. The only root of $\mathcal{T}^{dc}$ is $\Omega = (0.1)^d$. Clearly, Algorithms 2 and 3 produce a *complete subtree* $\mathcal{T}$ of $\mathcal{T}^{dc}$ in the sense that for any node in $\mathcal{T}$ its parent and all siblings are also in $\mathcal{T}$. The corresponding partition Δ consists of all leaves of $\mathcal{T}$. If we set $e(\omega) = \|f - f_\omega\|^p_{L_p(\omega)}$, then $E(\mathcal{T}) := \sum_{\omega \in \Delta} e(\omega) = \|f - s_\Delta(f)\|^p_p$. It is easy to see that $|\Delta| = 1 + (2^d - 1)n(\mathcal{T})$, where $n(\mathcal{T})$ denotes the number of subdivisions used to create $\mathcal{T}$. The quantity $n(\mathcal{T})$ measures the complexity of a tree, and $E_n := \inf_{n(\mathcal{T}) \leq n} E(\mathcal{T})$ gives the optimal error achievable by a tree of a given complexity. It is natural to look for optimal or near optimal trees. The concept of *tree approximation* was introduced in [8] in the context of n-term wavelet approximation. General results applicable in particular to the piecewise constant approximations on dyadic partitions are given in [4]. The idea is that replacing $\|f - f_\omega\|_{L_p(\omega)} > \varepsilon$ in Algorithm 3 by a more sophisticated refinement criterion leads to an algorithm that produces a near optimal tree.

Algorithm 4 ([4])

Assume $f \in L_p(\Omega)$, $\Omega = (0,1)^d$, for some $1 \le p < \infty$ and choose $\varepsilon > 0$. Generate a sequence of complete subtrees $\mathcal{T}_k$ of $\mathcal{T}^{dc}$ as follows. Set $\mathcal{T}_0 = \{\Omega\}$ and $\alpha(\Omega) = 0$. For $k = 0, 1, \ldots$, obtain $\mathcal{T}_{k+1}$ from $\mathcal{T}_k$ by the dyadic subdivision of those leaves ω of $\mathcal{T}_k$ for which

$$e(\omega) := \|f - f_\omega\|^p_{L_p(\omega)} > \varepsilon + \alpha(\omega),$$

and define $\alpha(\omega_i)$ for all children $\omega_1, \ldots, \omega_{2^d}$ of ω by

$$\alpha(\omega_i) = \frac{e(\omega_i)}{\sigma(\omega)}\Big[\alpha(\omega) + (\varepsilon - e(\omega) - \sigma(\omega))_+\Big], \quad \sigma(\omega) := \sum_j e(\omega_j),$$

assuming that $\sigma(\omega) \ne 0$. The algorithm terminates at some tree $\mathcal{T} = \mathcal{T}_m$ since $\|f - f_\omega\|_{L_p(\omega)} \to 0$ as $|\omega| \to 0$. The resulting piecewise constant approximation $s_\Delta(f)$ of f is given by (2), where Δ is the dyadic partition defined by the leaves of $\mathcal{T}$.

Theorem 5 ([4]). *The tree $\mathcal{T}$ produced by Algorithm 4 is near optimal as it satisfies*

$$E(\mathcal{T}) \le 2(2^d + 1)E_{[n/2]}, \quad n = n(\mathcal{T}).$$

Further results on tree approximation are reviewed in [15].

The above algorithms generate piecewise constant approximations by constructing an appropriate partition of Ω. A different approach is to look for an approximation as linear combination of a fixed set of piecewise constant 'basis functions', for example characteristic functions of certain subsets of Ω.

More general, let $\mathcal{D} \subset L_p(\Omega)$ be a set of functions, called *dictionary*, such that the finite linear combinations of the elements in $\mathcal{D}$ are dense in $L_p(\Omega)$. Note that the set $\mathcal{D}$ does not have to be linearly independent. Given $f \in L_p(\Omega)$, the error of the *best n-term approximation* is defined by

$$\sigma_n(f, \mathcal{D})_p = \inf_{s \in \Sigma_n} \|f - s\|_p,$$

where $\Sigma_n = \Sigma_n(\mathcal{D})$ is the set of all linear combinations of at most n elements of $\mathcal{D}$,

$$\Sigma_n(\mathcal{D}) := \Big\{\sum_{g \in D} c_g g : D \subset \mathcal{D},\ |D| \le n,\ c_g \in \mathbb{R}\Big\}.$$

If the functions in $\mathcal{D}$ are piecewise constants, then the approximants in $\Sigma_n(\mathcal{D})$ are piecewise constants as well. If each element $g \in \mathcal{D}$ can be described using a bounded number of parameters, then $s = \sum_{g \in D} c_g g \in \Sigma_n(\mathcal{D})$ requires $\mathcal{O}(n)$ parameters even though the number of cells in the partition Δ such that $s \in S(\Delta)$ may in general grow superlinearly (even exponentially) with n.

Piecewise constant approximation $s_\Delta(f)$ produced by Algorithm 2 or 3 belongs to $\Sigma_n(\mathcal{D}^c)$, with $n = |\Delta|$, where the dictionary $\mathcal{D}^c$ consists of the characteristic functions χ_ω of all dyadic cubes $\omega \subseteq (0,1)^d$. Therefore Theorems 3 and 4 imply that

$$\sigma_n(f,\mathcal{D}^c)_p \le \begin{cases} C(d,p,q)n^{-\alpha/d}|f|_{W^\alpha_q(\Omega)} & \text{if } f \in W^\alpha_q(\Omega),\ \alpha = 1, \\ C(d,p,q)n^{-\alpha/d}|f|_{B^\alpha_{q,\sigma}(\Omega)} & \text{if } f \in B^\alpha_{q,\sigma}(\Omega),\ 0<\alpha<1. \end{cases}$$

as soon as $q > \frac{d}{\alpha+d/p}$.

Clearly, $\Sigma_n(\mathcal{D}^c)$ includes many piecewise constants with more then n dyadic cells, for example, $\chi_{(0,1)^2} - \chi_{(0,1/2^m)^2} \in \Sigma_2(\mathcal{D}^c)$ is piecewise constant with respect to a partition of $(0,1)^2$ into $3m+1$ dyadic squares. A larger dictionary $\mathcal{D}^r$ consisting of the characteristic functions of all 'dyadic rings' (differences between pairs of embedded dyadic cubes) has been considered in [9, 20]. An adaptive algorithm proposed in [9] produces a piecewise constant approximation $s_\Delta(f)$ of any function f of bounded variation, where Δ is a partition of $(0,1)^2$ into N dyadic rings, such that

$$\|f - s_\Delta(f)\|_2 \le 18\sqrt{3}N^{-1/2}|f|_{BV((0,1)^2)},$$

where $|f|_{BV(\Omega)}$ is the variation of f over Ω. Recall that the space $BV(\Omega)$ coincides with $\mathrm{Lip}(1, L_1(\Omega))$ and contains the Besov space $B^1_{1,1}(\Omega)$. In [20] this result is generalized to certain spaces of bounded variation with respect to dyadic rings, and to the Besov spaces, leading in particular to the estimate

$$\sigma_n(f,\mathcal{D}^r)_p \le C(d,p)n^{-\alpha/d}|f|_{B^\alpha_{\tau,\tau}(\Omega)}, \quad f \in B^\alpha_{\tau,\tau}(\Omega),\ 0<\alpha<1,$$

where $\tau = \frac{d}{\alpha+d/p}$. Recall for comparison that $q > \tau$ in Theorems 3 and 4.

An important 'piecewise constant' dictionary $\mathcal{D}^h$ is given by the multivariate Haar wavelets. Let $\psi^0 = \chi_{(0,1)}$ and $\psi^1 = \chi_{(0,1/2)} - \chi_{(1/2,1)}$. For any $e = (e_1,\dots,e_d) \in V^d$, where V^d is the set consisting of the nonzero vertices of the cube $(0,1)^d$, let

$$\psi^e(x_1,\dots,x_d) := \psi^{e_1}(x_1)\cdots\psi^{e_d}(x_d).$$

Then the set

$$\Psi^h = \{\psi^e_{j,k} := 2^{jd/2}\psi^e(2^j(\cdot - k)) : j \in \mathbb{Z},\ k \in \mathbb{Z}^d,\ e \in V^d\}$$

of *Haar wavelets* $\psi^e_{j,k}$ is an orthonormal basis for $L_2(\mathbb{R}^d)$. Therefore every $f \in L_2(\mathbb{R}^d)$ has an L_2-convergent Haar wavelet expansion

$$f = \sum_{j,k,e}\langle f, \psi^e_{j,k}\rangle\psi^e_{j,k}, \quad \langle f, \psi^e_{j,k}\rangle := \int_{\mathbb{R}^d} f(x)\psi^e_{j,k}(x)\,dx.$$

If $f \in L_2(\Omega)$, $\Omega = (0,1)^d$, then $f - f_\Omega$ has zero mean on $(0,1)^d$ and hence by extending it to $\mathbb{R}^d$ by zero and taking a Haar wavelet decomposition, we obtain an L_2-convergent series

$$f = f_\Omega + \sum_{(j,k,e)\in\Lambda_\Omega} f_{j,k,e}\psi^e_{j,k}, \tag{10}$$

where Λ_Ω denotes the set of indices (j,k,e) such that $\operatorname{supp}\psi^e_{j,k} \subseteq \Omega$, and $f_{j,k,e}$ are the ***Haar wavelet coefficients*** of f,

$$f_{j,k,e} = \int_{(0,1)^d} f(x)\psi^e_{j,k}(x)\,dx.$$

Clearly, the Haar wavelet coefficients are well defined for any function $f \in L_1(\Omega)$. The series (10) converges unconditionally in L_p-norm if $f \in L_p(\Omega)$, $1 < p < \infty$. This implies in particular that every subset of $\{\|f_{j,k,e}\psi^e_{j,k}\|_p : (j,k,e) \in \Lambda_\Omega\}$ has a largest element. The dictionary of Haar wavelets on $\Omega = (0,1)^d$ is given by

$$\mathcal{D}^h = \{\psi^e_{j,k} : (j,k,e) \in \Lambda_\Omega\}.$$

A standard approximation method for this dictionary is *thresholding*, also called *greedy approximation*.

Algorithm 5 Haar wavelet thresholding.

Assume $f \in L_p(\Omega)$, $\Omega = (0,1)^d$, for some $1 < p < \infty$. Let $\int_\Omega f(x)\,dx = 0$. (Otherwise, replace f by $f - f_\Omega$.) Given $n \in \mathbb{N}$, choose n largest elements in the sequence $\{\|f_{j,k,e}\psi^e_{j,k}\|_p : (j,k,e) \in \Lambda_\Omega\}$ and denote the set of their indices by Λ^n_Ω. The resulting approximation of f is given by

$$G_n(f) = \sum_{(j,k,e)\in\Lambda^n_\Omega} f_{j,k,e}\psi^e_{j,k}.$$

If $p = 2$, then clearly $G_n(f)$ is the best n-term approximation of f with respect to the dictionary $\mathcal{D}^h$. The following theorem gives an error bound in this case.

Theorem 6 ([9]). *Let $f \in BV(\Omega)$, $\Omega = (0,1)^2$, and $\int_\Omega f(x)\,dx = 0$. Then the approximation $G_n(f)$ produced by Algorithm 5satisfies*

$$\|f - G_n(f)\|_2 \le C\,n^{-1/2}|f|_{BV(\Omega)},$$

where $C = 36(480\sqrt{5} + 168\sqrt{3})$.

It turns out that $G_n(f)$ is also near best for any $1 < p < \infty$.

Theorem 7 ([24]). *Let $f \in L_p(\Omega)$, $\Omega = (0,1)^d$, for some $1 < p < \infty$, and $\int_\Omega f(x)\,dx = 0$. The approximation $G_n(f)$ produced by Algorithm 5satisfies*

$$\|f - G_n(f)\|_p \le C(d,p)\,\sigma_n(f,\mathcal{D}^h)_p.$$

An estimate for $\sigma_n(f,\mathcal{D}^h)_p$ follows from the results of [16] by using the extension theorems for functions in Besov spaces, see [14, Section 7.7].

Theorem 8 ([16]). *Let $1<p<\infty$, $0<\alpha<1/p$, $\tau=\frac{d}{\alpha+d/p}$. If $f\in B^\alpha_{\tau,\tau}(\Omega)$, $\Omega=(0,1)^d$, and $\int_\Omega f(x)\,dx=0$, then*

$$\sigma_n(f,\mathcal{D}^h)_p \le C(d,p)n^{-\alpha/d}|f|_{B^\alpha_{\tau,\tau}(\Omega)}.$$

The best n-term approximation by piecewise constants (and by piecewise polynomials of any degree) on hierarchical partitions of $\mathbb{R}^d$ or $(0,1)^d$ into anisotropic dyadic boxes of the form

$$\Big(\frac{k_1}{2^{j_1d}},\frac{k_1+1}{2^{j_1d}}\Big)\times\cdots\times\Big(\frac{k_d}{2^{j_dd}},\frac{k_d+1}{2^{j_dd}}\Big),\quad j_s,k_s\in\mathbb{Z},$$

has been studied in [23]. Here, the smoothness of the target function is expressed in terms of certain Besov-type spaces defined with respect to a given hierarchical partition. In [21], results of the same type are obtained for even more flexible anisotropic hierarchical triangulations. Let $\mathcal{T}=\cup_{m\in\mathbb{Z}}\Delta_m$, where each Δ_m is a locally finite triangulation of $\mathbb{R}^2$ such that Δ_{m+1} is obtained from Δ_m by splitting each triangle $\omega\in\Delta_m$ into at least two and at most M subtriangles (children). The hierarchical triangulation $\mathcal{T}$ is called *weak locally regular* if there are constants $0<r<\rho<1$ ($r\le\frac14$), such that for any $\omega\in\mathcal{T}$ it holds $r|\omega'|\le|\omega|\le\rho|\omega'|$, where $\omega'\in\mathcal{T}$ is the parent triangle of ω. Clearly, the triangles in $\mathcal{T}$ may have arbitrarily small angles. The *skinny B-space* $B^{\alpha,k}_q(\mathcal{T})$, $0<q<\infty$, $\alpha>0$, $k\in\mathbb{N}$, is the set of all $f\in L^{\mathrm{loc}}_q(\mathbb{R}^2)$ such that

$$|f|_{B^{\alpha,k}_q(\mathcal{T})}:=\Big(\sum_{\omega\in\mathcal{T}}|\omega|^{-\alpha q}w_k(f,\omega)^q_q\Big)^{1/q},$$

where

$$w_k(f,\omega)_q:=\sup_{h\in\mathbb{R}^2}\|\delta^k_h(f)\|_{L_q(\omega)},$$

$\delta^k_h(f)$ being the k-th finite difference of f, in particular

$$\delta^1_h(f,x):=\begin{cases} f(x+h)-f(x), & \text{if } [x,x+h]\subset\omega,\\ 0, & \text{otherwise.}\end{cases}$$

It is shown in [21] that if $\mathcal{T}$ is *regular*, i.e. there is a positive lower bound for the minimum angles of all triangles in $\mathcal{T}$, then $B^{\alpha,k}_q(\mathcal{T})=B^{2\alpha}_{q,q}(\mathbb{R}^2)$ with equivalent norms whenever $0<2\alpha<\min\{1/q,k\}$.

Consider the dictionary $\mathcal{D}^{\mathcal{T}}=\{\chi_\omega:\omega\in\mathcal{T}\}$.

Theorem 9 ([21]). *Let $0<p<\infty$, $\alpha>0$, $\tau=\frac{2}{\alpha+2/p}$. If $f\in B^{\frac{\alpha}{2},1}_\tau(\mathcal{T})\cap L_p(\mathbb{R}^2)$, then*

$$\sigma_n(f,\mathcal{D}^{\mathcal{T}})_p\le C(p,\alpha,\rho,r)\,n^{-\alpha/2}|f|_{B^{\frac{\alpha}{2},1}_\tau(\mathcal{T})}.$$

Note that certain Haar type bases can be introduced on the anisotropic dyadic partitions and on hierarchical triangulations obtained by a special refinement rule, see [21, 23] for their definition and approximation properties. An extension of Theorem 9 to $\mathbb{R}^d$ with $d > 2$ is given in [13].

4 Anisotropic Partitions

We have seen in Theorem 2 that piecewise constants on isotropic partitions cannot approximate nontrivial smooth functions with order better than $N^{-1/d}$. We now turn to the question what approximation order can be achieved on anisotropic partitions. An argument similar to that in the proof of Theorem 2 shows that it is not better than $N^{-2/(d+1)}$ if we assume that the partition is convex, i.e. all its cells are convex sets.

Theorem 10 ([10]). *Assume that $f \in C^2(\Omega)$ and the Hessian of f is positive definite at a point $\hat{x} \in \Omega$. Then there is a constant C depending only on f and d such that for any convex partition Δ of Ω,*

$$E(f, \Delta)_\infty \geq C|\Delta|^{-2/(d+1)}.$$

The order of piecewise constant approximation on anisotropic partitions in two dimensions has been investigated in [19]. It is shown that for any $f \in C^2([0,1]^2)$ there is a sequence of partitions Δ_N of $(0,1)^2$ into polygons with the cell boundaries consisting of totally $\mathcal{O}(N)$ straight line segments, such that $\|f - s_{\Delta_N}(f)\|_\infty = \mathcal{O}(N^{-2/3})$. Moreover, the approximation order $N^{-2/3}$ cannot be improved on such partitions. Note that by triangulating each polygonal cell of Δ_N one obtains a convex partition with $\mathcal{O}(N)$ triangular cells, so that Theorem 10 also applies, giving the same saturation order $N^{-2/3}$. Another result of [19] is that for any $f \in C^3([0,1]^2)$ there is a sequence of partitions Δ_N of $(0,1)^2$ into cells with piecewise parabolic boundaries defined by a total of $\mathcal{O}(N)$ parabolic segments (pieces of graphs of univariate quadratic polynomials) such that $\|f - s_{\Delta_N}(f)\|_\infty = \mathcal{O}(N^{-3/4})$.

The following algorithm achieves the approximation order $N^{-2/(d+1)}$ on convex polyhedral partitions with totally $\mathcal{O}(N)$ facets.

Algorithm 6 ([10])

Assume $f \in W_1^1(\Omega)$, $\Omega = (0,1)^d$. Split Ω into $N_1 = m^d$ cubes $\omega_1, \ldots, \omega_{N_1}$ of edge length $h = 1/m$. Then split each ω_i into N_2 slices ω_{ij}, $j = 1, \ldots, N_2$, by equidistant hyperplanes orthogonal to the average gradient $g_i := |\omega_i|^{-1} \int_{\omega_i} \nabla f$ on ω_i. Set $\Delta = \{\omega_{ij} : i = 1, \ldots, N_1,\ j = 1, \ldots, N_2\}$, and define the piecewise constant approximation $s_\Delta(f)$ by (2). Clearly, $|\Delta| = N_1 N_2$ and each ω_{ij} is a convex polyhedron with at most $2(d+1)$ facets.

This algorithm is illustrated in Figure 3.

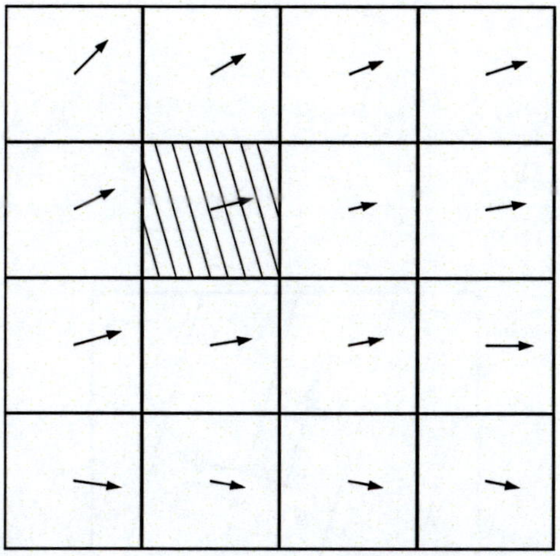

Fig. 3. Algorithm 6($d = 2$, $m = 4$). The arrows stand for the average gradients g_i on the cubes ω_i. The cells ω_{ij} are shown only for one cube.

Theorem 11 ([10]). *Assume that $f \in W_p^2(\Omega)$, $\Omega = (0,1)^d$, for some $1 \leq p \leq \infty$. For any $m = 1, 2, \ldots$, generate the partition Δ_m by using Algorithm 6 with $N_1 = m^d$ and $N_2 = m$. Then*

$$\|f - s_{\Delta_m}(f)\|_p \leq C(d,p) N^{-2/(d+1)} (|f|_{W_p^1(\Omega)} + |f|_{W_p^2(\Omega)}), \tag{11}$$

where $N = |\Delta_m| = m^{d+1}$.

Proof. For simplicity we assume $d = 2$ and $p = \infty$. (The general case is treated in [10].) Let us estimate the error of the best approximation of f by constants on ω_{ij},

$$E(f)_{L_\infty(\omega_{ij})} = \frac{1}{2}\Big(\max_{x \in \omega_{ij}} f(x) - \min_{x \in \omega_{ij}} f(x) \Big).$$

Let σ_i be a unit vector orthogonal to g_i. Since ∇f is continuous, there is $\tilde{x} \in \omega_i$ such that $g_i = \nabla f(\tilde{x})$. Then $D_{\sigma_i} f(\tilde{x}) = 0$ and hence $\|D_{\sigma_i} f\|_{L_\infty(\omega_i)} \leq c_1 h |f|_{W_\infty^2(\omega_i)}$. Given $x, y \in \omega_{ij}$, choose a point $x' \in \omega_{ij}$ such that $y - x'$ and $x' - x$ are collinear with g_i and σ_i, respectively. (This is always possible if we swap the roles of x and y when necessary, see Figure 4.)

Hence, denoting by $\hat{h}$ the distance between the hyperplanes that split ω_i, we obtain

$$\begin{aligned}
|f(y) - f(x)| &\leq |f(y) - f(x')| + |f(x') - f(x)| \\
&\leq \hat{h} \|\nabla f\|_{L_\infty(\omega_{ij})} + c_2 h \|D_\sigma f\|_{L_\infty(\omega_{ij})} \\
&\leq c_3 (\hat{h} |f|_{W_\infty^1(\omega_{ij})} + h^2 |f|_{W_\infty^2(\omega_i)}) \\
&\leq c_4 m^{-2} (|f|_{W_\infty^1(\omega_{ij})} + |f|_{W_\infty^2(\omega_i)}).
\end{aligned}$$

Thus,

$$\|f - f_{\omega_{ij}}\|_{L_\infty(\omega_{ij})} \le 2E(f)_{L_\infty(\omega_{ij})} \le c_4 m^{-2}(|f|_{W^1_\infty(\Omega)} + |f|_{W^2_\infty(\Omega)}),$$

and (11) follows. □

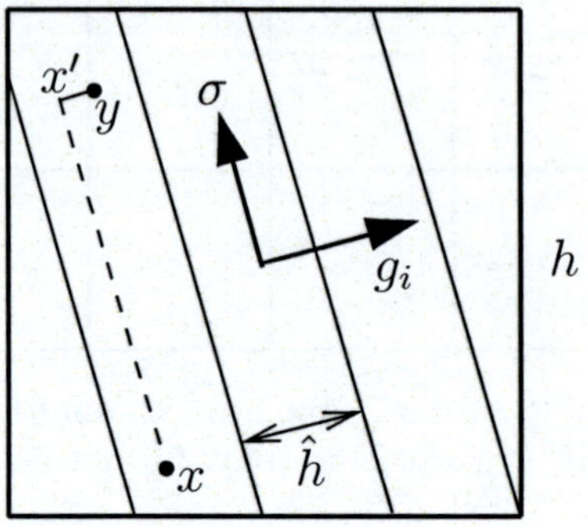

Fig. 4. Illustration of the proof of Theorem 11, showing a single ω_i.

The improvement of the approximation order by piecewise constants from $N^{-1/d}$ on isotropic partitions to $N^{-2/(d+1)}$ on convex partitions does not extend to higher degree piecewise polynomials. Given a partition Δ, let $E_1(f,\Delta)_p$ denote the best error of (discontinuous) piecewise linear approximation in L_p-norm. Then the approximation order on isotropic partitions is $N^{-2/d}$ for sufficiently smooth functions, and it cannot be improved in general on any convex partitions.

Theorem 12 ([10]). *Assume that $f \in C^2(\Omega)$ and the Hessian of f is positive definite at a point $\hat{x} \in \Omega$. Then there is a constant C depending only on f and d such that for any convex partition Δ of Ω,*

$$E_1(f,\Delta)_\infty \ge C|\Delta|^{-2/d}.$$

References

1. G. Acosta and R.G. Durán: An optimal Poincaré inequality in L^1 for convex domains. Proc. Amer. Math. Soc. **132**, 2004, 195–202.
2. R. Arcangeli and J.L. Gout: Sur l'evaluation de l'erreur d'interpolation de Lagrange dans un ouvert de $\mathbb{R}^n$. R.A.I.R.O. Analyse Numerique **10**, 1976, 5–27.
3. M. Bebendorf: A note on the Poincaré inequality for convex domains. J. Anal. Appl. **22**, 2003, 751–756.
4. P. Binev and R. DeVore: Fast computation in adaptive tree approximation. Numer. Math. **97**, 2004, 193–217.

5. M.S. Birman and M.Z. Solomyak: Piecewise polynomial approximation of functions of the classes W_p^α. Mat. Sb. **73**(115), no. 3, 1967, 331–355 (in Russian). English translation in Math. USSR-Sb. **2**, no. 3, 1967, 295–317.
6. S. Brenner and L.R. Scott: *The Mathematical Theory of Finite Element Methods*. Springer-Verlag, Berlin, 1994.
7. S. Buckley and P. Koskela: Sobolev-Poincaré implies John. Math. Res. Lett. **2**, 1995, 577–594.
8. A. Cohen, W. Dahmen, I. Daubechies, and R. DeVore: Tree approximation and optimal encoding. Appl. Comp. Harm. Anal. **11**, 2001, 192–226.
9. A. Cohen, R.A. DeVore, P. Petrushev, and H. Xu: Nonlinear approximation and the space $BV(\mathbb{R}^2)$. Amer. J. Math. **121**, 1999, 587–628.
10. O. Davydov: Approximation by piecewise constants on convex partitions. In preparation.
11. L.T. Dechevski and E. Quak: On the Bramble-Hilbert lemma. Numer. Funct. Anal. Optim. **11**, 1990, 485–495.
12. S. Dekel and D. Leviatan: The Bramble-Hilbert lemma for convex domains. SIAM J. Math. Anal. **35**, 2004, 1203–1212.
13. S. Dekel and D. Leviatan: Whitney estimates for convex domains with applications to multivariate piecewise polynomial approximation. Found. Comput. Math. **4**, 2004, 345–368.
14. R.A. DeVore: Nonlinear approximation. Acta Numerica **7**, 1998, 51–150.
15. R.A. DeVore: Nonlinear approximation and its applications. In *Multiscale, Nonlinear and Adaptive Approximation*, R.A. DeVore and A. Kunoth (eds.), Springer-Verlag, Berlin, 2009, 169–201.
16. R.A. DeVore, B. Jawerth, and V. Popov: Compression of wavelet decompositions. Amer. J. Math. **114**, 1992, 737–785.
17. R.A. DeVore and X.M. Yu: Degree of adaptive approximation. Math. Comp. **55**, 1990, 625–635.
18. P. Hajłasz: Sobolev inequalities, truncation method, and John domains. In *Papers on Analysis: A Volume Dedicated to Olli Martio on the Occasion of his 60th Birthday*. Report. Univ. Jyväskylä **83**, 2001, 109–126.
19. A.S. Kochurov: Approximation by piecewise constant functions on the square. East J. Approx. **1**, 1995, 463–478.
20. Y. Hu, K. Kopotun, and X. Yu: On multivariate adaptive approximation. Constr. Approx. **16**, 2000, 449–474.
21. B. Karaivanov and P. Petrushev: Nonlinear piecewise polynomial approximation beyond Besov spaces. Appl. Comput. Harmon. Anal. **15**(3), 2003, 177–223.
22. L.E. Payne and H.F. Weinberger: An optimal Poincaré inequality for convex domains. Arch. Rational Mech. Anal. **5**, 1960, 286–292.
23. P. Petrushev: Multivariate n-term rational and piecewise polynomial approximation. J. Approx. Theory **121**, 2003, 158–197.
24. V. Temlyakov: The best m-term approximation and greedy algorithms. Adv. Comput. Math. **8**, 1998, 249–265.
25. S. Waldron: Minimally supported error representations and approximation by the constants. Numer. Math. **85**, 2000, 469–484.

Anisotropic Triangulation Methods in Adaptive Image Approximation

Laurent Demaret[1] and Armin Iske[2]

[1] HelmholtzZentrum münchen, Neuherberg, Germany,
laurent.demaret@helmholtz-zentrum.de
[2] Department of Mathematics, University of Hamburg, D-20146 Hamburg, Germany, iske@math.uni-hamburg.de

Summary. Anisotropic triangulations are utilized in recent methods for sparse representation and adaptive approximation of image data. This article first addresses selected computational aspects concerning image approximation on triangular meshes, before four recent image approximation algorithms, each relying on anisotropic triangulations, are discussed. The discussion includes generic triangulations obtained by simulated annealing, adaptive thinning on Delaunay triangulations, anisotropic geodesic triangulations, and greedy triangle bisections. Numerical examples are presented for illustration.

1 Introduction

This article surveys recent triangulation methods for adaptive approximations and sparse representations of images. The main purpose is to give an elementary insight into recently developed methods that were particularly designed for the construction of suitable triangulations adapted to the specific features of images, especially to their geometrical contents. In this context, the use of anisotropic triangulations appears to be a very productive paradigm. Their construction, however, leads to many interesting and open questions. To better understand the problems being addressed in current research, selected aspects concerning adaptive approximations on triangulations are discussed. In particular, relevant algorithmic and computational issues are addressed, where special emphasis is placed on the representation of geometrical information contained in images.

Finite element methods (FEM) are among the most popular classical techniques for the numerical solution of partial differential equations, enjoying a large variety of relevant applications in computational science and engineering. FEM are relying on a partitioning of the computational domain, where triangulations are commonly used. In fact, FEM on triangulations provide very flexible and efficient computational methods, which are easy to implement.

E.H. Georgoulis, A. Iske, J. Levesley (eds.), *Approximation Algorithms for Complex Systems*, Springer Proceedings in Mathematics 3, DOI: 10.1007/978-3-642-16876-5_3,
© Springer-Verlag Berlin Heidelberg 2011

Moreover, from a theoretical viewpoint, the convergence and stability properties of FEM on regular meshes are well understood.

Current research concerning different classes of mesh-based approximation methods, including FEM, is focused on the design of suitable *adaptive* meshes to improve their convergence and stability properties over previous methods. Adaptivity is, for instance, particularly relevant for the numerical simulation of multiscale phenomena in time-dependent nonlinear evolution processes. More specific examples are convection-diffusion processes with space-dependent diffusivity, or the simulation of shock front behaviour in hyperbolic problems. In this case, the construction of *well-shaped* adaptive triangulations with *good* anisotropic properties (by the shape and alignment of their locally adapted triangles) is a key issue for the methods' numerical stability (see e.g. [14, 23] for further details).

As regards the central topic of this survey, *adaptive image approximation by anisotropic triangulations*, the basic problem is

- the construction of anisotropic triangular meshes that are *locally adapted* to the geometrical contents of the input image data

in combination with

- the selection of a reconstruction method leading to an *optimal* or *nearby optimal* image approximation scheme.

Note that the above problem formulation is rather general and informal. In fact, to further discuss this, we essentially require a more formal definition of the terms *"locally adaptive"* and *"nearby optimal"*. Although a comprehensive discussion on the approximation theoretical background of this problem is far beyond the aims and scope of this survey, we shall be more specific on the relevant basics later in Section 2.

The image approximation viewpoint taken here is very similar to that of terrain modelling, as investigated in our previous work [7, 12]. In that particular application, greyscale images may be considered as elevation fields, where the terrain's surface is replaced by locally adapted surface patches (e.g. polynomials) over triangulations. But this is only one related application example. In a more general context, we are concerned with the approximation of (certain classes of) bivariate functions by using suitable models for locally adapted approximations that are piecewise defined over anisotropic triangulations.

The rich variety of contemporary applications for mesh-based image approximation methods lead to different requirements for the construction of the utilized (triangular) meshes. This has provided a diversity of approximation methods, which, however, are not yet gathered in a unified theory. Despite of this apparent variety of different methods, we observe that their construction relies on merely a few common principles, two of which are as follows.

One construction principle is based on a very simple idea: sharp edges between objects visible in an image correspond to crucial information. The abstract mathematical modelling leads to measures of regularity which take into account singularities along curves. But triangulations are only one possible method for representing geometrical singularities of images. In fact, different methods were proposed for the rendering of contours in images. For a recent account of these methods, we refer to [13].

Another common principle requires the image representation to be sparse. In our particular situation, this means that the triangulation should be as small as possible. Note that the requirement of sparsity is not necessarily reflected by the number of vertices (or edges or triangles) in the triangulation, but it can also be characterized by some suitable entropy measure related to a compression scheme.

The construction of sparse triangulations for edge-adapted image approximation, according to the above construction principles, leads to an abstract approximation problem, whose general framework is briefly introduced in Section 2. In Section 3, we present four conceptually different approximation methods to solve the abstract approximation problem of Section 2.

This leads to four different image approximation algorithms, each of which achieves to combine the following desirable properties:

- good approximation behaviour;
- edge-preservation;
- efficient (sparse) image representation;
- small computational complexity.

Selected computational aspects concerning the implementation of the presented image approximation algorithms are discussed in Section 3. Supporting graphical illustrations and numerical simulations are presented in Section 3 and in Section 4. A short conclusion and directions for future work are finally provided in Section 5.

2 Image Approximation on Triangulations

2.1 Triangulations and Function Spaces

We first fix some notation and introduce basic definitions concerning triangulations and their associated function spaces.

Let $\Omega \subset \mathbb{R}^2$ denote a compact planar domain with polygonal boundary. Although we consider keeping this introduction more general, most of the following discussion assumes $\Omega = [0,1]^2$ for the (continuous) computational image domain.

Definition 1. *A* `continuous image` *is a bounded and measurable function* $f : \Omega \to [0, \infty)$*, so that* f *lies in the* L^∞*-class of measurable functions, i.e.,* $f \in L^\infty(\Omega)$*.*

Although images are bounded (i.e., are lying in $L^\infty(\Omega) \subset L^p(\Omega)$), we distinguish between the different function spaces $L^p(\Omega)$, corresponding to different norms $\|\cdot\|_{L^p(\Omega)}$ for measuring the reconstruction error on Ω. We further remark that a natural image can always be represented by a bounded function. However, the converse is (trivially) not true: for any fixed $p \in [1, \infty]$, functions in $L^p(\Omega)$ do often *not* correspond to natural images. One of the main tasks of functional analysis methods in image processing is to define function classes, being given by some suitable regularity conditions, which are as small as possible but contain relevant images. In the context of triangulation methods, this immediately leads us to one central question: which image classes may be well-recovered by approximation methods relying on triangular meshes? Later in this section, we shall briefly discuss this question and give pointers to the relevant literature.

But let us first define triangulations and their associated function spaces.

Definition 2. *A* `triangulation` $\mathcal{T}$ *of the domain* Ω *is a finite set* $\{T\}_{T\in\mathcal{T}}$ *of closed triangles* $T \in \mathbb{R}^2$ *satisfying the following conditions.*

(a) The union of the triangles in $\mathcal{T}$ *covers the domain* Ω*, i.e.,*

$$\Omega = \bigcup_{T\in\mathcal{T}} T,$$

(b) for any pair $T, T' \in \mathcal{T}$ *of two distinct triangles,* $T \neq T'$*, the intersection of their interior is empty, i.e.,*

$$\mathring{T} \cap \mathring{T}' = \emptyset \qquad \text{for } T \neq T'.$$

We denote the set of triangulations of Ω *by* $\mathscr{T}(\Omega)$*.*

Note that condition (b) in the above definition disallows overlaps between different triangles in $\mathcal{T}$. Nevertheless, according to Definition 2, a triangulation $\mathcal{T}$ may contain *hanging vertices*, where a hanging vertex is a vertex of a triangle in $\mathcal{T}$ which is lying on the interior of an adjacent triangle's edge. Triangulations without hanging nodes are *conform*, which leads us to the following definition.

Definition 3. *A triangulation* $\mathcal{T}$ *of* Ω *is a* `conforming triangulation` *of* Ω*, if any pair of two distinct triangles in* $\mathcal{T}$ *intersect at most at one common vertex or along one common edge. We denote the set of conforming triangulations of* Ω *by* $\mathscr{T}_c(\Omega)$*.*

For the purpose of illustration, Figure 1 shows one non-conforming triangulation and one conforming triangulation.

On given planar domain Ω, there are many different (conforming) triangulations of Ω. We remark that many relevant applications rely on triangulations, where long and thin triangles need to be avoided, e.g. in finite element methods for the sake of numerical stability. In this case, *Delaunay triangulations* are a popular choice.

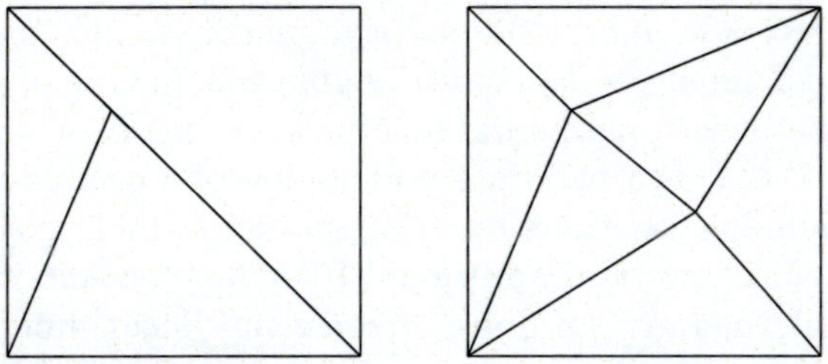

Fig. 1. **(a)** a non-conforming triangulation with three triangles and five vertices, where one vertex is a hanging vertex; **(b)** a conforming triangulation with six triangles and six vertices.

Definition 4. *A* `Delaunay triangulation` $\mathcal{D}$ *of* Ω *is a conforming triangulation of* Ω*, such that for any triangle in* $\mathcal{D}$ *its circumcircle does not contain any vertex from* $\mathcal{D}$ *in its interior.*

We remark that the Delaunay triangulation $\mathcal{D}$ of Ω with vertices $X \subset \Omega$ maximizes the minimal angle among all possible triangulations of Ω with vertices X. In this sense, Delaunay triangulations are *optimal* triangulations. Moreover, the Delaunay triangulation $\mathcal{D} \equiv \mathcal{D}(X)$ is unique among all triangulations with vertices X, provided that no four points in X are co-circular [20]. Figure 2 shows one Delaunay triangulation for the square domain $\Omega = [0,1]^2$ with $|X| = 30$ vertices.

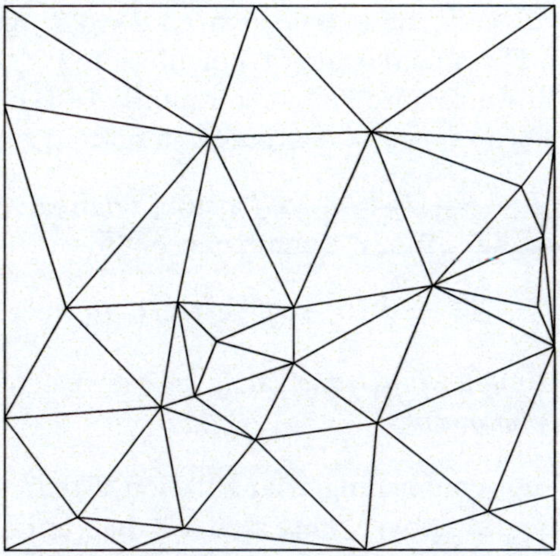

Fig. 2. Delaunay triangulation of a square domain with 30 vertices.

Now let us turn to functions on triangulations, i.e., bivariate functions $f : \Omega \to \mathbb{R}$ to be defined over a fixed triangulation $\mathcal{T}$. One possibility for doing so is by using piecewise polynomial functions. In this case, the restriction $f\big|_T$ to any triangle $T \in \mathcal{T}$ is a bivariate polynomial of a certain degree. To define piecewise polynomial functions on triangulations, their maximum degree is usually fixed beforehand and additional boundary or smoothness conditions are utilized. This then gives a finite dimensional linear function space.

In the situation of image approximation, we prefer to work with piecewise *linear* polynomial functions f. This is due to the simple representation of f. Moreover, we do not require any global smoothness conditions for f apart from global continuity. Since natural images are typically discontinuous, it also makes sense to refrain from assuming global continuity for f. This leads us to the following definition of two suitable function spaces of piecewise linear polynomials, one with requiring global continuity, the other without requiring global continuity. In this definition, $\mathcal{P}_1$ denotes the linear space of bivariate linear polynomials.

Definition 5. *Let $\mathcal{T} \in \mathscr{T}(\Omega)$ be a triangulation of Ω. The set of* `piecewise linear functions` *on $\mathcal{T}$,*

$$\mathcal{S}_{\mathcal{T}} := \left\{ f : \Omega \to \mathbb{R} : f\big|_{\mathring{T}} \in \mathcal{P}_1 \right\}$$

is given by all functions whose restriction to the interior $\mathring{T}$ of any triangle $T \in \mathcal{T}$ is a linear polynomial.

Note that in the above definition, the restriction $f\big|_{\mathring{T}}$ of f may, for any individual triangle $T \in \mathcal{T}$, be extended from $\mathring{T}$ to T, in which case f may not be well-defined. For a conforming triangulation $\mathcal{T}(\Omega)$, however, f will be well-defined on Ω, if we require global continuity. In the following definition, $\mathscr{C}(\Omega)$ denotes the linear space of continuous functions on Ω.

Definition 6. *Let $\mathcal{T} \in \mathscr{T}_c(\Omega)$ be a conforming triangulation of Ω. The set of* `continuous piecewise linear functions` *on $\mathcal{T}$,*

$$\mathcal{S}^0_{\mathcal{T}} := \left\{ f \in \mathscr{C}(\Omega) : f\big|_T \in \mathcal{P}_1 \right\},$$

is given by all continuous functions on Ω whose restriction to any triangle $T \in \mathcal{T}$ is a linear polynomial.

Note that (for any conforming triangulation $\mathcal{T}$) $\mathcal{S}^0_{\mathcal{T}}$ is a linear subspace of $\mathcal{S}_{\mathcal{T}}$, i.e., $\mathcal{S}^0_{\mathcal{T}} \subset \mathcal{S}_{\mathcal{T}}$. Moreover, note that $\mathcal{S}^0_{\mathcal{T}}$ has finite dimension. Indeed, the (finite) set of *Courant elements* $\varphi_v \in \mathcal{S}^0_{\mathcal{T}}$, for v in $\mathcal{T}$, each of which being uniquely defined by

$$\varphi_v(x) = \begin{cases} 1 & \text{for } x = v; \\ 0 & \text{for } x \neq v; \end{cases} \qquad \text{for any vertex } x \text{ in } \mathcal{T},$$

is a basis of $\mathcal{S}_\mathcal{T}^0$. Therefore, the dimension of $\mathcal{S}_\mathcal{T}^0$ is equal to the number of vertices in $\mathcal{T}$. Similarly, the linear function space $\mathcal{S}_\mathcal{T}$ has dimension $3|\mathcal{T}|$, where $|\mathcal{T}|$ denotes the number of triangles in $\mathcal{T}$.

We remark that one important differences between the two approximation spaces, $\mathcal{S}_\mathcal{T}^0$ and $\mathcal{S}_\mathcal{T}$, is that $\mathcal{S}_\mathcal{T}^0$ requires conforming triangulations, whereas for $\mathcal{S}_\mathcal{T}$ the triangulation $\mathcal{T}$ may contain hanging vertices. Note that this leads to different approximation schemes, as illustrated in Figure 3.

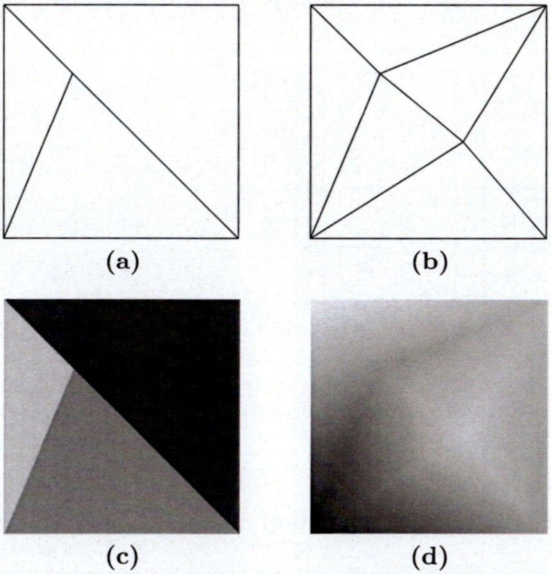

Fig. 3. **(a)** a non-conforming triangulation $\mathcal{T}$; **(b)** a conforming triangulation $\mathcal{T}'$ (cf. Figure 1); **(c)** a piecewise constant function $f \in \mathcal{S}_\mathcal{T}$; **(d)** a continuous piecewise linear function $g \in \mathcal{S}_{\mathcal{T}'}^0$.

2.2 Isotropic and Anisotropic Approximation Methods

When approximating an image f by (best approximating) functions from $\mathcal{S}_\mathcal{T}^0$ or $\mathcal{S}_\mathcal{T}$, the resulting approximation quality heavily depends on the *quality* of the triangulation $\mathcal{T}$, in particular on the shape of the triangles in $\mathcal{T}$. Several alternative quality measures for triangulations are proposed in [23].

In this subsection, we discuss relevant principles concerning the construction of well-adapted triangulations, leading to adaptive image approximation methods. In this construction, it is essential to take (possible) anisotropic features of the image data into account. Second derivatives of the image (if they exist) are not necessarily of comparable amplitude (i.e., magnitude) in all directions. Furthermore, the ratio between the maximal and the minimal amplitudes may vary quite significantly. Due to typical such heterogeneity (and other related multiscale phenomena) in image data, this gives rise to prefer *anisotropic* triangulations.

To be more specific, image approximation methods by triangulations are split into non-adaptive methods (relying on uniform triangulations) and adaptive methods (relying on non-uniform triangulations). Another distinction between image approximation methods (and their underlying triangulations) is by *isotropic* and *anisotropic* methods. For illustration, Figure 4 shows examples for uniform vs non-uniform and isotropic vs anisotropic triangulations.

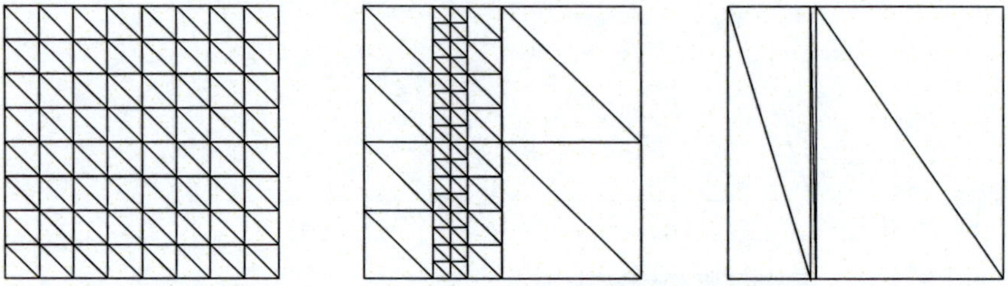

(a) **(b)** **(c)**

Fig. 4. **(a)** a uniform triangulation; **(b)** a non-uniform isotropic (non-conforming) triangulation; **(c)** an anisotropic (conforming) triangulation.

As regards characteristic features of different triangulation types, Figure 4,

(a) triangles in a *uniform triangulation* have comparable sizes and shapes, so that the shape of each triangle is similar to that of an equilateral triangle.
(b) *non-uniform isotropic* triangulations comprise triangles of varying sizes, but the triangles have similar shapes.
(c) *anisotropic* triangulations comprise triangles of varying sizes and shapes.

We remark that non-adaptive image approximations (relying on uniform triangulations) are, in terms of their approximation quality, inferior to adaptive methods. In adaptive image approximation methods, however, the construction of their underlying non-uniform triangulations requires particular care. While non-uniform isotropic triangulations are suitable in situations, where the target image has point singularities at isolated points, anisotropic triangulations are particularly useful to locally adapt singularities of the image along curves, or other features of the image that may be reflected by a heterogeneous distribution of directional derivatives of first and second order. Therefore, anisotropic triangulations are usually preferred, especially when it comes to design suitable triangulations for adaptive image approximation methods.

2.3 Techniques for Proving Approximation Rates

Let us finally discuss available techniques for proving error estimates for image approximation. The following discussion includes classical results from finite elements as well as more recent results concerning Besov spaces and the related smoothness spaces, relying on the Mumford-Shah model. We keep the discussion of this section rather short, but we give pointers to the relevant literature. Further details are deferred to a follow-up paper.

The Bramble-Hilbert Lemma. Let us first recall classical error estimates from finite element methods. To obtain estimates for the global approximation error of a function by a piecewise function on a triangulation $\mathcal{T}$, standard analysis on finite element methods provides estimates for a single triangle $T \in \mathcal{T}$ (see [2]), where the key estimate is derived from the Bramble-Hilbert lemma [3]. The basic error estimate of the Bramble-Hilbert lemma leads to

$$\|f - \Pi_{\mathcal{S}^0_{\mathcal{T}_n}} f\|_{L^2(\Omega)} \leq \frac{1}{n}|f|_{W^{2,2}(\Omega)} \text{ for } f \in W^{2,2}(\Omega),$$

where $\Pi_{\mathcal{S}^0_{\mathcal{T}_n}} f$ is the orthogonal L^2-projection of f onto $\mathcal{S}^0_{\mathcal{T}_n}$, and where $\mathcal{T}_n$ is a uniform triangulation of Ω with n vertices.

Slim and Skinny Besov Spaces. Although classical isotropic Besov spaces offer a more suitable framework for adaptive approximation schemes, they usually fail to represent approximation classes relying on anisotropic triangulations. Just recently, a more flexible concept was proposed to remedy this problem.

In [15], Karavainov and Petrushev introduced two different classes of anisotropic Besov spaces, *slim* and *skinny* Besov spaces. The construction of such Besov spaces relies on subdivision schemes leading to a family of nested triangulations. The approach taken in [15] is rather technical, but their main results can loosely be explained as follows.

The set of functions belonging to a slim Besov space are the functions which can be approximated at a given convergence rate by piecewise linear and globally continuous functions on a specific subdivision scheme. Skinny Besov spaces are obtained by using similar construction principles for the special case of piecewise linear (not necessarily continuous) approximations. The quality of the resulting image approximation, however, heavily relies on the properties of the utilized subdivision scheme. The construction of suitable subdivision schemes remains a rather critical and challenging task. For further details, we refer to [15].

Bivariate Smoothness Spaces. Cartoon models lead to a large family of approximation methods which are based on the celebrated Mumford-Shah model [19]. These methods essentially take sharp edges of images into account, and their basic idea is to regard images as piecewise regular functions being separated by piecewise smooth curves. A generalization of the Mumford-Shah model has been proposed by Dekel, Leviatan & Sharir in [6]. In their work [6], smoothness spaces are defined by using a combination of two distinct notions of smoothness, one for the inner pieces (away from the edge singularities), the other for the singularity supporting curves.

The corresponding smoothness spaces, $\mathcal{B}$-spaces, are in [6] defined in a similar way as by the standard interpolation techniques used for classical isotropic Besov spaces. In the present setting, we are interested in the use of smoothness spaces for the characterization of the approximation spaces

$$\mathcal{A}^\alpha := \left\{ f \in L^2(\Omega) : \inf_{|T|=n, T \in \mathcal{T}(\Omega)} \|f - \Pi_{S_T} f\|_{L^2(\Omega)} \leq \frac{C}{n^\alpha} \text{ for some } C > 0 \right\}.$$

Further in this context, we consider the smoothness space $\mathcal{B}_q^{\alpha, r_1, r_2}(L^p(\Omega))$ in [6, Definition 1.3] with $p = 2$, $r_1 = r_2 = 2$ and $q = \infty$. This combination of parameters is in [6] used for the special case of piecewise affine functions ($r_1 = 1$) on triangles ($r_2 = 2$) by measuring the error in $L^2(\Omega)$ ($p = 2$), and the approximation rates in the L^∞-norm ($q = \infty$).

Useful error estimates for $\mathcal{B}$-spaces are proven in [6], where also a suitable characterization for the relevant approximation classes $\mathcal{A}^\alpha \equiv \mathcal{A}^\alpha(\Omega)$ is given. In fact, the main result in [6, Theorem 1.9] states that the inclusion $\mathcal{A}^\alpha \subset \mathcal{B}^\alpha$ holds for any planar domain $\Omega \subset \mathbb{R}^2$, and, conversely, we have the inclusion

$$\mathcal{B}^\alpha(\Omega) \cap L^\infty(\Omega) \subset \mathcal{A}^\alpha(\Omega),$$

which provides an almost sharp characterization of the approximation class $\mathcal{A}^\alpha$. For further details, we refer to [6].

3 Four Algorithms for Adaptive Image Approximation

In this section we discuss four different algorithmic concepts for adaptive image approximation, which were proposed during the last five years. Each of the resulting approximation algorithms, to be discussed in Subsections 3.1-3.4, aims at the construction of suitable anisotropic triangulations to obtain sufficiently accurate image approximations. Moreover, their utilized adaptation strategies achieve a well-balanced trade-off between essential requirements concerning computational costs, approximation properties, and information redundancy. Computational details and key features of the five different concepts are explained in the following Subsections 3.1-3.4.

3.1 Generic Triangulations and Simulated Annealing

A naive approach to solve the image approximation problem of the previous Section 2 would consist in finding an *optimal* triangulation (among all possible triangulations of equal size) to obtain a *best* image approximation. However, the problem of finding such an optimal triangulation is clearly intractable. In fact, the set of all possible triangulations is huge, and so it would be far too costly to traverse all triangulations to locate an optimal one.

An alternative way for (approximately) solving this basic approximation problem is to traverse a smaller set of generic triangulations, according to a suitable set of traversing rules, to compute only a *suboptimal* triangulation but at much smaller computational complexity.

Recently, Lehner, Umlauf & Hamann [18] have introduced such a method for traversing a set of generic triangulations, where their basic algorithm relies on simulated annealing. The triangulations output by the method in [18] are very sparse, i.e., for a target approximation error, the output triangulation (whose resulting approximation error is below the given tolerance), requires only a small number of vertices (cf. the numerical comparison in Figure 5).

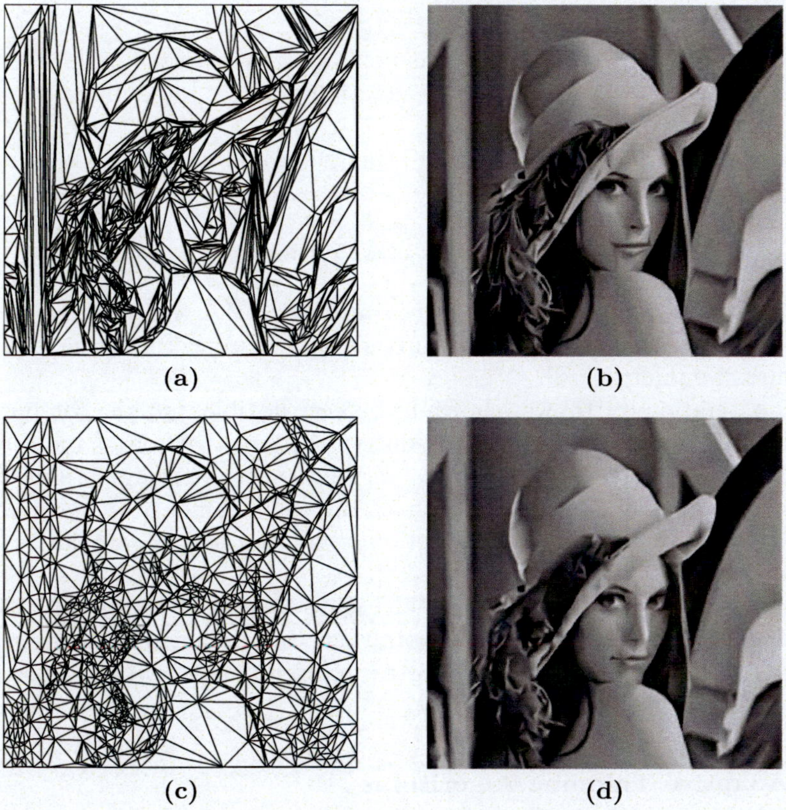

Fig. 5. **(a)** Triangulation and **(b)** image reconstruction obtained by simulated annealing [18]; **(c)** triangulation and **(d)** image reconstruction obtained by adaptive thinning [8] (Section 3.2). In either test example, the triangulation has 979 vertices.

The generic algorithm of [18] is iterative and can briefly be explained as follows. On given initial triangulation $\mathcal{T}_0$, local modifications are performed, which yields a sequence $\{\mathcal{T}_n\}_n$ of triangulations. Any local modification on a current triangulation $\mathcal{T}_k$ is accomplished by using one of the following three basic operations (edge flip, local and global vertex move), yielding the subsequent triangulation $\mathcal{T}_{k+1}$.

- For two triangles sharing a common edge in a convex quadrilateral, an *edge flip* replaces the diagonal of the quadrilateral by the opposite diagonal;
- *local vertex move:* a vertex is moved to a new position in its neighbourhood;
- *global vertex move:* a vertex is moved and its cell is retriangulated.

To each of the three elementary operations, corresponding probabilities p_e, p_ℓ, p_g, satisfying

$$p_e + p_\ell + p_g = 1,$$

are assigned, according to which the next operation (edge flip, local or global vertex move) is performed. But the selection of edges to be flipped or vertices to be moved is by random.

To avoid local optima, the next triangulation is either taken or rejected, according another probability,

$$p_n = p(\Delta E_n, n),$$

where ΔE_n is the difference between approximation errors induced by the triangulations $\mathcal{T}_n$ and $\mathcal{T}_{n+1}$. For further details concerning the probability measures we refer to [18].

The iteration of [18] was shown to be very flexible, but the computational efficiency is rather critical. This is due to the construction of the greedy algorithm, which navigates through a very large set of triangulations. Further improvements may be used to speed-up the convergence of the simulated annealing procedure, where the reduction of computational complexity is done by working with *local* probabilities [18].

Finally, the numerical example in Figure 5 (a)-(b) shows the efficacy of the simulated annealing method. Note that, especially from the viewpoint of anisotropy, the triangles in Figure 5 (a) are well-aligned with the sharp edges in the test image.

3.2 Adaptive Thinning Algorithms

Adaptive thinning algorithms [12] are a class of greedy point removal schemes for bivariate scattered data. In our recent work [8, 9, 10], adaptive thinning algorithms were applied to image data, and more recently, also to video data [11] to obtain an adaptive approximation scheme for images and videos. The resulting compression methods were shown to be competitive with JPEG2000 (for images) and MPEG4-H264 (for videos).

To explain the basic ideas of adaptive thinning for image data, let X denote the set of image pixels, whose corresponding pixel positions are lying on a rectangular two-dimensional grid. This, in combination with the pixels' luminance values defines a bivariate discrete function $f : X \to \mathbb{R}$. Now the aim of the adaptive thinning algorithm is to select a small subset $Y \subset X$ of *significant pixels*, whose corresponding Delaunay triangulation $\mathcal{D} \equiv \mathcal{D}(Y)$ gives a suitable finite-dimensional ansatz space $\mathcal{S}_{\mathcal{D}}^0$ of globally continuous piecewise linear functions.

The image approximation $s : [X] \to \mathbb{R}$ to f is then given by the *best approximation* $s^* \in \mathcal{S}_{\mathcal{D}}^0$ in the least squares sense, i.e. s^* minimizes the ℓ^2-error

$$\|s - f\|_2^2 := \sum_{x \in X} |s(x) - f(x)|^2$$

among all functions $s \in \mathcal{S}_{\mathcal{D}}^0$. Note that s^* is unique and can be computed efficiently by standard least squares approximation.

The challenge of this particular approximation method is to determine a *good* adaptive spline space $s \in \mathcal{S}_{\mathcal{D}}^0$, by the selection of a suitable subset $Y \subset X$, such that the resulting least squares error

$$\eta \equiv \eta(Y) = \|s^* - f\|_2^2$$

is as small as possible. Ideally, one wishes to compute an optimal $Y^* \subset X$ which minimizes $\eta(Y)$ among all subsets $Y \subset X$ of equal size. But the problem of computing Y^* is NP-complete. This requires greedy approximation algorithms for the selection of suitable (sub-optimal) solutions $Y \subset X$.

In greedy adaptive thinning algorithms, a subset $Y \subset X$ of significant pixels is computed by the recursive removal of pixel points, one after the other. The generic formulation of the adaptive thinning algorithm is as follows.

Algorithm (Adaptive Thinning)

INPUT: set of pixel positions X and luminance values $\{f(x)\}_{x \in X}$.

(1) Let $X_N = X$;
(2) **FOR** $k = 1, \ldots, N - n$
  (2a) Find a `least significant` pixel $x \in X_{N-k+1}$;
  (2b) Let $X_{N-k} = X_{N-k+1} \setminus x$;

OUTPUT: subset $Y = X_n \subset X$ of significant pixels.

To implement an adaptive thinning algorithm, it remains to give, for any $Y \subset X$, a definition for *least significant* pixels in Y. To this end, several different significance measures were proposed in [7, 8, 9, 10, 12]. Each of the utilized significance measures are relying on (an estimate) of the *anticipated error* that is incurred by the removal of the pixel point. The anticipated error for a pixel y is a local measure $\sigma(y)$ for the incurred ℓ^2-error due to its removal. In the greedy implementation of adaptive thinning, a pixel y^* is least significant (in any step of the algorithm), whenever its anticipated error $\sigma(y^*)$ is smallest among all points in the current subset $Y \subset X$. Since $\sigma(y)$ can be computed and updated in constant time, this allows for an efficient implementation of adaptive thinning in only $\mathcal{O}(N \ \log(N))$ operations, where $N = |X|$.

3.3 Anisotropic Geodesic Triangulations

The anisotropic meshing problem, as pointed out in [1], can be interpreted as the search for a criterion based on a locally modified metric, according to which triangulations are then constructed. To connect this viewpoint with the concept of anisotropic triangulations, the Euclidean metric corresponds to a uniform triangulation, whereas metrics whose unit balls are disks of varying sizes lead to isotropic adaptive triangulations. Finally, anisotropic triangulations can be generated by using metrics whose unit balls are ellipses of varying sizes, shapes and directions. A suitable triangulation algorithm is then required to produce triangulations which are *aligned* with this modified metric, i.e., each triangle should be included in such an ellipse. Moreover, direction and ratio between the major and minor radii are directed by the local structure of data, while the size of the ellipse (i.e., the radius of the ball in the modified metric) is a parameter depending on the global target reconstruction quality.

The local metric is commonly defined by a positive definite tensor field, i.e., a mapping which associates to each point $x \in \Omega$ a symmetric positive definite tensor matrix $H(x) \in \mathbb{R}^{2\times 2}$. For any point $x_0 \in \Omega$, the local metric $H(x_0)$ is then defined by the distance between a point $x \in \Omega$ and x_0,

$$\|x - x_0\|_{H(x_0)} = \sqrt{(x - x_0)^t H(x_0)(x - x_0)}.$$

Note that this concept enables us to define the length of a piecewise smooth curve $\gamma : [0,1] \to \Omega$ w.r.t. metric H by

$$L_H(\gamma) = \int_0^1 \|\gamma'(t)\|_{H(\gamma(t))} dt$$

and so the *geodesic distance* between two points $x, y \in \Omega$ by

$$d_H(x,y) = \min_{\gamma} L_H(\gamma),$$

where the minimum is taken over all piecewise smooth curves joining x and y.

A quite natural choice for H seems to be the Hessian. One construction of anisotropic triangulations for image approximation, based on an anisotropic geodesic metric, has recently been proposed in [1]. Instead of taking the Hessian matrix as tensor structure matrix, a regularized version of the gradient tensor is used in [1]. Regularization is then performed by the convolution with a Gaussian kernel; the role of this regularization is to ensure a robust estimation in the presence of noise. Let us denote this regularized gradient tensor by $T(x)$ and assume that $T(x)$ can be diagonalized in an orthonormal basis, i.e., for a suitable basis we have (with λ_1, λ_2 depending on x):

$$T(x) = \begin{bmatrix} \lambda_1 & 0 \\ 0 & \lambda_2 \end{bmatrix}.$$

The geodesic metric is then defined as follows: the matrix T is slightly perturbed and then set to a power α, which is an ad hoc parameter controlling the anisotropy of the triangulation. This leads to

$$H(x) = \begin{bmatrix} (\lambda_1 + \varepsilon)^\alpha & 0 \\ 0 & (\lambda_2 + \varepsilon)^\alpha \end{bmatrix}.$$

Using this particular definition for a locally modified metric, the following difficult problem needs to be solved: determine a small as possible finite set of vertices $V \subset \Omega$ satisfying

$$\inf_{y \in V} d_H(x, y) \leq \delta \text{ for some } \delta > 0.$$

Here, δ is a global parameter which controls the reconstruction quality. The meshing of this set of points is then based on the use of the anisotropic Voronoi diagram, where the Voronoi cells are defined by the modified metric d_H rather than by the Euclidean one. Since the dual of a so obtained Voronoi diagram is not necessarily a Delaunay triangulation, some effort is necessary to construct a valid triangulation. This can be achieved by greedy insertion algorithms, where at each step the farthest point (w.r.t. the modified metric) to the current set of vertices is added to this set. This strategy is coupled with a suitable triangulation technique. In Figure 6, it is shown how this method works for a smooth image with steep gradients in an area around a regular curve. The result is an anisotropic triangulation. For more details, see [1, 16, 17].

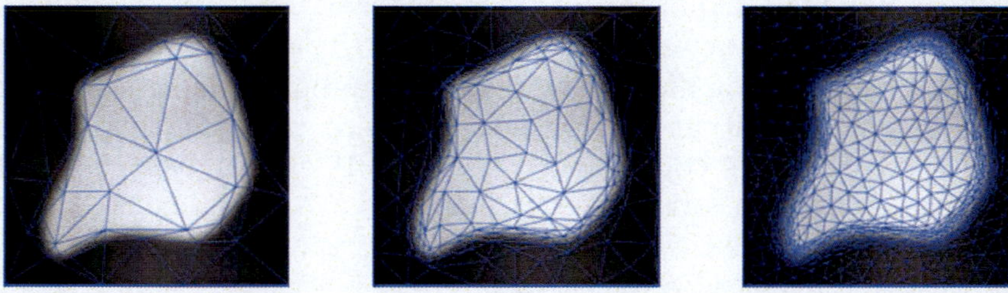

(a) (b) (c)

Fig. 6. Geodesic triangulation with **(a)** 50, **(b)** 200, and **(c)** 800 vertices.

3.4 Greedy Triangle Bisections

In [4], Cohen, Dyn, Hecht & Mirebeau propose a greedy algorithm which is based on a very simple but effective rule for recursive subdivisions of triangles. The main operations in their method are *bisections* of triangles, where a bisection of a triangle T is given by a subdivision of T in two smaller triangles, obtained by the insertion of an edge which connects one vertex in T with the midpoint of the opposite edge. Therefore, for any triangle there are three possible bisections, with the (inserted) edges e_1,e_2 and e_3, say. An example of the two first steps of such a recursive subdivision is shown in Figure 7. Note that this method produces non-conforming triangulations. Therefore, the approximation is performed w.r.t. the reconstruction space $\mathcal{S}_\mathcal{T}$.

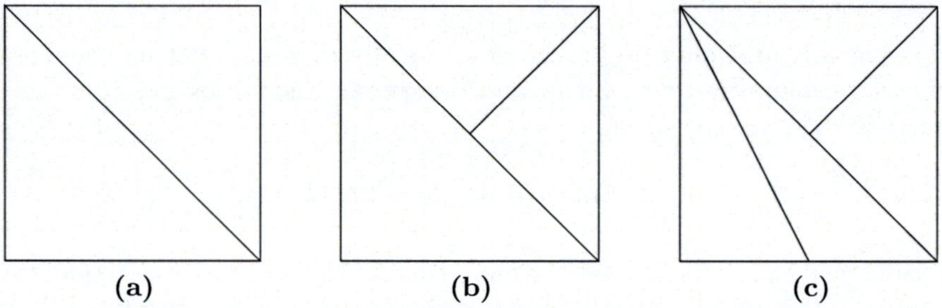

Fig. 7. **(a)** Initial triangulation; **(b)**,**(c)** triangulations by greedy bisection [4].

To derive a refinement algorithm from this bisection rule, a suitable criterion is required for selecting the triangle to be subdivided along with the bisection rule. The criterion proposed in [4] is straight forward: it takes one triangle with maximal approximation error, along with a bisection whose resulting approximation error is minimal. Therefore, an edge e^* corresponding to an optimal bisection of a triangle T is given by

$$e^* = \underset{e \in \{e_1,e_2,e_3\}}{\operatorname{argmin}} \left(\|f - \Pi_{\mathcal{S}_{T_1(e)}}\|^2_{L^2(T_1(e))} + \|f - \Pi_{\mathcal{S}_{T_2(e)}}\|^2_{L^2(T_2(e))} \right),$$

where $T_1(e)$ and $T_2(e)$ are the triangles resulting from the bisection of T by e.

This procedure outputs a sequence of refined triangulations, which we denote by $\mathcal{T}_{b,n}$ (b for bisection). In [5, Theorem 5.1], optimality of the bisection algorithm is proven for strictly convex functions. We can formulate the result in the relevant L^2-setting as follows. For a strictly convex function $f \in \mathscr{C}^2(\Omega)$, there exists a constant $C > 0$ satisfying

$$\|f - \Pi_{\mathcal{S}_{\mathcal{T}_{b,n}}} f\|_{L^2(\Omega)} \le \frac{C}{n} \|\sqrt{\det(D^2 f)}\|_{L^\tau(\Omega)}, \text{ with } \tau = \frac{2}{3},$$

where $\mathcal{T}_{b,n}$ is the sequence of triangulations produced by greedy bisection in combination with an L^1-based selection criterion [5].

4 Numerical Simulations

In this final section, we wish to demonstrate the utility of anisotropic triangulation methods for image approximation. To this end, we apply adaptive thinning of Subsection 3.2 to four different images of size 512×512, as shown in Figure 8: **(a)** one artificial image generated by a piecewise quadratic function, `PQuad`, and three natural images, **(b)** `Aerial`, **(c)** `Game`, and **(d)** `Boats`.

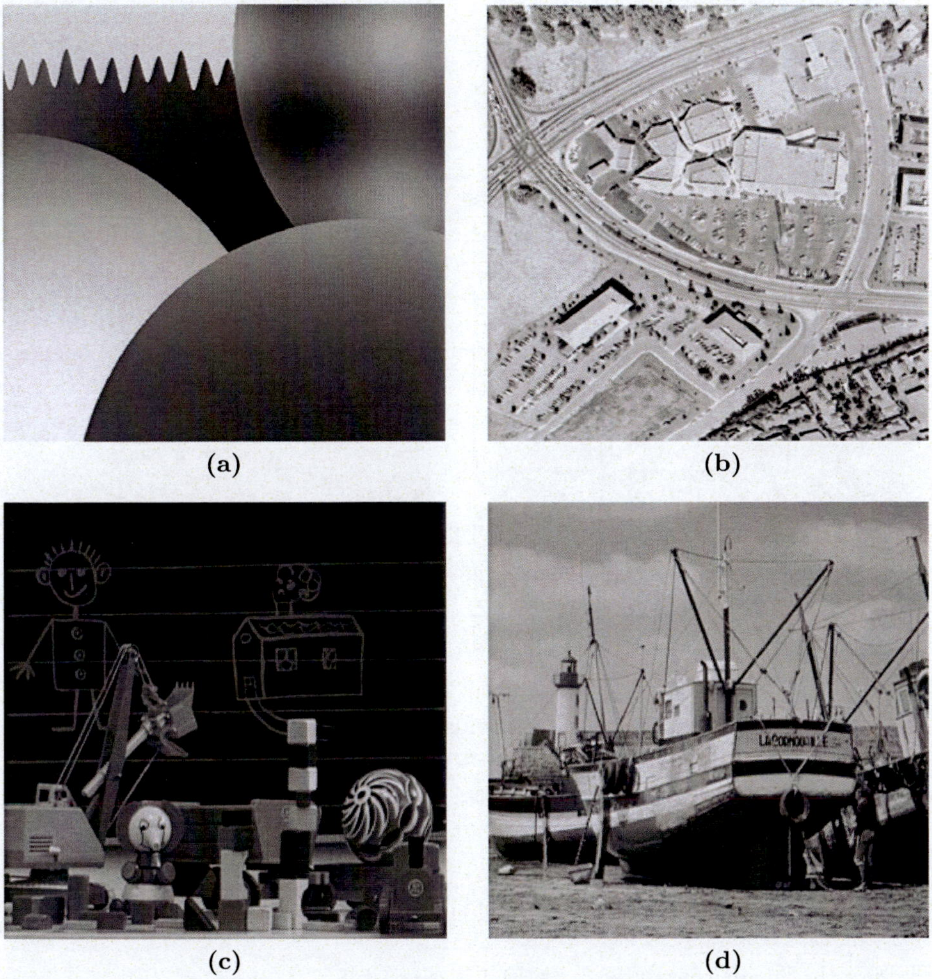

Fig. 8. Four images of size 512×512: **(a)** `PQuad`; **(b)** `Aerial`; **(c)** `Game`; **(d)** `Boats`.

The Delaunay triangulations of the significant pixels, output by adaptive thinning, are shown in Figure 9. The quality of the image approximations, shown in Figure 10, is measured in `dB` (decibel) by the *peak signal to noise*

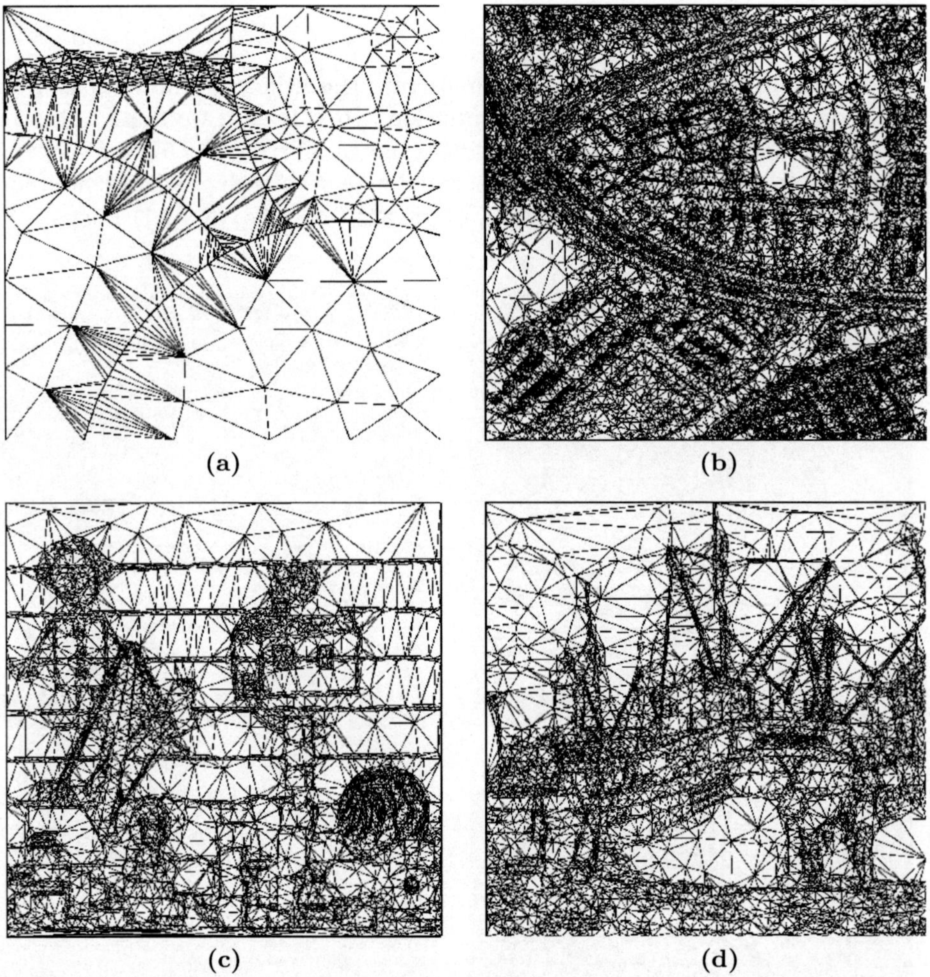

Fig. 9. Anisotropic Delaunay triangulations. **(a)** `PQuad` with 800 vertices **(b)** `Aerial`: 16000 vertices **(c)** `Game`: 6000 vertices **(d)** `Boats`: 7000 vertices.

ratio,

$$\mathrm{PSNR} = 10 * \log_{10}\left(\frac{2^r \times 2^r}{\bar{\eta}^2(Y,X)}\right),$$

where the *mean square error* (MSE) is given by

$$\bar{\eta}^2(Y,X) = \frac{1}{|X|}\sum_{x\in X}|s(x) - f(x)|^2.$$

Note that for each test case the anisotropic triangulations achieve to capture the image geometry fairly well. This results in image approximations

(a)

(b)

(c)

(d)

Fig. 10. Image approximation by adaptive thinning. **(a)** `PQuad` at PSNR `42.85 dB` **(b)** `Aerial`: PSNR `30.33 dB` **(c)** `Game`: PSNR `36.54 dB` **(d)** `Boats`: PSNR `31.83 dB`.

where the key features of the images (e.g. sharp edges) and finer details are recovered very well, at reasonable coding costs, as reflected the small number of significant pixels. In fact, the efficient distribution of sparse significant pixels is one key feature of adaptive thinning, which yields a very competitive compression method [8, 9, 11].

Note that the test case `PQuad` reflects the behaviour of adaptive thinning for an artificial cartoon image: inside on the smooth parts, the triangles are uniform and close to equilateral, whereas long and thin triangles are obtained along the discontinuities. As regards the two test cases `Game` and `Boats`, it is shown how anisotropic triangulations help concentrate the representing triangles in the content-rich areas of these natural images.

Finally, the performance of adaptive thinning for the test case `Aerial` shows the potential of anisotropic triangulations yet once more, but now in a context where much more information are available. The anisotropy of the triangles vary according to the kind of feature they represent, and so the corresponding triangulation helps reproduce the geometrical properties of the underlying features very well, in particular the roads and buildings.

5 Final Remarks and Future Work

One of the practical issues of anisotropic meshing is related to the high computational costs induced by the search for nearby optimal approximating triangulations. In this article, we have considered methods which are too slow for applied fields where real-time computational costs are indispensable. Recently, very fast methods, coming from a slightly different world, closer to the preoccupations of engineering applications, have been developed [21, 22, 24]. These methods rely on some heuristic intended to find an adapted sampling set of pixels together with corresponding meshing and reconstruction techniques. In [22], a quite detailed comparison of these methods in terms of number of triangles versus quality and in terms of computational costs is provided, including comparisons with adaptive thinning. In comparison with the methods discussed in the present survey (in Subsections 3.1-3.4), the number of vertices required in the methods of [21, 22, 24] is much higher for a given target quality, but they allow for very fast implementations. One of the most challenging tasks for future research remains to bridge the gap between these rather pragmatic but highly efficient methods and the mathematically well-motivated but much slower methods of Subsections 3.1-3.4.

Acknowledgement

We gratefully thank Jean-Marie Mirebeau and Gabriel Peyré for several fruitful discussions concerning image approximation. The graphical illustrations in Subsection 3.1 were provided by Burkhard Lehner, those in Subsection 3.3 were provided by Gabriel Peyré.

References

1. S. Bougleux, G. Peyré, and L. Cohen: Image compression with anisotropic geodesic triangulations. Proceedings of ICCV'09, Oct. 2009.
2. D. Braess: *Finite Elements. Theory, Fast Solvers and Applications in Solid Mechanics.* 3rd edition, Cambridge University Press, Cambridge, UK, 2007.
3. J. Bramble and S. Hilbert: Estimation of linear functionals on Sobolev spaces with application to Fourier transforms and spline interpolation. SIAM J. Numer. Anal. **7**(1), 1970.
4. A. Cohen, N. Dyn, F. Hecht, and J.-M. Mirebeau: Adaptive multiresolution analysis based on anisotropic triangulations. Preprint.
5. A. Cohen and J.-M. Mirebeau: Greedy bisection generates optimally adapted triangulations. Preprint.
6. S. Dekel, D. Leviatan, and M. Sharir: On bivariate smoothness spaces associated with nonlinear approximation. Constructive Approximation **20**, 2004, 625–646.
7. L. Demaret, N. Dyn, M.S. Floater, and A. Iske: Adaptive thinning for terrain modelling and image compression. In: *Advances in Multiresolution for Geometric Modelling*, N.A. Dodgson, M.S. Floater, and M.A. Sabin (eds.), Springer-Verlag, Berlin, 2004, 321–340.
8. L. Demaret, N. Dyn, and A. Iske: Image compression by linear splines over adaptive triangulations. Signal Processing **86**(7), July 2006, 1604–1616.
9. L. Demaret and A. Iske: Adaptive image approximation by linear splines over locally optimal Delaunay triangulations. IEEE Signal Processing Letters **13**(5), May 2006, 281–284.
10. L. Demaret and A. Iske: Scattered data coding in digital image compression. In: *Curve and Surface Fitting: Saint-Malo 2002*, A. Cohen, J.-L. Merrien, and L.L. Schumaker (eds.), Nashboro Press, Brentwood, 2003, 107–117.
11. L. Demaret, A. Iske, and W. Khachabi: Sparse representation of video data by adaptive tetrahedralisations. In: *Locally Adaptive Filters in Signal and Image Processing*, L. Florack, R. Duits, G. Jongbloed, M.-C. van Lieshout, and L. Davies (eds.) Springer, Dordrecht, 2010, 197–220.
12. N. Dyn, M.S. Floater, and A. Iske: Adaptive thinning for bivariate scattered data. Journal of Computational and Applied Mathematics **145**(2), 2002, 505–517.
13. H. Führ, L. Demaret, and F. Friedrich: Beyond wavelets: new image representation paradigms. In: *Document and Image Compression*, M. Barni and F. Bartolini (eds.), 2006, 179–206.
14. P.-L. George and H. Borouchaki: *Delaunay Triangulations and Meshing: Application to Finite Elements.* Hermes, Paris, 1998.
15. B. Karavainov and P. Petrushev: Nonlinear piecewise polynomial approximation beyond Besov spaces. Appl. Comput. Harmon. Anal. **15**(3), 2003, 177–223.
16. F. Labelle and J. Shewchuk: Anisotropic Voronoi diagrams and guaranteed-quality anisotropic mesh generation. Proc. 19th Annual Symp. on Computational Geometry, ACM, 2003, 191–200.
17. G. Leibon and D. Letscher: Delaunay triangulations and Voronoi diagrams for Riemannian manifolds. Proc. 16th Annual Symp. on Computational Geometry, ACM, 2009, 341–349.
18. B. Lehner, G. Umlauf, and B. Hamann: Image compression using data-dependent triangulations. In: *Advances in Visual Computing*, G. Bebis et al. (eds.), Springer, LNCS **4841**, 2007, 351–362.

19. D. Mumford and J. Shah: Optimal approximation of piecewise smooth functions and associated variational problems. Comm. in Pure and Appl. Math. **43**, 1989, 577–685.
20. F.P. Preparata and M.I. Shamos: *Computational Geometry*. Springer, New York, 1988.
21. G. Ramponi and S. Carrato: An adaptive sampling algorithm and its application to image coding. Image and Vision Computing **19**(7), 2001, 451–460.
22. M. Sarkis and K. Diepold: Content adaptive mesh representation of images using binary space partitions. IEEE Transactions on Image Processing **18**(5), 2009, 1069–1079.
23. J. Shewchuk: What is a good linear finite element? Interpolation, conditioning, anisotropy and quality measures. Preprint, December 2002.
24. Y. Yang, M. Wernick, and J. Brankov: A fast approach for accurate content-adaptive mesh generation. IEEE Transaction on Image Processing **12**(8), 2003, 866–880.

Form Assessment in Coordinate Metrology

Alistair B. Forbes and Hoang D. Minh

National Physical Laboratory, Teddington TW11 0LW, UK

Summary. A major activity in quality control in manufacturing engineering involves comparing a manufactured part with its design specification. The design specification usually has two components, the first defining the ideal geometry for the part and the second placing limits on how far a manufactured artefact can depart from ideal geometry and still be fit for purpose. The departure from ideal geometry is known as form assessment. Traditionally, the assessment of fitness for purpose was achieved using hard gauges in which the manufactured part was physically compared with the gauge. Increasingly, part assessment is done on the basis of coordinate data gathered by a coordinate measuring machine and involves fitting geometric surfaces to the data. Two fitting criteria are commonly used, least squares and Chebyshev, with the former being far more popular. Often the ideal geometry is specified in terms of standard geometric elements: planes, spheres, cylinders, etc. However, many modern engineering surfaces such as turbine blades, aircraft wings, etc., are so-called 'free form geometries', complex shapes often represented in computer-aided design packages by parametric surfaces such as nonuniform rational B-splines. In this paper, we consider i) computational approaches to form assessment according to least squares and Chebyshev criteria, highlighting issues that arise when free form geometries are involved, and ii) how reference data can be generated to test the performance of form assessment software.

1 Introduction

The National Physical Laboratory is concerned with the accuracy and traceability of measurements. In coordinate metrology, an artefact is represented by the measured coordinates of data points lying on the artefact's surface. The data is then analysed to evaluate how the manufactured artefact relates to its design specification. In particular, form assessment relates to the departure from ideal shape, e.g., how close is a shaft to an ideal cylinder. In terms of traceability of measurements, two aspects are important i) the accuracy of the coordinate measurements and ii) the reliability of the algorithms and software analysing the data. This paper is concerned with the second aspect. Through

E.H. Georgoulis, A. Iske, J. Levesley (eds.), *Approximation Algorithms for Complex Systems*, Springer Proceedings in Mathematics 3, DOI: 10.1007/978-3-642-16876-5_4,
© Springer-Verlag Berlin Heidelberg 2011

various international activities in the 1990s, e.g., [20, 21, 36, 37] a testing methodology for least squares analysis for standard geometric shapes such as planes and cylinders [24] has been developed and applied, allowing software developers to demonstrate the correctness of their algorithms and there are no real concerns about the state of the art for such calculations. However, many modern engineering surfaces such as turbine blades, aircraft wings, etc., are so-called 'free form geometries'. Standards such as ISO 1101 [38] call for assessment methods based on minimising the maximum form error: Chebyshev form assessment. In this paper, we discuss how testing methodologies can be developed to cover these more general and difficult form assessment problems.

In section 2 we overview the problem of the computer-aided inspection of manufactured parts. Section 3 concerns how surface geometries are parametrized in terms of position, size and shape and discusses problems that can arise for free form surfaces. Sections 4 and 5 look at form assessment according to least squares and Chebyshev criteria and the generation of test data. Our concluding remarks are given in section 6.

2 Computer-Aided Inspection of Manufactured Parts

2.1 Specification of Ideal Geometry

The specification of the ideal shape of an artefact are typically defined in terms of i) geometric elements: planes, spheres, cylinders, cones, tori, ii) (more general) surfaces of revolution: aspherics, paraboloids, hyperboloids, and iii) freeform parametric surfaces: parametric splines, nonuniform rational B-splines (NURBS). In general, the geometry can be defined as a mathematical surface $\mathbf{u} \mapsto \mathbf{f}(\mathbf{u}, \mathbf{b})$, where $\mathbf{u} = (u, v)^T$ are the *patch* or *footpoint parameters*, and $\mathbf{b}$ are parameters that define the position in some fixed frame of reference, size and shape of the surface.

2.2 Coordinate Measuring Machines

A standard coordinate measuring machine (CMM) can be thought of as a robot that can move a probe along three nominally orthogonal axes, each of which is associated with a line scale that indicates the distance along the axis. The probe usually has a spherical tip and when the tip contacts an artefact, the position of the probe tip is determined from the scale readings. In this way, the CMM gathers points $\mathbf{x}_i$, $i = 1, \ldots, m$, on (or related to) the surface of the artefact. Form assessment involves matching $\mathbf{x}_i$ to $\mathbf{f}(\mathbf{u}, \mathbf{b})$ to compare the manufactured part, as represented by the coordinate data, to the design.

2.3 Orthogonal Distances

Let $\mathbf{x}$ be a data point reasonably close to the surface $\mathbf{u} \mapsto \mathbf{f}(\mathbf{u}, \mathbf{b})$, and let $\mathbf{u}^*$ solve the *footpoint problem*

$$\min_{\mathbf{u}} \ (\mathbf{x} - \mathbf{f}(\mathbf{u}, \mathbf{b}))^T (\mathbf{x} - \mathbf{f}(\mathbf{u}, \mathbf{b})), \tag{1}$$

so that $\mathbf{u}^* = \mathbf{u}^*(\mathbf{b})$ specifies the point $\mathbf{f}^* = \mathbf{f}(\mathbf{u}^*, \mathbf{b})$ on $\mathbf{f}(\mathbf{u}, \mathbf{b})$ closest to $\mathbf{x}$. The optimality conditions associated with (1) implicitly define $\mathbf{u}^*$ as a function of $\mathbf{b}$ through the equations

$$(\mathbf{x} - \mathbf{f}(\mathbf{u}, \mathbf{b}))^T \mathbf{f}_u = 0, \quad (\mathbf{x} - \mathbf{f}(\mathbf{u}, \mathbf{b}))^T \mathbf{f}_v = 0,$$

where $\mathbf{f}_u = \partial \mathbf{f}/\partial u$ and $\mathbf{f}_v = \partial \mathbf{f}/\partial v$. Let $\mathbf{n} = \mathbf{n}(\mathbf{b})$ be the normal to the surface at $\mathbf{f}^*$, likewise a function of $\mathbf{b}$, and set

$$d(\mathbf{x}, \mathbf{b}) = (\mathbf{x} - \mathbf{f}^*)^T \mathbf{n} = (\mathbf{x} - \mathbf{f}(\mathbf{u}^*(\mathbf{b}), \mathbf{b}))^T \mathbf{n}(\mathbf{b}). \tag{2}$$

Then $d(\mathbf{x}, \mathbf{b})$ is the signed distance of $\mathbf{x}$ from $\mathbf{f}(\mathbf{u}, \mathbf{b})$, where the sign is consistent with the convention for choosing the surface normal. Furthermore,

$$\frac{\partial d}{\partial b_k} = -\left(\frac{\partial \mathbf{f}}{\partial b_k}\right)^T \mathbf{n}, \quad k = 1, \ldots, n, \tag{3}$$

where all terms on the lefthand side are evaluated at $\mathbf{u}^*$. Although $\mathbf{u}^* = \mathbf{u}^*(\mathbf{b})$ and $\mathbf{n} = \mathbf{n}(\mathbf{b})$ are both functions of $\mathbf{b}$, there is no need to calculate their derivatives with respect to $\mathbf{b}$ in order to evaluate the derivatives of the distance function. This is because their contribution is always as linear combinations of vectors orthogonal to $\mathbf{n}$ or $\mathbf{x} - \mathbf{f}$. For standard geometric elements, the distance function $d(\mathbf{x}, \mathbf{b})$ can be defined as an explicit function of the parameters but for free form surfaces, the optimal footpoint parameters $\mathbf{u}^*$ have to be determined using numerical techniques [2, 7, 28, 35, 39].

3 Surface Parametrization

3.1 Separating Position, Size and Shape

It is often (but not always) convenient to separate out parameters that specify position, size and shape. In this approach, we parametrize a surface in standard position, $\mathbf{u} \mapsto \mathbf{f}(\mathbf{u}, \mathbf{s})$ with only the size and shape parameters $\mathbf{s}$ to be adjusted and regard the position parameters $\mathbf{t}$ as applying to the point coordinates. We can parametrize a rigid body transformation $T(\mathbf{x}, \mathbf{t})$ in terms of six parameters $\mathbf{t}^T = (\mathbf{x}_0^T, \boldsymbol{\alpha}^T)$, involving three translation parameters $\mathbf{x}_0 = (x_0, y_0, z_0)^T$ and three rotation angles $\boldsymbol{\alpha} = (\alpha, \beta, \gamma)^T$. One such parametrization is specified by

$$\hat{\mathbf{x}} = T(\mathbf{x}, \mathbf{t}) = R(\boldsymbol{\alpha})(\mathbf{x} - \mathbf{x}_0), \quad R(\boldsymbol{\alpha}) = R_z(\gamma) R_y(\beta) R_x(\alpha) R_0, \tag{4}$$

where R_0 is a fixed rotation matrix and $R_x(\alpha)$, $R_y(\beta)$ and $R_z(\gamma)$ are plane rotations about the x-, y- and z-axes. With this separation, the distance function $d(\mathbf{x}, \mathbf{b})$ is calculated as $d(\hat{\mathbf{x}}, \mathbf{s})$, $\hat{\mathbf{x}} = T(\mathbf{x}, \mathbf{t})$, with

$$d(\mathbf{x},\mathbf{b}) = (\hat{\mathbf{x}} - \mathbf{f}^*)^T \mathbf{n}, \quad \frac{\partial d}{\partial t_k} = \left(\frac{\partial \hat{\mathbf{x}}}{\partial t_k}\right)^T \mathbf{n}, \quad \frac{\partial d}{\partial s_j} = -\left(\frac{\partial \mathbf{f}}{\partial s_j}\right)^T \mathbf{n},$$

where $\mathbf{f}^*$, $\mathbf{n}$ and the derivatives are evaluated at $\mathbf{u}^*$, the solution of the footpoint problem (1) for the data point $\hat{\mathbf{x}}$ and surface $\mathbf{u} \mapsto \mathbf{f}(\mathbf{u},\mathbf{s})$. Evaluated at $\mathbf{t} = \mathbf{0}$ and $R_0 = I$, the derivatives with respect to the transformation parameters are given by

$$\frac{\partial d}{\partial \mathbf{t}^T} = \begin{bmatrix} -\mathbf{n} \\ \mathbf{x}\times\mathbf{n} \end{bmatrix}. \tag{5}$$

It is often possible to further separate size from shape so that $\mathbf{u} \mapsto \mathbf{f}(\mathbf{u},\mathbf{s})$ can be written as $\mathbf{u} \mapsto s\tilde{\mathbf{f}}(\mathbf{u},\tilde{\mathbf{s}})$ where s represents a global scale parameter and $\tilde{\mathbf{s}}$ shape parameters. If $\mathbf{s}$ incorporates a global scale parameter s, then

$$\frac{\partial d}{\partial s} = -\left(\tilde{\mathbf{f}}^*\right)^T \mathbf{n}, \quad \tilde{\mathbf{f}}^* = \tilde{\mathbf{f}}(\mathbf{u}^*,\tilde{\mathbf{s}}). \tag{6}$$

For example, a cylinder in standard position can be parametrized as $(u,v)^T \mapsto s(\cos u, \sin u, v)^T$, involving one global scale parameter s, its radius. Given a data point $\mathbf{x} = (x,y,z)^T$, the point on the cylinder closest to $\mathbf{x}$ is specified by $u^* = \tan^{-1}(y/x)$ and $v^* = z/s$, $\tilde{\mathbf{f}}^* = (\cos u^*, \sin u^*, v^*)^T$ and the corresponding normal is $(\cos u^*, \sin u^*, 0)^T$. For this case, (5) and (6) give

$$\frac{\partial d}{\partial \mathbf{t}^T} = (\cos u^*, \sin u^*, 0, -sv^* \sin u^*, sv^* \cos u^*, 0)^T, \quad \frac{\partial d}{\partial s} = -1.$$

Note that the derivatives corresponding to a translation in the z direction and rotation about the z-axis are zero, as expected from the symmetries associated with a cylinder.

The separation of shape from size and position is generally only a local separation. Consider an ellipse in 2 dimensions. In standard position, that is, aligned with the x- and y-axes, it has a size parameter s_0, the length of the semi-axis aligned with the x-axis, and a shape parameter, s, that jointly specify the length ss_0 of the other semi-axis, $s > 0$. (We allow for the possibility that $s > 1$.) Three rigid body transformation parameters act on the ellipse, two translation parameters x_0 and y_0, and one rotation angle γ. The ellipses specified by $(x_0, y_0, \gamma + \pi/2, s_0, s)^T$ is the same as that specified by $(x_0, y_0, \gamma, s_0 s, 1/s)^T$. Problems occur with this parametrization when $s = 1$, for in this case the rotation angle γ is not defined since the associated ellipse is a circle. Alternatively, an ellipse can be specified by equation

$$b_1 x^2 + b_2 y^2 + b_3 xy + b_4 x + b_5 y + b_6 = 0.$$

Adding a single constraint to the coefficients $\mathbf{b}$ defines a parametrization of the ellipse. For example, setting $b_6 = -1$ parametrizes ellipses that do not pass through the origin. Such a parametrization does not become singular for circles ($b_1 = b_2$, $b_3 = 0$) [23, 30, 50].

3.2 Parametrization of Geometric Elements

Even for the case of standard geometric elements such as spheres and cylinders, element parametrization is not straightforward [3, 26]. Let $\mathcal{E}$ be the space of geometric elements, e.g., cylinders. A parametrization $\mathbf{b} \mapsto \{\mathbf{u} \mapsto \mathbf{f}(\mathbf{u}, \mathbf{b})\}$ is a locally one-to-one and onto mapping $\mathcal{R}^n \longrightarrow \mathcal{E}$. Locally one-to-one means that if $\mathbf{f}(\mathbf{u}, \mathbf{b}_1) \equiv \mathbf{f}(\mathbf{u}, \mathbf{b}_2)$ and $\mathbf{b}_1$ is close enough to $\mathbf{b}_2$ then $\mathbf{b}_1 = \mathbf{b}_2$. Parametrizations are not necessarily globally one-to-one: the cylinder with axis normal $\mathbf{n}$ is the same as that defined by $-\mathbf{n}$. Two elements are equal if $d(\mathbf{x}, \mathbf{b}_1) = d(\mathbf{x}, \mathbf{b}_2)$, $\forall \mathbf{x} \in D$, where D is some open, nonempty domain in $\mathcal{R}^3$. Locally onto means that if E **is sufficiently near** $F = \mathbf{f}(\mathbf{u}, \mathbf{b}_1)$, then $\exists \mathbf{b}_2$ such that $E = \mathbf{f}(\mathbf{u}, \mathbf{b}_2)$.

Parametrizations are not unique. For example, a cone is defined by six parameters. Standard parametrizations of a cone whose axis is approximately aligned with the z-axis are

- Cone vertex (3), direction of the axis (2), e.g., angles of rotation about the x- and y-axes, cone angle, i.e., the angle between the cone generator and its axis,
- Intersection of cone axis with the plane $z = 0$ (2), direction of the axis (2), radii (2) at two distances h_1 and h_2 along the cone axis from the point of intersection with $z = 0$.

These two parametrizations are not equivalent. The first parametrization breaks down when the cone angle approaches zero while the second parametrization copes with any cone angle less than $\pi/2$.

3.3 Topology of the Space of Elements

The condition that a parametrization is locally onto cannot, in general, be strengthened to being globally onto. The reason for this is that the topology of $\mathcal{E}$ need not be flat like $\mathcal{R}^n$. For example, the space $\mathcal{N}$ of cylinder axis direction vectors $\mathbf{n}$ is a sphere in $\mathcal{R}^3$ with $\mathbf{n}$ identified with $-\mathbf{n}$ and has a nontrivial topology. For element spaces with non-flat topologies, any parametrization $\mathcal{R}^n \mapsto \mathcal{E}$ has at least one singularity. This has implications for developers of element fitting algorithms since the algorithms may need to change the parametrization as the optimisation algorithm progresses. Any implementation that uses only one parametrization will (likely) breakdown for data representing an element at (or close to) a singularity for that particular parametrization; parametrizations in general only have local domains of validity.

Usually, it is possible to determine a family of parametrizations that can cover all eventualities. For example, the parametrization of a rigid body transformation given in (4) is a family with the member of the family specified by the fixed rotation matrix R_0. For a particular application, R_0 should be chosen so that the rotation angles $\boldsymbol{\alpha}$ are close to zero and avoid singularities at $\pm\pi/2$.

3.4 Condition of a Parametrization

If there is more than one candidate parametrization of a geometric element, how do we choose which one to use? Let D be a finite region of a surface $\mathbf{u} \mapsto \mathbf{f}(\mathbf{u}, \mathbf{b})$. For example, D might be the section of a cylindrical shaft. We define the distance function $d(\mathbf{x}, \mathbf{b})$, $\mathbf{x} \in D$, and the $n \times n$ matrix

$$H_{jk}(\mathbf{b}) = \int_D \frac{\partial d}{\partial b_j}(\mathbf{x}, \mathbf{b}) \frac{\partial d}{\partial b_k}(\mathbf{x}, \mathbf{b}) d\mathbf{x}.$$

The square root of the condition of the matrix $H(\mathbf{b})$ is a measure of the condition of the parametrization at $\mathbf{b}$ [3, 26]. In practice, if $X = \{\mathbf{x}_i\}_1^m$ is a random and representative scatter of points in D, the condition can be estimated by that of J, $J_{ij} = \partial d_i / \partial b_j$, where $d_i = d(\mathbf{x}_i, \mathbf{b})$. The condition of this matrix relates to the ratio of maximal rate of change of the surface to the minimal rate of change with respect to $\mathbf{b}$, as measured by the distance functions d_i.

3.5 Parametrization of NURBS Surfaces

The problem of parametrization becomes more difficult for free form surfaces defined by parametric B-splines or the more general nonuniform rational B-splines (NURBS) [49]. In a parametric B-spline surface, each coordinate $\mathbf{f}(\mathbf{u}, \mathbf{b})$ is represented as a tensor product spline on the same knot set:

$$\mathbf{f}(\mathbf{u}, \mathbf{b}) = \sum_{i=1}^{n} \sum_{j=1}^{m} N_i(u) N_j(v) \mathbf{p}_{ij}.$$

where $N_i(u)$, etc., are the B-spline basis functions for the knot set, $\mathbf{p}_{ij} = (p_{ij}, q_{ij}, r_{ij})^T$ are control points, the coefficients of the basis functions, and $\mathbf{b}$ is the vector of control points. The control points $\mathbf{p}_{ij}$ form a discrete approximation to the surface and changing $\mathbf{p}_{ij}$ influences the shape of the surface local to $\mathbf{p}_{ij}$. Together, the control points define the position, size and shape of the surface. Derivatives with respect to $\mathbf{u}$ can be calculated in terms of lower order tensor product splines.

NURBS generalise parametric splines so that standard geometric elements can be represented also by NURBS. Consider $f(u) = (1 - u^2)/(1 + u^2)$ and $g(u) = 2u/(1 + u^2)$ which parametrically defines an arc of circle in terms of ratios of quadratics involving the same denominator. This motivates the NURBS definition

$$\mathbf{f}_{\mathbf{w}}(\mathbf{u}, \mathbf{b}) = \frac{\sum_{i=1}^{n} \sum_{j=1}^{m} N_i(u) N_j(v) w_{ij} \mathbf{p}_{ij}}{\sum_{i=1}^{n} \sum_{j=1}^{m} w_{ij} N_i(u) N_j(v)}.$$

The numerator is a parametric tensor product B-spline surface and the denominator is a tensor product spline surface. Weights w_{ij} specify the relative influence of the control points $\mathbf{p}_{ij}$. For a NURBS surface, the position, size and shape of the surface are determined by the control point $\mathbf{p}_{ij}$ and the weights w_{ij}. Derivatives with respect to $\mathbf{u}$ can again be expressed in terms of parametric B-spline surfaces.

3.6 Position and Shape of a Parametric Surface

For these parametric spline surfaces, there is a strong geometrical relationship between the control points and the surface itself. For example, applying a rigid body transformation to the control points results in the same transformation of the surface; scaling the control points scales the surface by the same amount. While the control points $\mathbf{b} = \{\mathbf{p}_{ij}\}$ (along with weights if we are dealing with a NURBS surface) specify the surface, the vector $\mathbf{b}$ does not in general represent a parametrization, in the sense of section 3.2, of parametric spline surfaces on the fixed knot set. However, it is generally possible to separate out shape from parameters determining position and size on the basis of the control points. Let $P = \{\mathbf{p}_i, i \in I\}$ be a set of points in $\mathcal{R}^3$, and $\mathbf{p}_I$ be the $3m$-vector of their coordinates. Let $\hat{\mathbf{p}}_i = T^{-1}(\mathbf{p}_i, \mathbf{t}) = \mathbf{x}_0 + R^T(\boldsymbol{\alpha})\mathbf{p}_i$, the transformed points under the inverse rigid body transformation (4) and let $\hat{\mathbf{p}}_I$ be the vector of transformed points. We let J be the $3m \times 6$ Jacobian matrix of partial derivatives of $\hat{\mathbf{p}}_I$ with respect to t_j, evaluated at $\mathbf{t} = \mathbf{0}$. If J has QR factorisation [32]

$$J = \begin{bmatrix} Q_1 \; Q_2 \end{bmatrix} \begin{bmatrix} R_1 \\ \mathbf{0} \end{bmatrix},$$

and $\mathbf{q}_I$ is near $\mathbf{p}_I$, there exist unique $\mathbf{t}$ and $(3m-6)$-vector $\tilde{\mathbf{p}}_I$ such that the ith data point in $\mathbf{q}_I$ is the image of a point $\check{\mathbf{q}}_i$, where $\check{\mathbf{q}}_I$ is a perturbation of $\mathbf{p}_I$ specified by $\tilde{\mathbf{p}}_I$:

$$\mathbf{q}_i = T^{-1}(\check{\mathbf{q}}_i, \mathbf{t}), \quad \check{\mathbf{q}}_I = \mathbf{p}_I + Q_2\tilde{\mathbf{p}}_I.$$

The pair $(\tilde{\mathbf{p}}_i^T, \mathbf{t}^T)$ represents and alternative parametrization of $\mathbf{p}_I$ that separates shape from position. The component $Q_2\tilde{\mathbf{p}}_I$ represents the local change in shape of $\mathbf{p}_I$, and $\mathbf{t}$ specifies its local change of position. Similarly, it is possible to separate out position and scale from shape for a set of points.

The main problem with the parametrization of parametric surfaces is that a change in the shape of the control points $\mathbf{p}_{ij}$ does not necessarily mean a change in the shape of the associated surface. A simple example is given by parametric spline curves of the form $\mathbf{f}(u, \mathbf{b}) = (f(u, \mathbf{p}), g(u, \mathbf{q}))^T$ on the same knot set. If $\mathbf{p} = \mathbf{q}$ the parametric spline curve specifies a section of the line $y = x$. Changing the coefficients associated with the interior knots but keeping the relationship $\mathbf{p} = \mathbf{q}$ does not change the line segment, only the speed with respect to u the point $\mathbf{f}(u, \mathbf{b})$ moves along the line segment. This suggests that we allow only those changes in shape of the control points $\mathbf{p}_{ij}$ that correspond to changes in the shape of the surface. For example, we can constrain the control points to lie on straight lines $\mathbf{p}_{ij} = \mathbf{p}_{ij}(t_{ij}) = \mathbf{p}_{ij,0} + t_{ij}\mathbf{n}_{ij,0}$, where $\mathbf{n}_{ij,0}$ is the normal vector to the surface at the point closest to $\mathbf{p}_{ij,0}$ [11].

More generally, we can perform a singular value decomposition analysis [26, 32] to ensure that a change in the shape of the control points induces a change in the surface shape, i.e., induces a change that has a component normal to the surface at some point. One approach is as follows. Let X^* be a moderately dense set of points on $\mathbf{f}(\mathbf{u}, \mathbf{b}_0)$. Calculate the Jacobian matrix J associated with $d(\mathbf{x}, \mathbf{b})$ for X^* and $\mathbf{b}_0$. Calculate the singular value decomposition $J = USV^T$ with $V = [\ V_1\ V_2\ ]$ where V_1 corresponds to singular values above a threshold. Constrain $\mathbf{b} = \mathbf{b}_0 + V_1\tilde{\mathbf{b}}$. The constrained Jacobian $\tilde{J} = JV_1$ should be full rank for fitting to data reasonably close to $\mathbf{f}(\mathbf{u}, \mathbf{b}_0)$. More general regularisation techniques can also be applied [33].

For form assessment in coordinate metrology, the problem of surface parametrization is not so acute as the design specification will define the ideal geometry $\mathbf{u} \mapsto \mathbf{f}(\mathbf{u}, \mathbf{s}) = s_0\tilde{\mathbf{f}}(\mathbf{u}, \tilde{\mathbf{s}}_0)$ in a fixed frame of reference. The surface fitting may simply involve finding the transformation (and scale) parameters that define the best match of the transformed data to the (scaled) design surface. For parametric surfaces defined in terms of control points, this means that the fitting can be performed in terms of the position and scale of the control points, usually a well-posed problem.

4 Least Squares Orthogonal Distance Regression

In least squares orthogonal distance regression (LSODR) [1, 7, 9, 10, 24, 25, 34, 35, 54, 55, 60, 61], the best fit surface $\mathbf{u} \mapsto \mathbf{f}(\mathbf{u}, \mathbf{b})$ to a set of data $X = \{\mathbf{x}_i, i = 1, \ldots, m\}$ is that which minimises the sum of squares of the orthogonal distances, i.e., solves

$$\min_{\mathbf{b}} \sum_i d^2(\mathbf{x}_i, \mathbf{b}).$$

The Gauss-Newton algorithm is usually employed to perform the optimisation. If

$$\mathbf{d} = (d(\mathbf{x}_1, \mathbf{b}), \ldots, d(\mathbf{x}_m, \mathbf{b}))^T, \quad J_{ij} = \frac{\partial d_i}{\partial b_j},$$

calculated according to (2) and (3), and J has QR factorisation

$$J = QR = \left[Q_1\ Q_2\right] \begin{bmatrix} R_1 \\ \mathbf{0} \end{bmatrix} = Q_1 R_1$$

then the update step $\mathbf{p}$ for $\mathbf{b}$ solves $R_1\mathbf{p} = -Q_1^T\mathbf{d}$ and $\mathbf{b}$ is updated according to $\mathbf{b} := \mathbf{b} + \mathbf{p}$. The advantage of the Gauss-Newton algorithm over more general optimisation approaches based on the Newton's algorithm is that only first order information is required [22, 31]. For standard geometric elements, the distance functions $d(\mathbf{x}, \mathbf{b})$ can be evaluated analytically. Otherwise, optimisation techniques are required to solve the footpoint problem (1). Since the footpoint problem is itself a least squares problem, the Gauss-Newton algorithm can also be applied for its solution. However, there is generally scope for speeding up the convergence by using second order information [2, 7, 28]. An alternative is to bring the footpoint parameters $\mathbf{u}_i$ explicitly into the optimisation process. The resulting Jacobian matrix has a block angular structure that can be exploited efficiently during the QR factorisation of the Jacobian matrix [8, 16, 27].

4.1 Validation of LSODR Software

One of the issues in coordinate metrology is ensuring that the surface fitting software used is giving correct answers. Numerical software with a well defined computational aim can be tested using the following general approach [12]: i) determine reference data sets (appropriate for the computational aim) and corresponding reference results, ii) apply the software under test to the reference data sets to produce test results, and iii) compare the test results with the reference results. The so-called nullspace data generation approach [18, 17, 29] starts with a statement of the optimality conditions associated with the computational aim and then generates data for which the optimality conditions are automatically satisfied. For the LSODR problem, the first order optimality conditions are given by $J^T\mathbf{d} = \mathbf{0}$, where $d_i = d(\mathbf{x}_i, \mathbf{b})$ and J is the Jacobian matrix of partial derivatives $J_{ij} = \partial d_i/\partial b_j$. If a data point $\mathbf{x}_i^*$ lies exactly on the surface $\mathbf{u} \mapsto \mathbf{f}(\mathbf{u}, \mathbf{b})$, $\mathbf{n}_i^*$ is the normal to the surface at $\mathbf{x}_i^*$ and $\mathbf{x}_i = \mathbf{x}_i^* + d_i\mathbf{n}_i^*$, then $d(\mathbf{x}_i, \mathbf{b}) = d_i$ and $\partial d(\mathbf{x}_i, \mathbf{b})/\partial b_j = \partial d(\mathbf{x}_i^*, \mathbf{b})/\partial b_j$. These facts can be used to generate reference data for the LSODR problem, as follows. Given points $X^* = \{\mathbf{x}_i^* = \mathbf{f}(\mathbf{u}_i^*, \mathbf{b}^*)\}$ lying exactly on the surface $\mathbf{f} = \mathbf{f}(\mathbf{u}, \mathbf{b})$, corresponding normals $N^* = \{\mathbf{n}_i^*\}$, $i = 1, \ldots, m$, optimal transformation parameters $\mathbf{t}^*$ and perturbation constant Δ:

I Determine Jacobian matrix J of partial derivatives of $d(\mathbf{x}, \mathbf{b})$ evaluated for X^* and $\mathbf{b}^*$.
II Determine the nullspace Z^* of J^T from its QR decomposition.
III Choose, at random, a non-zero $m-n$ vector $\boldsymbol{\nu}$ normalised so that $\|\boldsymbol{\nu}\| = \Delta$.
IV Set $\mathbf{d}^* = Z^*\boldsymbol{\nu}$ and for each $i = 1, \ldots, m$, $\check{\mathbf{x}}_i = \mathbf{x}_i^* + d_i^*\mathbf{n}_i^*$.
V Set $X = \{\mathbf{x}_i : \mathbf{x}_i = T^{-1}(\check{\mathbf{x}}_i, \mathbf{t}^*)\}$.

If Δ is sufficiently small (smaller than the radius of curvature of the surface at any $\mathbf{x}_i^*$), the best-fit surface to X is given by $\mathbf{b}^*$ and $\mathbf{t}^*$.

For geometric elements, the normals $\mathbf{n}_i^*$ are easy to calculate. For more general surfaces they can be calculated in terms of $\partial\mathbf{f}/\partial u \times \partial\mathbf{f}/\partial v$. If the surface fitting only involves position and scale parameters then, from (5) and (6), the ith row of the Jacobian matrix in step I is given by

$$\left[-\mathbf{x}_i^*,\, \mathbf{x}_i^* \times \mathbf{n}_i^*,\, -(\mathbf{x}_i^*)^T\mathbf{n}_i^*\right], \tag{7}$$

which means that the data generator only requires points $\mathbf{x}_i^*$ on the surface and the corresponding normals $\mathbf{n}_i^*$; no other shape information is required, nor yet is the parametrization of the surface involved.

5 Chebyshev Orthogonal Distance Regression

In Chebyshev orthogonal distance regression (ChODR), the best fit surface $\mathbf{u} \mapsto \mathbf{f}(\mathbf{u}, \mathbf{b})$ to a set of data $X = \{\mathbf{x}_i, i = 1, \ldots, m\}$ is that which minimises the maximum orthogonal distance i.e., solves

$$\min_{\mathbf{b}} \max_{i=1}^{m} |d(\mathbf{x}_i, \mathbf{b})|. \tag{8}$$

The term 'minimum zone' is used in coordinate metrology for Chebyshev best fits. The term relates to the concept of a surface lying inside a tolerance zone of a given width and the minimum zone fit defines a tolerance zone of minimum width. For example ISO 1101 [38] defines a tolerance zone for a surface as

> The tolerance zone is limited by two surfaces enveloping spheres of diameter t, the centres of which are situated on a surface having the theoretically exact geometric form.

The 'enveloping spheres' of diameter t relates to orthogonal distances of at most $t/2$. However, the concept of minimum zone can be interpreted differently. For example, the earlier Carr and Ferreira [13, Definition DF.1.8] have

> The minimum zone solution is the minimum distance between two similar perfect-form features [surfaces] so that the two features maintain some relative location and/or orientation requirement and all data points are between the two features.

For example, suppose the perfect form is an ellipse with a fixed eccentricity e. The Carr and Ferreira definition specifies two ellipses with the same centre, orientation and eccentricity of different semi-axes lengths that just enclose the data points. According to the ISO 1101 definition, the enclosing profiles are orthogonally offset from an ellipse and will not be ellipses unless $e = 1$. In this paper, we use the ISO 1101 definition.

By introducing the parameter $e = \max_i |d(\mathbf{x}_i, \mathbf{a})|$, (8) can be reformulated as

$$\min_{\mathbf{a},e} e \quad \text{subject to} \quad e - d(\mathbf{x}_i, \mathbf{b}) \geq 0, \quad e - d(\mathbf{x}_i, \mathbf{b}) \geq 0, \quad i = 1, \ldots, m,$$

i.e., as a constrained optimisation problem involving parameters $\mathbf{b}$ and e.

5.1 Optimisation Subject to Nonlinear Inequality Constraints

We consider the more general constrained optimisation problem [22, 31, 46]

$$\min_{\mathbf{a}} F(\mathbf{a}) \quad \text{subject to} \quad c_i(\mathbf{a}) \geq 0, \quad i \in I, \tag{9}$$

involving n parameters $\mathbf{a} = (a_1, \ldots, a_n)^T$. A point $\mathbf{a}$ is said to be *feasible* if all the constraints are satisfied: $c_i(\mathbf{a}) \geq 0, i \in I$. At a feasible point $\mathbf{a}$, we distinguish between those constraints that are satisfied with equality, and those that are not. A constraint j for which $c_j(\mathbf{a}) = 0$ is said to be *active* at $\mathbf{a}$, otherwise it is *inactive*.

At a point $\mathbf{a}$, let I^* be the set of indices corresponding to the constraints that are active, and define the associated *Lagrangian* function by $\mathcal{L}_{I^*}(\mathbf{a}, \boldsymbol{\lambda})$ defined by

$$\mathcal{L}_{I^*}(\mathbf{a}, \boldsymbol{\lambda}) = F(\mathbf{a}) - \sum_{i \in I^*} \lambda_i c_i(\mathbf{a}).$$

Let $\mathbf{g}(\mathbf{a}) = \nabla_{\mathbf{a}} F(\mathbf{a})$, $W = \nabla^2_{\mathbf{a}} \mathcal{L}(\mathbf{a}, \boldsymbol{\lambda})$, $C^* = C^*(\mathbf{a})$ be the matrix of gradients $\nabla_{\mathbf{a}} c_i$, $i \in I^*$. The *constraint qualification* condition holds if the columns of C^* are linearly independent. If the number p of active constraints is less than n, let $Z = Z(\mathbf{a})$ be the $n \times (n-p)$ orthogonal complement to C^*, so that $Z^T C^* = \mathbf{0}$. The conditions for the Lagrangian to have zero gradient with respect to both $\mathbf{a}$ and $\boldsymbol{\lambda}$ are known as the Kuhn-Tucker equations:

$$\mathbf{g}(\mathbf{a}) = \sum_{i \in I^*} \lambda_i \nabla c_i(\mathbf{a}), \quad c_i(\mathbf{a}) = 0, \quad i \in I^*, \tag{10}$$

an $(n+p) \times (n+p)$ system of equations.

5.2 Optimality Conditions

If the constraint qualification holds, the following are *necessary* conditions for $\mathbf{a}^*$ to be a local minimizer for the problem defined by (9).

- N1 *Feasibility:* $c_i(\mathbf{a}^*) \geq 0, \ i \in I$.
- N2 *First order:* If I^* denotes the set of constraints active at $\mathbf{a}^*$, there exist Lagrange multipliers $\boldsymbol{\lambda}^*$ for which
  (a) $\mathbf{g}(\mathbf{a}^*) = \sum_{i \in I^*} \lambda_i^* \nabla c_i(\mathbf{a}^*)$, and
  (b) $\lambda_i^* \geq 0, \ i \in I^*$.
- N3 *Second order:* If $p < n$, the matrix $Z(\mathbf{a}^*)^{\mathrm{T}} W(\mathbf{a}^*, \boldsymbol{\lambda}^*) Z(\mathbf{a}^*)$ is positive semi-definite.

Under constraint qualification, the following are *sufficient* conditions for $\mathbf{a}^*$ to be local minimizer:

- S1 *Feasibility:* $c_i(\mathbf{a}^*) \geq 0, \ i \in I$.
- S2 *First order:* If I^* denotes the set of constraints active at $\mathbf{a}^*$, there exist Lagrange multipliers $\boldsymbol{\lambda}^*$ for which

(a) $\mathbf{g}(\mathbf{a}^*) = \sum_{i \in I^*} \lambda_i^* \nabla c_i(\mathbf{a}^*)$, and
(b) $\lambda_i^* > 0$, $i \in I^*$.

S3 *Second order:* If $p < n$, the matrix $Z(\mathbf{a}^*)^{\mathrm{T}} W(\mathbf{a}^*, \boldsymbol{\lambda}^*) Z(\mathbf{a}^*)$ is positive definite.

The sufficient conditions replace $\geq$ with $>$ in N2, b) and N3. The sufficient conditions can be modified to allow for zero Lagrange multipliers by including additional restrictions on the Hessian of the objective function F.

Mathematical programming approaches to solving the constrained optimisation problems at each iteration determine a set of working constraints, an estimate of the constraints active at the solution, and solve for the Lagrange multipliers $\boldsymbol{\lambda}$ by solving the Kuhn-Tucker equations (10). If any Lagrange multiplier is negative (and the constraint qualification holds) it means it is possible to move away from the constraint in a direction $\mathbf{p}$ that reduces F. A step is taken in such a direction until another constraint becomes active or F attains a minimum along that step. If all the Lagrange multipliers are positive, a second order test for optimality is made. If the second order optimality condition fails, then there exists a $\mathbf{p}$ with $\mathbf{p}^T Z(\mathbf{a})^{\mathrm{T}} W(\mathbf{a}, \boldsymbol{\lambda}) Z(\mathbf{a}) \mathbf{p} < 0$ and moving along $\mathbf{p}$ reduces F while maintaining (approximately) the working constraints. Such algorithms solve a sequence of quadratic programming problems.

For linearly constrained problems, managing the working set is generally straightforward since it is possible to determine step directions that keep a subset of the constraints active. For nonlinearly constrained problems, a balance has to be sought between reducing the objective function and solving $c_i(\mathbf{a}) = 0$ for the constraints judged to be active.

For both linearly and nonlinearly constrained problems, special steps have to be taken if the active constraint gradient matrix is rank deficient. In this case, the Kuhn-Tucker equations will have a space of solutions so that tests on the positivity of the Lagrange multipliers have to be implemented carefully.

The general theory of optimisation subject to nonlinear inequality constraints can be adapted to Chebyshev (sometimes minimax) optimisation to take into account its special features, mainly that the objective function is linear and the constraints appear in pairs; see, e.g., [15, 41, 45, 58, 59].

5.3 Chebyshev Optimisation for Surface Fitting

For the surface fitting problems with which we are concerned, the vector of optimisation parameters is $\mathbf{a}^T = (\mathbf{b}^T, e)$. Given $\mathbf{b}$, setting $e = \max_i d(\mathbf{x}_i, \mathbf{b})$ determines a feasible $\mathbf{a}$. An active constraint defines a point that is e away from the surface determined by $\mathbf{b}$. A point associated with an active constraint is known as a *contacting point.* We set $I^+ = \{i \in I : e = d(\mathbf{x}_i, \mathbf{b})\}$ and $I^- = \{i \in I : e = -d(\mathbf{x}_i, \mathbf{b})\}$. At a local minimum the Kuhn-Tucker equations can be formulated as $\lambda_i \geq 0$ and

$$\sum_{i \in I^+ \cup I^-} \lambda_i = 1, \quad \left(\sum_{i \in I^+} \lambda_i \nabla d_i(\mathbf{x}_i, \mathbf{b}) \right) - \left(\sum_{i \in I^-} \lambda_i \nabla d(\mathbf{x}_i, \mathbf{b}) \right) = \mathbf{0}. \quad (11)$$

Geometrically these conditions mean the convex hull of $\{\pm \nabla d_i, i \in I^+ \cup I^-\}$, contains the origin, where the sign is determined from by which of the two constraints is active. If the fitting problem only involves position parameters, the second condition reads as

$$\sum_{i \in I^+} \lambda_i \mathbf{n}_i = \sum_{i \in I^-} \lambda_i \mathbf{n}_i, \quad \sum_{i \in I^+} \lambda_i \mathbf{x}_i \times \mathbf{n}_i = \sum_{i \in I^-} \lambda_i \mathbf{x}_i \times \mathbf{n}_i.$$

If a global scale parameter is also included, the corresponding additional relation is

$$\sum_{i \in I^+} \lambda_i \mathbf{x}_i^T \mathbf{n}_i = \sum_{i \in I^-} \lambda_i \mathbf{x}_i^T \mathbf{n}_i.$$

For the geometric elements lines, circles and more general conic sections in 2D, planes, spheres and quadratic surfaces, the optimisation problem can be posed as minimising a nonlinear function subject to linear constraints, a considerable simplification. Geometrically, the constraints define a region of space bounded by hyperplanes. The solution can be at a vertex, in which case the number of constraints equals the number of parameters, or on an edge or on a face, etc. For some problems, the solution is known to be at a vertex, so that an algorithm can be developed purely on the basis of first order information. For other fitting problems, the problem is one of minimising a linear function, $F(\mathbf{a}) = e$, subject to nonlinear constraints. The geometry is similar in that the constraints define a region of space bounded by curvilinear hyper-surfaces and the solution can be at a vertex, on a edge or face, etc. Since the objective function is linear, a non-vertex solution relies on the curvature of the constraints.

Because of the general complexity of nonlinear Chebyshev approximation, much effort has gone into defining algorithms specifically defined for a type of geometric element. In the engineering literature, considerable attention has been paid to straightness, flatness, circularity, sphericity and cylindricity; see e.g. [13, 14, 51, 52, 56, 57]. All but the last can be formulated as linearly constrained problems and it can be shown for these that the solutions are all vertex solutions, i.e., if the geometric element is defined in terms of n parameters, at least $n + 1$ constraints are active at the solution. The global solution can therefore be determined by brute force, examining all solutions defined by $n + 1$ data points for feasibility and optimality. For even modest numbers of data points, this procedure is computationally expensive. Much more efficient are the active set algorithms that start from a feasible solution and descend to a local minimum exchanging points in and out of the active set. A technical issue to be catered for in such algorithms is the presence of degenerate solutions in which more than $n + 1$ constraints are active, but no subset of $n + 1$ constraints (data points) define a local minimum. Such an example is given in Fig. 1 where six contacting points define a local minimum but no subset of four defines a local minimum.

A second issue is that an active set descent strategy will only deliver a local minimum in general. However, the ChODR problem for a set of data X has the property that if a solution defined by $\mathbf{a}$ is the global optimum for $Y \subset X$ and feasible for X then it is also the global solution for X; any better solution for X would also be better for Y. This property motivates the following algorithm for the linearly constrained problems [19, 44]. Start with any $n+1$ points $E_1 \subset X$ and set $k = 1$.

I Find the global solution for E_k. If the global solution $\mathbf{a}_k$ for E_k is also feasible for X then it is also a global solution for X and the algorithms stop. Otherwise go to II.

II Expand: choose a point that violates a constraint for $\mathbf{a}_k$ and add it to E_k, forming E_k^+. Find the global solution $\mathbf{a}_k^+$ for E_k^+.

III Contract if possible: for each $\mathbf{x}_i \in E_k$, form the global solution $\mathbf{a}_i$ for $E_k^+ \setminus \{\mathbf{x}_i\}$; if $\mathbf{a}_i = \mathbf{a}_k^+$, remove $\mathbf{x}_i$ from E_k^+. Set $E_{k+1} = E_k^+$ (after all possible points have been removed) and go to I.

The algorithm terminates with the global solution $\mathbf{a}^*$ and a minimal subset $E^* \subset X$, for which the global solution for E^* is the same as that for X. E^* is known as the *essential subset* for X [19]. In fact, the algorithm will determine all global solutions for X. The algorithm employs an active set strategy but the active set contains not only the contacting points corresponding to active constraints but also points necessary to define the global solution at the kth stage. For example, in Fig. 2, the global solution for the four contacting points is not feasible for the fifth point. The essential subset contains all five points. The algorithm uses a brute force strategy but only applied to the active/essential set and so long as the active set does not become large, the algorithm is efficient and delivers all global solutions. It also copes well with degeneracy.

The essential subset approach works for the linearly constrained geometric elements because a brute force approach can be applied to the essential subsets: all possible solutions are determined by solving a number of sets of linear equations. For the nonlinear elements, all possible solutions for a set of points is much more difficult to determine. Furthermore, there is no guarantee that the global solution is a vertex solution.

For the calculations of cylindricity involving nonlinear constraints, many algorithms use an active set strategy to find a vertex solution; see e.g. [14, 57]. Very few such algorithms cope well with nonvertex solutions. A more general approach is based on the Osborne-Watson algorithm [48, 58, 59], similar to the Gauss-Newton algorithm for least squares fitting except that at each iteration, the update step $\mathbf{p}$ solves a linear Chebyshev problem [4, 5, 6] involving the Jacobian matrix J and function values $\mathbf{d}$. The algorithm is easy to implement and software to solve the linear Chebyshev problem is in the public domain [4]. In practice, Osborne-Watson algorithm works well if the solution is a vertex

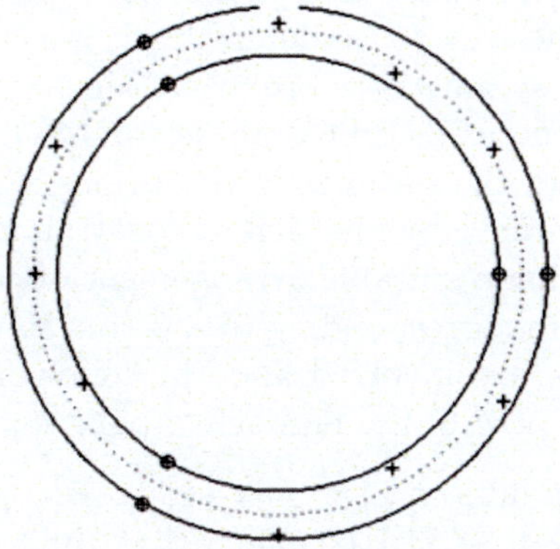

Fig. 1. Degenerate solution for the ChODR circle defined by six contacting points 'o'.

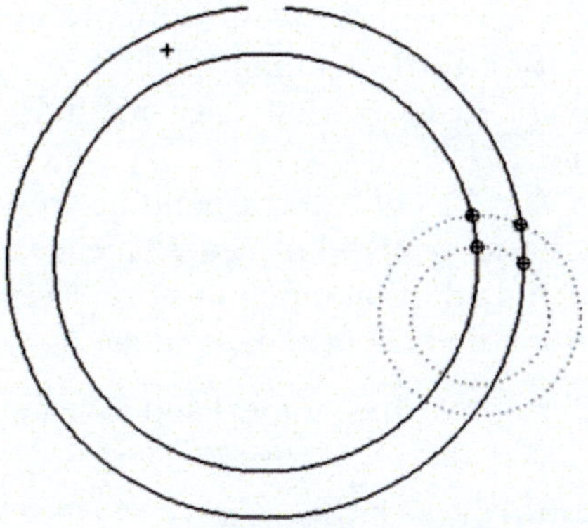

Fig. 2. The global minimum for the four contacting points (dotted circles) is not feasible for the fifth point; the essential subset contains all five points.

solution; for non-vertex solutions, convergence can be very slow as second order information is required to converge to a minimum quickly [42]. Non-vertex solutions do occur in practice. For example, in 1000 randomly generated datasets involving four data points on each of four parallel circles on a cylinder, a five point local minimum was detected in about 16 % of the datasets.

5.4 Validation of Chebyshev ODR Software

Because of the technical difficulty in implementing ChODR algorithms, there is a need to validate software claiming to provide Chebyshev fits. We limit ourselves to the case of fitting fixed shapes to data, so that only the position and scale of the surface are to be determined. For a surface with no translational nor rotational symmetry there are seven free parameters. We use the same approach to generating test data as for LSODR, i.e., generating data in a way derived from the optimality conditions. Since the local optimality conditions are determined by the active constraints, the problem is largely confined to determining a suitable arrangement of contacting points. For the case of generating data corresponding to a vertex solution, only the first order optimality conditions are involved and the following simple scheme can be implemented. Assign $m \geq 8$, the number of points, and $e > 0$, the form error.

I Determine eight points $\mathbf{x}_i^*$, $i \in I = \{1, \ldots, 8\}$, lying exactly on the ideal (design) geometry $\mathbf{u} \mapsto \mathbf{f}(\mathbf{u}, \mathbf{b}^*)$ such that the 8×7 Jacobian matrix determined from (7) is of full rank. Note that only the $\mathbf{x}_i^*$ and corresponding normals $\mathbf{n}_i$ are involved; no other aspect of the surface geometry is involved.

II For all partitions $I = I^+ \cup I^-$, representing the assignment of the points to the outer or inner enveloping surfaces, solve the Kuhn-Tucker equations (11) for $\boldsymbol{\lambda}$. There will be two, mutually anti-symmetric partitions for which the solution Lagrange multipliers are positive. (In practice, we assign the first point to I^+ so that only one half of the possibilities have to be examined.)

III For the partition $I = I^+ \cup I^-$ corresponding to positive Lagrange multipliers and $e > 0$, set $\mathbf{x}_i = \mathbf{x}_i^* + e\mathbf{n}_i$, $i \in I^+$, $\mathbf{x}_i = \mathbf{x}_i^* - e\mathbf{n}_i$, $i \in I^-$.

IV Generate, at random, additional points $\mathbf{x}_i^*$ on the surface, form errors $\delta_i \in [-e, e]$ and corresponding normals $\mathbf{n}_i$ and set $\mathbf{x}_i = \mathbf{x}_i^* + \delta_i \mathbf{n}_i$, $i = 9, \ldots m$.

For sufficiently small e, $\mathbf{b}^*$ defines a local solution for the ChODR problem.

5.5 Non-Vertex Solutions: Cylindricity

The vertex solution depends on finding a partition such that the convex hull of $(-1)^{\pi_i} \nabla_{\mathbf{b}} d(\mathbf{x}_i, \mathbf{b}^*)$, $\pi_i = 0$ or 1, contains the origin. A non-vertex solution corresponds to a face (or edge) of the convex hull passing through the origin. For particular geometric shapes, it may be straightforward to choose the data points to correspond to a non-vertex solution. The following scheme can be used to generate data sets such that the best fit Chebyshev cylinder has only five contacting points (a vertex solution has six). Recall that for a cylinder in standard position, $\nabla_{\mathbf{b}} d(\mathbf{x}, \mathbf{b}) = (\cos\theta, \sin\theta, -z\sin\theta, z\cos\theta, -1)^T$ in cylindrical coordinates.

I Generate points $(x_i^*, y_i^*) = (r_0 \cos\theta_i, r_0 \sin\theta_i)$, $i = 1, 2, 3, 4, 5$, on the circle $x^2 + y^2 = r_0^2$ in increasing polar angle order with $\boldsymbol{\pi} = (1, 0, 1, 0, 1)^T$; going around the circle, the points are assigned alternatively to the outer and inner surfaces. This arrangement ensures that the optimality conditions for a Chebyshev circle fit are met. Let $\boldsymbol{\lambda}_0 = (1, 1, 1, 1, 1)^T/5$,

$$C_1 = \begin{bmatrix} 1 & 1 & 1 & 1 & 1 \\ \cos\theta_1 & -\cos\theta_2 & \cos\theta_3 & -\cos\theta_4 & \cos\theta_5 \\ \sin\theta_1 & -\sin\theta_2 & \sin\theta_3 & -\sin\theta_4 & \sin\theta_5 \\ 1 & -1 & 1 & -1 & 1 \end{bmatrix}, \quad \mathbf{g} = (1, 0, 0, 0, 0)^T.$$

and solve

$$\min_{\boldsymbol{\lambda}} (\boldsymbol{\lambda} - \boldsymbol{\lambda}_0)^T (\boldsymbol{\lambda} - \boldsymbol{\lambda}_0) \quad \text{subject to} \quad C_1 \boldsymbol{\lambda} = \mathbf{g}.$$

The resulting Lagrange multipliers $\boldsymbol{\lambda} = (\lambda_1, \lambda_2, \lambda_3, \lambda_4, \lambda_5)^T$ will be positive. Alternatively, we can solve $\max \min_i \lambda_i$ subject to $C\boldsymbol{\lambda} = \mathbf{b}$.

II Generate at random $\mathbf{z} = (z_1, z_2, z_3, z_4, z_5)^T$ that solves $C_2 \mathbf{z} = \mathbf{0}$, where

$$C_2 = \begin{bmatrix} 1 & 1 & 1 & 1 & 1 \\ -\lambda_1 \sin\theta_1 & \lambda_2 \sin\theta_2 & -\lambda_3 \sin\theta_3 & \lambda_4 \sin\theta_4 & -\lambda_5 \sin\theta_5 \\ \lambda_1 \cos\theta_1 & -\lambda_2 \cos\theta_2 & \lambda_3 \cos\theta_3 & -\lambda_4 \cos\theta_4 & \lambda_5 \cos\theta_5 \end{bmatrix}.$$

The first constraint simply ensures that $\sum z_i = 0$, while the other two constraints are the optimality conditions corresponding to the axes direction parameters.

III Form the full 6×5 constraint matrix C with columns $(1, \pm\nabla_{\mathbf{b}^T} d(\mathbf{x}_i, \mathbf{b}^*))^T$ and a nonzero null space vector $\mathbf{p}$ satisfying $C^T \mathbf{p} = \mathbf{0}$. A step in the direction $\mathbf{p}$ from $\mathbf{b}^*$ keeps all five constraints active to first order. Let $\mathbf{p}_\mathbf{b}$ be the latter five elements of $\mathbf{p}$, corresponding to the cylinder parameters $\mathbf{b}$.

IV Form the matrix

$$W_\mathbf{b} = \sum_{i=1}^{5} (-1)^{\pi_i} \lambda_i \nabla^2_\mathbf{b} d(\mathbf{x}_i, \mathbf{b}^*).$$

If $\mathbf{p}_\mathbf{b}^T W_\mathbf{b} \mathbf{p}_\mathbf{b} > 0$ the points (x_i, y_i, z_i), $i = 1, \ldots, 5$ define active constraints corresponding to a local minimum for the Chebyshev cylinder problem. Otherwise go to step II.

In step IV, since there is only one degree of freedom for five contacting points it is possible to check if $\mathbf{z}$ corresponds to a local minimum by evaluating the distances $\max_i |d(\mathbf{x}_i, \mathbf{b}^* \pm \delta\mathbf{p})|$ for a small perturbation δ. This obviates the need to calculate the second order information $W_\mathbf{b}$.

These data generation schemes have been tested on Chebyshev fitting algorithms using the NAG library routine E04UCF [47] called from within Matlab [43]. The data generation scheme guarantees only that $\mathbf{b}^*$ defines a local minimum and does not rule out the possibility of a better local minimum nearby. Using a large number of starting points sampled at random near $\mathbf{b}^*$ the optimisation software can be used to test for other local minima. Other global optimisation techniques such as as genetic algorithms and simulated annealing have also been suggested [40, 53].

6 Concluding Remarks

This paper has considered the problem of form assessment – the departure from ideal geometry – from coordinate data according to two criteria: least squares and Chebyshev in which the L_2 norm, respectively, L_∞ norm of the vector of orthogonal distances is minimised. For standard geometric elements, the orthogonal distances can be calculated as analytical functions of the surface parameters. For more general surfaces these functions have to be evaluated numerically; nevertheless, the same overall approach applies in both cases.

Even for standard geometric elements such as a cylinder, the question of parametrization of the space of elements is not straightforward since the latter space need not be topologically trivial. This means in practice that a family of parametrizations is needed, and an appropriate member of the family chosen (adaptively) for a particular application. In general, surface parameters specify position, size and shape and it is usually useful to separate out these three components. For free form surfaces such as NURBS defined in terms of control points, the position and size of the surface is specified by the position and size of the control points. However, a change in shape of the control points does not necessarily mean a change in shape of the surface so that surface fitting with the control points as unconstrained parameters is likely to be ill-posed and some additional, functionally relevant constraints are required. Form assessment can be thought of in terms of departure from an ideal, fixed shape, with only position and (usually) size regarded as free parameters and usually gives rise to well-posed optimisation problems.

Least squares form assessment is reasonably straightforward and algorithms such as the Gauss-Newton algorithm usually perform well. The only complication for free form surfaces is that the orthogonal distance functions have to be evaluated numerically and involves solving for the footpoint parameters. While the Gauss-Newton algorithm relies only on first order information, there are likely to be advantages (in terms of convergence and speed) in using second order information to solve for the footpoints. The generation of test data exploiting the optimality conditions is very straightforward.

Chebyshev form assessment is considerably more involved, particularly for surfaces that involve nonlinear constraints. For problems that can be posed in terms of linear constraints the essential subset algorithm is attractive as it delivers the global minimum. For nonlinear problems, many proposed algorithms work well if the solution happens to be defined by a vertex and can be found using first order information. For non-vertex solutions, (approximations to) second order information are required in order to converge efficiently to the solution and to confirm that the solution is a local minimum. The generation of test data for vertex solutions is straightforward. For special cases, non-vertex solutions can be constructed easily.

Many form assessment problems involve checking that the form error is within a tolerance. If the form error determined from a least squares solution is within the tolerance there is no need to find the Chebyshev solution. The grey area is when the least squares solution indicates that the tolerance criteria have not been met but the Chebyshev solution does lead to conformance to specification. However, two other factors have to be considered. The first is that the form assessment is performed on a discrete set of data which may under-represent the true surface, leading to an optimistic estimate of the form error. The second is that the coordinate measuring machine producing the data is not perfectly accurate and systematic and random effects associated with the measuring system will tend to inflate the estimate of the form error. These two factors in practice may have a bigger effect than the choice of criterion so that inferences based on a Chebyshev assessment will be neither more nor less valid than those using the much more tractable least squares criterion.

References

1. S.J. Ahn, E. Westkämper, and W. Rauh: Orthogonal distance fitting of parametric curves and surfaces. In: *Algorithms for Approximation IV*, J. Levesley, I.J. Anderson, and J.C. Mason (eds.), University of Huddersfield, 2002, 122–129.
2. I.J. Anderson, M.G. Cox, A.B. Forbes, J.C. Mason, and D.A. Turner: An efficient and robust algorithm for solving the footpoint problem. In: *Mathematical Methods for Curves and Surfaces II*, M. Daehlen, T. Lyche, and L.L. Schumaker (eds.), Vanderbilt University Press, Nashville TN, 1998, 9–16.
3. G.T. Anthony, H.M. Anthony, M.G. Cox, and A.B. Forbes: *The Parametrization of Fundamental Geometric Form*. Report EUR 13517 EN, Commission of the European Communities (BCR Information), Luxembourg, 1991.
4. I. Barrodale and C. Phillips: Algorithm 495: Solution of an overdetermined system of linear equations in the Chebyshev norm. Transactions of Mathematical Software, 1975, 264–270.
5. R. Bartels and A.R. Conn: A programme for linearly constrained discrete ℓ_1 problems. ACM Trans. Math. Soft. **6**(4), 1980, 609–614.
6. R. Bartels and G.H. Golub: Chebyshev solution to an overdetermined linear system. Comm. ACM **11**(6), 1968, 428–430.
7. M. Bartholomew-Biggs, B.P. Butler, and A.B. Forbes: Optimisation algorithms for generalised regression on metrology. In: *Advanced Mathematical and Computational Tools in Metrology IV*, P. Ciarlini, A.B. Forbes, F. Pavese, and D. Richter (eds.), World Scientific, Singapore, 2000, 21–31.
8. A. Björck: *Numerical Methods for Least Squares Problems*. SIAM, Philadelphia, 1996.
9. P.T. Boggs, R.H. Byrd, and R.B. Schnabel: A stable and efficient algorithm for nonlinear orthogonal distance regression. SIAM Journal of Scientific and Statistical Computing **8**(6), 1987, 1052–1078.
10. P.T. Boggs, J.R. Donaldson, R.H. Byrd, and R.B. Schnabel: ODRPACK: software for weighted orthogonal distance regression. ACM Trans. Math. Soft. **15**(4), 1989, 348–364.
11. B.P. Butler, M.G. Cox, and A.B. Forbes: The reconstruction of workpiece surfaces from probe coordinate data. In: *Design and Application of Curves and*

Surfaces, R.B. Fisher (ed.), IMA Conference Series, Oxford University Press, 1994, 99–116.
12. B.P. Butler, M.G. Cox, A.B. Forbes, S.A. Hannaby, and P.M. Harris: A methodology for testing the numerical correctness of approximation and optimisation software. In: *The Quality of Numerical Software: Assessment and Enhancement*, R. Boisvert (ed.), Chapman and Hall, 1997, 138–151.
13. K. Carr and P. Ferreira: Verification of form tolerances part I: basic issues, flatness and straightness. Precision Engineering **17**(2), 1995, 131–143.
14. K. Carr and P. Ferreira: Verification of form tolerances part II: cylindricity and straightness of a median line. Precision Engineering **17**(2), 1995, 144–156.
15. A.R. Conn and Y. Li: *An Efficient Algorithm for Nonlinear Minimax Problems.* Report CS-88-41, University of Waterloo Computer Science Department, Waterloo, Ontario, Canada, November 1989.
16. M.G. Cox: The least-squares solution of linear equations with block-angular observation matrix. In: *Advances in Reliable Numerical Computation*, M.G. Cox and S. Hammarling (eds.), Oxford University Press, 1989, 227–240.
17. M.G. Cox, M.P. Dainton, A.B. Forbes, and P.M. Harris: Validation of CMM form and tolerance assessment software. In: *Laser Metrology and Machine Performance V*, G.N. Peggs (ed.), WIT Press, Southampton, 2001, 367–376.
18. M.G. Cox and A.B. Forbes: *Strategies for Testing Form Assessment Software.* Report DITC 211/92, National Physical Laboratory, Teddington, December 1992.
19. R. Drieschner: Chebyshev approximation to data by geometric elements. Numerical Algorithms **5**, 1993, 509–522.
20. R. Drieschner, B. Bittner, R. Elligsen, and F. Wäldele: *Testing Coordinate Measuring Machine Algorithms, Phase II.* Report EUR 13417 EN, Commission of the European Communities (BCR Information), Luxembourg, 1991.
21. S.C. Feng and T.H. Hopp: *A Review of Current Geometric Tolerancing Theories and Inspection Data Analysis Algorithms.* Report NISTIR 4509, National Institute of Standards and Technology, U.S., 1991.
22. R. Fletcher: *Practical Methods of Optimization.* 2nd edition, John Wiley and Sons, Chichester, 1987.
23. A.B. Forbes: *Fitting an Ellipse to Data.* Report DITC 95/87, National Physical Laboratory, Teddington, 1987.
24. A.B. Forbes: *Least-Squares Best-Fit Geometric Elements.* Report DITC 140/89, National Physical Laboratory, Teddington, 1989.
25. A.B. Forbes: Least squares best fit geometric elements. In: *Algorithms for Approximation II*, J.C. Mason and M.G. Cox (eds.), Chapman & Hall, London, 1990, 311–319.
26. A.B. Forbes: Model parametrization. In: *Advanced Mathematical Tools for Metrology*, P. Ciarlini, M.G. Cox, F. Pavese, and D. Richter (eds.), World Scientific, Singapore, 1996, 29–47.
27. A.B. Forbes: Efficient algorithms for structured self-calibration problems. In: *Algorithms for Approximation IV*, J. Levesley, I. Anderson, and J.C. Mason (eds.), University of Huddersfield, 2002, 146–153.
28. A.B. Forbes: Structured nonlinear Gauss-Markov problems. In: *Algorithms for Approximation V*, A. Iske and J. Levesley (eds.), Springer, Berlin, 2007, 167–186.

29. A.B. Forbes, P.M. Harris, and I.M. Smith: Correctness of free form surface fitting software. In: *Laser Metrology and Machine Performance VI*, D.G. Ford (ed.), WIT Press, Southampton, 2003, 263–272.
30. W. Gander, G.H. Golub, and R. Strebel: Least squares fitting of circles and ellipses. BIT **34**, 1994.
31. P.E. Gill, W. Murray, and M.H. Wright: *Practical Optimization*. Academic Press, London, 1981.
32. G.H. Golub and C.F. Van Loan: *Matrix Computations*. 3rd edition, John Hopkins University Press, Baltimore, 1996.
33. P. Hansen: Regularization tools: a Matlab package for analysis and solution of discrete ill-posed problems. Numerical Algorithms **6**, 1994, 1–35.
34. H.-P. Helfrich and D. Zwick: A trust region method for implicit orthogonal distance regression. Numerical Algorithms **5**, 1993, 535–544.
35. H.-P. Helfrich and D. Zwick: Trust region algorithms for the nonlinear distance problem. Numerical Algorithms **9**, 1995, 171–179.
36. T.H. Hopp: *Least-Squares Fitting Algorithms of the NIST Algorithm Testing System*. Technical report, National Institute of Standards and Technology, 1992.
37. ISO: *ISO 10360-6:2001 Geometrical Product Specifications (GPS) – Acceptance and Reverification Tests for Coordinate Measuring Machines (CMM) – Part 6: Estimation of Errors in Computing Gaussian Associated Features*. International Organization for Standardization, Geneva, 2001.
38. ISO: *BS EN ISO 1101: Geometric Product Specifications (GPS) – Geometric Tolerancing – Tolerances of Form, Orientation, Location and Run Out*. International Organization for Standardization, Geneva, 2005.
39. X. Jiang, X. Zhang, and P.J. Scott: Template matching of freeform surfaces based on orthogonal distance fitting for precision metrology. Measurement Science and Technology **21**(4), 2010.
40. H.-Y. Lai, W.-Y. Jywe, C.-K. Chen, and C.-H. Liu: Precision modeling of form errors for cylindricity evaluation using genetic algorithms. Precision Engineering **24**, 2000, 310–319.
41. K. Madsen: An algorithm for minimax solution of overdetermined systems of non-linear equations. J. Inst. Math. Appl. **16**, 1975, 321–328.
42. K. Madsen and J. Hald: A 2-stage algorithm for minimax optimization. In: *Lecture Notes in Control and Information Sciences 14*, Springer-Verlag, 1979, 225–239.
43. MathWorks, Inc., Natick, Mass, `http://www.mathworks.com`.
44. G. Moroni and S. Petrò: Geometric tolerance evaluation: a discussion on minimum zone fitting algorithms. Precision Engineering **32**, 2008, 232–237.
45. W. Murray and M.L. Overton: A projected Lagrangian algorithm for nonlinear minimax optimization. SIAM Journal for Scientific and Statistical Computing **1**(3), 1980, 345–370.
46. G.L. Nemhhauser, A.H.G. Rinnooy Kan, and M.J. Todd (eds.): *Handbooks in Operations Research and Management Science, Volume 1: Optimization*. North-Holland, Amsterdam, 1989.
47. The Numerical Algorithms Group: *The NAG Fortran Library, Mark 22, Introductory Guide*, 2009. `http://www.nag.co.uk/`.
48. M.R. Osborne and G.A. Watson. An algorithm for minimax approximation in the nonlinear case. Computer Journal **12**, 1969, 63–68.
49. L. Piegl and W. Tiller: *The NURBS Book*. 2nd edition, Springer-Verlag, New York, 1996.

50. V. Pratt: Direct least-squares fitting of algebraic surfaces. Computer Graphics **21**(4), July 1987, 145–152.
51. G.L. Samuel and M.S. Shunmugam: Evaluation of circularity from coordinate data and form data using computational geometric techniques. Precision Engineering **24**, 2000, 251–263.
52. G.L. Samuel and M.S. Shunmugam: Evaluation of sphericity error from form data using computational geometric techniques. Int. J. Mach. Tools & Manuf. **42**, 2002, 405–416.
53. C.M. Shakarji and A. Clement: Reference algorithms for Chebyshev and one-sided data fitting for coordinate metrology. CIRP Annals - Manufacturing Technology **53**, 2004, 439–442.
54. D. Sourlier and W. Gander: A new method and software tool for the exact solution of complex dimensional measurement problems. In: *Advanced Mathematical Tools in Metrology II*, P. Ciarlini, M.G. Cox, F. Pavese, and D. Richter (eds.), World Scientific, Singapore, 1996, 224–237.
55. D.A. Turner, I.J. Anderson, J.C. Mason, M.G. Cox, and A.B. Forbes: An efficient separation-of-variables approach to parametric orthogonal distance regression. In: *Advanced Mathematical Tools in Metrology IV*, P. Ciarlini, A.B. Forbes, F. Pavese, and R. Richter (eds.), World Scientific, Singapore, 2000, 246–255.
56. N. Venkaiah and M.S. Shunmugam: Evaluation of form data using computational geometric techniques - part I: circularity error. Int. J. Mach. Tools & Manuf. **47**, 2007, 1229–1236.
57. N. Venkaiah and M.S. Shunmugam: Evaluation of form data using computational geometric techniques - part II: cylindricity error. Int. J. Mach. Tools & Manuf. **47**, 2007, 1237–1245.
58. G.A. Watson: The minimax solution of an overdetermined system of non-linear equations. Journal of the Institute of Mathematics and its Applications **23**, 1979, 167–180.
59. G.A. Watson: *Approximation Theory and Numerical Methods.* John Wiley & Sons, Chichester, 1980.
60. G.A. Watson: Some robust methods for fitting parametrically defined curves or surfaces to measured data. In: *Advanced Mathematical and Computational Tools in Metrology IV*, P. Ciarlini, A.B. Forbes, F. Pavese, and D. Richter (eds.), World Scientific, Singapore, 2000, 256–272.
61. D. Zwick: Algorithms for orthogonal fitting of lines and planes: a survey. In: *Advanced Mathematical Tools in Metrology II*, P. Ciarlini, M.G. Cox, F. Pavese, and D. Richter (eds.), World Scientific, Singapore, 1996, 272–283.

Discontinuous Galerkin Methods for Linear Problems: An Introduction

Emmanuil H. Georgoulis

Department of Mathematics, University of Leicester, LE1 7RH, UK

Summary. Discontinuous Galerkin (dG) methods for the numerical solution of partial differential equations (PDE) have enjoyed substantial development in recent years. Possible reasons for this are the flexibility in local approximation they offer, together with their good stability properties when approximating convection-dominated problems. Owing to their interpretation both as Galerkin projections onto suitable energy (native) spaces and, simultaneously, as high order versions of classical upwind finite volume schemes, they offer a range of attractive properties for the numerical solution of various classes of PDE problems where classical finite element methods under-perform, or even fail. These notes aim to be a gentle introduction to the subject.

1 Introduction

Finite element methods (FEM) have been proven to be extremely useful in the numerical approximation of solutions to self-adjoint or "nearly" self-adjoint elliptic PDE problems and related indefinite PDE systems (e.g., Darcy's equations, Stokes' system, elasticity models), or to their parabolic counterparts.

Possible reasons for the success of FEM are their applicability in very general computational geometries of interest and the availability of tools for their rigorous error analysis. The error analysis is usually based on the variational interpretation of the FEM as a minimisation problem over finite-dimensional sets (or gradient flows of such, in the case of parabolic PDEs). The variational structure is inherited by the corresponding variational interpretation of the underlying PDE problems, thereby facilitating the use of tools from PDE theory for the error analysis of the FEM.

However, the use of (classical) FEM for the numerical solution of hyperbolic (or "nearly" hyperbolic) problems and other strongly non-self-adjoint PDE problems is, generally speaking, not satisfactory. These problems do not arise naturally in a variational setting. Indeed, the use of FEM for such problems has been mainly of academic interest in the 1970's and 1980's and for

E.H. Georgoulis, A. Iske, J. Levesley (eds.), *Approximation Algorithms for Complex Systems*, Springer Proceedings in Mathematics 3, DOI: 10.1007/978-3-642-16876-5_5,
© Springer-Verlag Berlin Heidelberg 2011

most of the 1990's. Instead, finite volume methods (FVM) have been predominantly used in industrial software packages for the numerical solution of hyperbolic (or "nearly" hyperbolic) systems, especially in the area of Computational Fluid Dynamics.

Nevertheless, in 1971 Reed and Hill [46] proposed a new class of FEM, namely the *discontinuous Galerkin finite element method* (dG method, for short) for the numerical solution of the nuclear transport PDE problem, which involves a linear first-order hyperbolic PDE. This methods was later analysed by LeSaint and Raviart [42] and by Johnson and Pitkäranta [39]. A significant volume of literature on dG methods for hyperbolic problems has since appeared in the literature; we suggest interested readers consult [17, 16, 14, 19, 24, 13, 8], the volume [15] and the references therein.

In the area of elliptic problems, Nitsche's seminal work on weak imposition of essential boundary conditions [44] for (classical) FEM, allowed for finite element solution spaces that do not satisfy the essential boundary conditions. This was followed up a few years later by Baker [6] who proposed the first modern discontinuous Galerkin method for elliptic problems, later followed by Wheeler [54], Arnold [3] and others. We also mention here the relevant finite element method with penalty of Babuška [5]. Since then, a plethora of DGFEMs have been proposed for a variety of PDE problems: we refer to [15, 34, 48] and the references therein for details.

DG methods exhibit attractive properties for the numerical approximation of problems of hyperbolic or nearly–hyperbolic type, compared to both classical FEM and FVM. Indeed, in contrast with classical FEM, but together with FVM, dG methods are, by construction, locally (or "nearly" locally) conservative with respect to the state variable; moreover, they exhibit enhanced stability properties in the vicinity of sharp gradients (e.g., boundary or interior layers) and/or discontinuities which are often present in the analytical solution of convection/transport dominated PDE problems. Additionally, dG methods offer advantages in the context of automatic local mesh and order adaptivity, such as increased flexibility in the mesh design (irregular grids are admissible) and the freedom to choose the elemental polynomial degrees without the need to enforce any conformity requirements. The implementation of genuinely (locally varying) high-order reconstruction techniques for FVM still remains a computationally difficult task, particularly on general unstructured hybrid grids.

Therefore, dG methods emerge as a very attractive class of arbitrary order methods for the numerical solution of various classes of PDE problems where classical FEM are not applicable and FVM produce typically low order approximations.

The rest of this work is structured as follows. In Section 2, we give a brief revision of the classical FEM for elliptic problems, along with its error analysis, and we discuss its limitations. An introduction to the philosophy of dG methods, along with some basic notation is given in Section 3. Section 4 deals with the construction of the popular interior-penalty dG method for linear elliptic problems, along with derivation of a priori and a posteriori error bounds. In Section 5, we present a dG method for first order linear hyperbolic problems, along with its a priori error analysis. In Section 7, two numerical experiments indicating the good performance of the dG method for PDE problems of mixed type, are given. Section 8 deals with the question of the efficient solution of the large linear systems arising from the discretization using dG methods, while Section 9 contains some final concluding remarks.

1.1 Sobolev Spaces

We start by recalling the notion of a Sobolev space, based on the Lebesgue space $L^p(\omega)$, $p \in [1,\infty]$, for some open domain $\omega \subset \mathbb{R}^d$, $d = 1,2,3$ (for more on Sobolev spaces see, e.g, [1]).

Definition 1. *For $k \in \mathbb{N} \cup \{0\}$, we define the* Sobolev space $W_p^k(\omega)$ *over an open domain $\omega \subset \mathbb{R}^d$, $d = 1,2,3$, by*

$$W_p^k(\omega) := \{u \in L^p(\omega) : D^\alpha u \in L^p(\kappa) \text{ for } |\alpha| \le k\},$$

with $\alpha = (\alpha_1, \dots, \alpha_d)$ being the standard multi-index notation. We also define the associated norm $\|\cdot\|_{W_p^k(\omega)}$ and seminorm $|\cdot|_{W_p^k(\omega)}$ by:

$$\|u\|_{W_p^k(\omega)} := \Big(\sum_{|\alpha|\le k} \|D^\alpha u\|_{L^p(\omega)}^p \Big)^{\frac{1}{p}}, \quad |u|_{W_p^k(\omega)} := \Big(\sum_{|\alpha|= k} \|D^\alpha u\|_{L^p(\omega)}^p \Big)^{\frac{1}{p}},$$

for $p \in [1,\infty)$, and

$$\|u\|_{W_\infty^k(\omega)} := \max_{|\alpha|\le s} \|D^\alpha u\|_{L^\infty(\omega)}, \quad |u|_{W_\infty^k(\omega)} := \max_{|\alpha|= k} \|D^\alpha u\|_{L^\infty(\omega)},$$

for $p = \infty$, respectively, for $k \in \mathbb{N}\cup\{0\}$. For $p = 2$, we shall use the abbreviated notation $W_2^k(\omega) \equiv H^s(\omega)$; equipped with the standard inner product, these spaces become Hilbert spaces. For $k = 0$, $p = 2$, we retrieve the standard $L^2(\omega)$ space, whose norm is abbreviated to $\|\cdot\|_\omega$, with associated inner product denoted by $\langle\cdot,\cdot\rangle_\omega$.

Negative and fractional order Sobolev spaces (i.e., where the *Sobolev index* $k \in \mathbb{R}$) are also defined by (standard) duality and function-space interpolation procedures, respectively, (for more on these techniques see, e.g., [1]). Also, we shall make use of Sobolev spaces on manifolds, as we are interested in the regularity properties of functions on boundaries of domains. These are defined in a standard fashion via diffeomorphisms and partition of unity arguments (see, e.g., [47] for a nice exposition). Finally, we shall denote by $H_0^1(\omega)$ the space

$$H_0^1(\omega) := \{v \in H^1(\omega) : v = 0 \text{ on } \partial\omega\}.$$

2 The Finite Element Method

We illustrate the classical Finite Element Method for linear elliptic problems by considering the Poisson problem with homogeneous Dirichlet boundary conditions over an open bounded polygonal domain $\Omega \subset \mathbb{R}^d$, $d = 2, 3$:

$$\begin{aligned} -\Delta u =& f, \qquad && \text{in } \Omega \\ u =& 0, && \text{on } \partial\Omega, \end{aligned} \tag{1}$$

where $f : \Omega \to \mathbb{R}$ is a known function.

To simplify the notation, we use the following abbreviations for the L^2-norm and the corresponding inner product when defined over the computational domain Ω: $\|\cdot\|_\Omega \equiv \|\cdot\|$ and $\langle\cdot,\cdot\rangle_\Omega \equiv \langle\cdot,\cdot\rangle$, respectively.

The first step in defining a finite element method is to rewrite the problem (1) in the so-called *weak form* or *variational form*. Let $V = H_0^1(\Omega)$ be the *solution space* and consider a function $v \in V$. Upon multiplication of the PDE in (1) by v (usually, referred to as the *test function*) and integration over the domain Ω, we obtain

$$-\int_\Omega \Delta u \, v \, \mathrm{d}x = \int_\Omega f v \, \mathrm{d}x.$$

Applying the divergence theorem to the integral on the left-hand side, and the fact that $v = 0$ on $\partial\Omega$ for all $v \in V$, we arrive at

$$\int_\Omega \nabla u \cdot \nabla v \, \mathrm{d}x = \int_\Omega f v \, \mathrm{d}x,$$

for all $v \in \mathcal{H}$. Hence, the Poisson problem with homogeneous Dirichlet boundary conditions can be transformed to the following problem in *weak form*:

$$\text{Find } u \in V \text{ such that } a(u, v) = \langle f, v\rangle, \quad \text{for all } v \in V, \tag{2}$$

with the *bilinear form* $a(\cdot,\cdot)$ defined by

$$a(u, v) := \int_\Omega \nabla u \cdot \nabla v \, \mathrm{d}x.$$

The second step is to consider an approximation to the problem (2). To this end, we restrict the (infinite-dimensional) space V of eligible solutions to a finite-dimensional subspace $V_h \subset V$ and we consider the approximation problem:

$$\text{Find } u_h \in V_h \text{ such that } a(u_h, v_h) = \langle f, v_h\rangle, \quad \text{for all } v_h \in V_h. \tag{3}$$

This procedure is usually referred to as the *Galerkin projection* (also known as the *Ritz projection* when $a(\cdot,\cdot)$ is a symmetric bilinear form, as is the case here).

Setting $v = v_h \in V_h$ in (2) and subtracting (3) from the resulting equation, we arrive at

$$a(u - u_h, v_h) = 0 \quad \text{for all} \quad v_h \in V_h. \tag{4}$$

The identity (4) is usually referred to as the *Galerkin orthogonality property.* Noting that, in this case, the bilinear form $a(\cdot, \cdot)$ satisfies the properties of an inner product on $H_0^1(\Omega)$, the Galerkin orthogonality property states that u_h is the best approximation of u in V_h, with respect to the inner product defined by the bilinear form $a(\cdot, \cdot)$.

We remark that, although $a(\cdot, \cdot)$ may not satisfy the properties of an inner product if it is not symmetric (e.g., as a result of writing a non-self-adjoint PDE problem in weak form), Galerkin orthogonality still holds by construction, as long as the approximation is *conforming*, that is as long as $V_h \subset V$.

From the analysis point of view, there is great flexibility in the choice of an appropriate approximation space. The conformity of the approximation space requires $V_h \subset V$. To investigate what other assumptions on V_h are sufficient for (3) to deliver a useful approximation, we consider a family of basis functions ψ_i, with $i = 1, 2, \ldots, N$, for some $N \in \mathbb{N}$, spanning V_h, viz., $V_h = \text{span}\{\psi_i : i = 1, 2, \ldots, N\}$. Due to linearity of the bilinear form, the approximation problem (3) is equivalent to the problem:

$$\text{Find } u_h \in V_h \text{ such that } a(u_h, \psi_i) = \langle f, \psi_i \rangle, \quad \text{for all } i = 1, 2, \ldots, N. \tag{5}$$

Since $u_h \in V_h$, there exist $U_j \in \mathbb{R}$, $j = 1, 2, \ldots, N$, so that $u_h = \sum_{j=1}^N U_j \psi_j$, which upon insertion into (5), leads to the linear system

$$A\mathbf{U} = \mathbf{F}, \tag{6}$$

with $A = [A_{ij}]_{i,j=1}^N$, $\mathbf{U} = (U_1, \ldots, U_N)^T$ and $\mathbf{F} = (F_1, \ldots, F_N)^T$, where

$$A_{ij} = \int_\Omega \nabla \psi_j \cdot \nabla \psi_i \, \mathrm{d}x, \quad \text{and} \quad F_i = \int_\Omega f \psi_i \, \mathrm{d}x.$$

Notice that the matrix A is symmetric. For the approximation u_h to be well defined, the linear system (6) should have a unique solution. It is, therefore, reasonable to consider a space V_h so that the matrix A is positive definite.

Further restrictions on the choices of "good" subspaces V_h become evident when considering the practical implementation of the Galerkin procedure. In particular, the supports of the basis functions ψ_i should be a covering the computational domain Ω, while being simultaneously relatively simple in shape so that the entries A_{ij} can be computed in an efficient fashion. Also, given that the linear system (6) can be quite large, it would be an advantage if A is a sparse matrix, to reduce the computational cost of solving (6).

The (classical) finite element method (FEM) is defined by the Galerkin procedure described above through a particular choice of the subspace V_h, which we now describe.

We begin by splitting the domain Ω into a covering $\mathcal{T}$, which will be referred to as the *triangulation* or the *mesh*, consisting of open triangles if $d = 2$ or open tetrahedra if $d = 3$, which we shall refer to as the *elements*, with the following properties:

(a) $\Omega = \cup_{T\in\mathcal{T}}\bar{T}$, with $\bar{\cdot}$ denoting the closure of a set in $\mathbb{R}^d$;
(b) for $T, S \in \mathcal{T}$, we only have the possibilities: either $T = S$, or $\bar{T} \cap \bar{S}$ is a common (whole) $d-k$-dimensional face, with $1 \leq k \leq d$ (i.e., face, edge or vertex, respectively).

In Figure 1, we illustrate a mesh for a domain $\Omega \subset \mathbb{R}^2$.

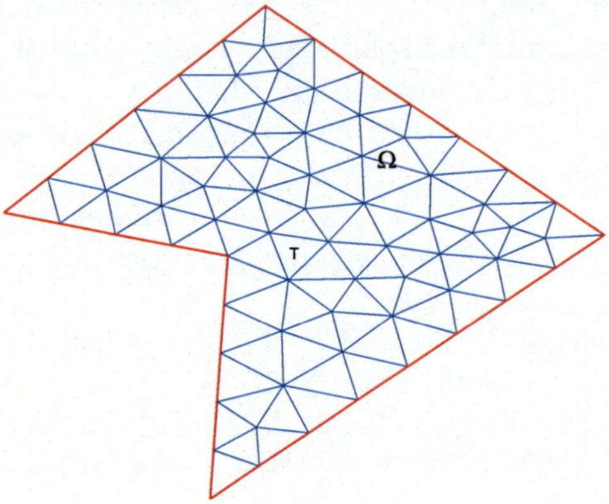

Fig. 1. A mesh in two dimensions

The finite element space V_h^p of degree p is then defined as the space of element-wise d-variate polynomials of degree at most p that are continuous across the inter-element boundaries, viz.,

$$V_h^p := \{w_h \in C(\Omega) : w_h|_T \in \mathcal{P}_p(T),\ T \in \mathcal{T} \text{ and } w_h|_{\partial\Omega} = 0\}$$

with $\mathcal{P}_p(T)$ denoting the space of d-variate polynomials of degree at most p. It is evident that $V_h^p \subset H_0^1(\Omega) = V$. The Galerkin procedure with the particular choice of $V_h = V_h^p$ is called the (classical) *finite element method.*

We observe that this choice of finite dimensional subspace is in line with the practical requirements that a "good" subspace should admit. Indeed, the element-wise polynomial functions over simple triangular or tetrahedral domains enable efficient quadrature calculations for the entries of the matrix A. Moreover, choosing carefully a basis for V_h^p (for instance, the *Lagrange elements*, see, e.g., [12, 9] for details) the resulting linear system becomes sparse. This is because the Lagrange elements have very small support consisting of an element together with only *some* of its immediate neighbours sharing a face or an edge or a vertex. Moreover, as we shall see next, the choice of V_h^p yields a positive definite matrix A, thereby it is uniquely solvable.

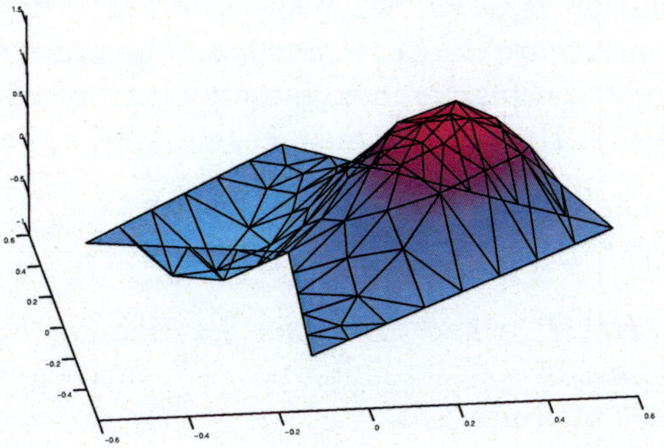

Fig. 2. Example 1. Approximation using the Finite Element Method

Example 1. We use the finite element method to approximate the solution to the Poisson problem with homogeneous Dirichlet boundary conditions:

$$-\Delta u = 100\sin(\pi x), \quad \text{in } \Omega, \quad u = 0, \quad \text{on } \partial\Omega,$$

where Ω is given by the domain in Figure 1, along with the mesh used in the approximation. The finite element approximation for using element-wise linear basis (i.e., $V_h = V_h^1$) is shown in Figure 2.

2.1 Error Analysis of the FEM

Let $\|\nabla w\| := \left(\int_\Omega |\nabla w|^2 \,\mathrm{d}x\right)^{1/2}$ for a weakly differentiable scalar function w, and note that $\|\nabla\cdot\|$ is a norm on $V = H_0^1(\Omega)$. It is evident that that $a(w, w) = \|\nabla w\|^2$ for $w \in V$, i.e., the bilinear form is *coercive in* V. This immediately implies that $a(\cdot,\cdot)$ is also coercive in the closed subspace V_h^p of V. Hence $\|\nabla\cdot\|$ is a norm on V_h^p also and, thus, the corresponding matrix A is positive definite, yielding unique solvability of (6) for $V_h = V_h^p$. The norm for which the bilinear form is coercive for is usually referred to as the *energy norm.*

Coercivity, Galerkin orthogonality (4) and the Cauchy-Schwarz inequality, respectively, imply

$$\begin{aligned}\|\nabla(u - u_h)\|^2 &= a(u - u_h, u - u_h) = a(u - u_h, u - v_h)\\ &\le \|\nabla(u - u_h)\|\,\|\nabla(u - v_h)\|,\end{aligned}$$

for any $v_h \in V_h^p$, which yields

$$\|\nabla(u - u_h)\| \leq \inf_{v_h \in V_h^p} \|\nabla(u - v_h)\|, \tag{7}$$

that is, the finite element method produces the best approximation of V_h^p for the exact solution $u \in V$ with respect to the energy norm. This result is known in the literature as *Cea's Lemma.* The error analysis of the FEM can be now completed using (7) in conjunction with Jackson-type inequalities (such as the Bramble-Hilbert Lemma, see, e.g., [9]) of the form

$$\inf_{v_h \in V_h^p} \|\nabla(u - v_h)\| \leq Ch^{\min\{p,r-1\}}|u|_{H^r(\Omega)}, \tag{8}$$

for $u \in H^r(\Omega) \cap H_0^1(\Omega)$, where $h = \max_{T \in \mathcal{T}} \operatorname{diam}(T)$ and the constant C is independent of h and of u. Combining now (7) with (8) results to standard a priori error bounds for the FEM:

$$\|\nabla(u - u_h)\| \leq Ch^{\min\{p,r-1\}}|u|_{H^r(\Omega)},$$

i.e., as $h \to 0$, the error decreases at an algebraic rate which depends on the local polynomial degree used and the regularity of the solution in the domain Ω.

3 Discontinuous Galerkin Methods

The restriction $V_h \subset V$ essentially dictates that the underlying space contains only functions of particular smoothness (e.g., when $V = H_0^1(\Omega)$, we choose $V_h \subset \{v \in C(\bar{\Omega}) : v|_{\partial\Omega} = 0\} \subset H_0^1(\Omega)$). Although the FEM is, generally speaking, well suited for PDE problems related to a variational/minimisation setting, it is well known that this restriction can have a degree of severity in the applicability of FEM for a larger class of PDE problems (e.g. first order hyperbolic PDE problems). Since the 1970s there has been a substantial amount of work in the literature on so-called *non-conforming* FEM, whereby $V_h \not\subset V$. The discontinuous Galerkin (dG) methods described below will admit finite element spaces with "severe" non-conformity, i.e., element-wise discontinuous polynomial spaces, viz.,

$$S_h^p := \{w_h \in L^2(\Omega) : w_h|_T \in \mathcal{P}_p(T),\ T \in \mathcal{T}\}.$$

Let us introduce some notation first. We denote by $\mathcal{T}$ a subdivision of Ω into (triangular or quadrilateral if $d = 2$ and tetrahedral or hexahedral if $d = 3$) elements T. We define $\Gamma := \cup_{T \in \mathcal{T}} \partial T$ the *skeleton* of the mesh (i.e., the union of all $(d-1)$-dimensional element faces) and let $\Gamma_{\text{int}} := \Gamma \backslash \partial\Omega$. Let also $\Gamma_{\text{int}} := \Gamma \backslash \partial\Omega$, so that $\Gamma = \partial\Omega \cup \Gamma_{\text{int}}$.

Let T^+, T^- be two (generic) elements sharing a face $e := T^+ \cap T^- \subset \Gamma_{\text{int}}$ with respective outward normal unit vectors $\boldsymbol{n}^+$ and $\boldsymbol{n}^-$ on e. For $q : \Omega \to \mathbb{R}$ and $\boldsymbol{\phi} : \Omega \to \mathbb{R}^d$, let $q^\pm := q|_{e \cap \partial T^\pm}$ and $\boldsymbol{\phi}^\pm := \boldsymbol{\phi}|_{e \cap \partial T^\pm}$, and set

$$\{q\}|_e := \frac{1}{2}(q^+ + q^-), \qquad \{\boldsymbol{\phi}\}|_e := \frac{1}{2}(\boldsymbol{\phi}^+ + \boldsymbol{\phi}^-),$$
$$[\![q]\!]|_e := q^+\boldsymbol{n}^+ + q^-\boldsymbol{n}^-, \quad [\![\boldsymbol{\phi}]\!]|_e := \boldsymbol{\phi}^+ \cdot \boldsymbol{n}^+ + \boldsymbol{\phi}^- \cdot \boldsymbol{n}^-;$$

if $e \subset \partial T \cap \partial\Omega$, we set $\{\boldsymbol{\phi}\}|_e := \boldsymbol{\phi}^+$ and $[\![q]\!]|_e := q^+\boldsymbol{n}^+$. Finally, we introduce the *meshsize* $h : \Omega \to \mathbb{R}$, defined by $h(x) = \operatorname{diam}(T)$, if $x \in T \backslash \partial T$ and $h(x) = \{h\}$, if $x \in \Gamma$. The subscript e in these definitions will be suppressed when no confusion is likely to occur.

4 Discontinuous Galerkin Methods for Elliptic Problems

Now we are ready to derive the weak form for the Poisson problem (1), which will lead to the discontinuous Galerkin (dG) finite element method.

Since the dG method will be non-conforming, we should work on an extended variational framework, making use of the space $\mathcal{S} := H_0^1(\Omega) + S_h^p$. Assuming for the moment that u is smooth enough, we multiply the equation by a test function $v \in \mathcal{S}$, we integrate over Ω and we split the integrals:

$$-\sum_{T \in \mathcal{T}} \int_T \Delta u v \,\mathrm{d}x = \sum_{T \in \mathcal{T}} \int_T f v \,\mathrm{d}x.$$

Using the divergence theorem on every elemental integral (as v is now element-wise discontinuous), using the anti-clockwise orientation, we have

$$\sum_{T \in \mathcal{T}} \int_T \nabla u \cdot \nabla v \,\mathrm{d}x - \sum_{T \in \mathcal{T}} \int_{\partial T} (\nabla u \cdot \boldsymbol{n}) v \,\mathrm{d}s = \int_\Omega f v \,\mathrm{d}x = \langle f, v \rangle,$$

where $\boldsymbol{n}$ denotes the outward normal to each element edge.

The second term on the left-hand side contains the integrals over the element faces. Thus, when the face is common to two adjacent element, we have two integrals over every interior face.

Now, from standard elliptic regularity estimates (see, e.g., Corollary 8.36 in Gilbarg & Trudinger [32]), we have that $u \in C^1(\Omega')$ for all $\Omega' \subset \Omega$ and, hence, ∇u is continuous across the interior element faces. Thus we can substitute ∇u by $\{\nabla u\}$ for all faces on the skeleton Γ, noting that this is just the definition of $\{\nabla u\}$ on the boundary $\partial\Omega$. Taking into account the orientation convention we have adopted, we can see that this sum can be rewritten as follows:

$$\sum_{T \in \mathcal{T}} \int_T \nabla u \cdot \nabla v \,\mathrm{d}x - \int_\Gamma \{\nabla u\} \cdot [\![v]\!] \,\mathrm{d}s = \langle f, v \rangle. \tag{9}$$

One may now be tempted to define a bilinear form and a linear form from the left- and right-hand sides of (9), respectively and attempt to solve the resulting variational problem. Such an endeavour would be deemed to have failed for the reason that the left-hand side does not give rise to a positive-definite operator, (not even a conditionally positive definite operator,) over S_h^p. In other words, there is no coercivity property for such a bilinear form in any relevant norm. There is also the somewhat philosophically discomforting issue that a symmetric variational problem (such as the Poisson problem in variational form) is approximated using a Galerkin procedure based on a *non-symmetric* bilinear form (such as the one stemming from the left-hand side of (9)). This issue may also have practical implications as solving non-symmetric linear systems is usually a far more computationally demanding procedure than solving a symmetric linear system.

To rectify the lack of positivity, we work as follows. We begin by noting that $[\![u]\!] = 0$ on Γ, due to elliptic regularity on $\Gamma_{\rm int}$ and due to the boundary conditions on $\partial\Omega$. Therefore, we have

$$\int_\Gamma \sigma [\![u]\!] \cdot [\![v]\!] \,\mathrm{d}s = 0, \tag{10}$$

for any positive function $\sigma : \Gamma \to \mathbb{R}$. Note that this term is symmetric with respect to the two arguments u and v and can be arbitrarily large (upon choosing σ an arbitrarily large positive function) upon replacing u with a function $v \in \mathcal{S}$. Adding (9) and (10) up, we arrive at

$$\sum_{T\in\mathcal{T}} \int_T \nabla u \cdot \nabla v \,\mathrm{d}x - \int_\Gamma (\{\nabla u\} \cdot [\![v]\!] - \sigma [\![u]\!] \cdot [\![v]\!]) \,\mathrm{d}s = \langle f, v \rangle. \tag{11}$$

Since the term of the left-hand side of (11), stemming from (10) gives rise to a positive-definite term in the bilinear form, which implies that there is a range of (large enough) σ that will render the resulting bilinear form coercive (at least over a finite dimensional subspace of $\mathcal{S}$) to give rise to a positive definite finite element matrix. The choice of the *discontinuity-penalisation parameter* σ, as it is often called in the literature, will arise from the error analysis of the method.

We note that the left-hand side of (11) is still non-symmetric with respect to the arguments u and v. To rectify this, we observe that also

$$\int_\Gamma \{\nabla v\} \cdot [\![u]\!] \,\mathrm{d}s = 0,$$

assuming that v is smooth enough, which can be subtracted from (11), resulting in

$$\sum_{T\in\mathcal{T}} \int_T \nabla u \cdot \nabla v \,\mathrm{d}x - \int_\Gamma (\{\nabla u\} \cdot [\![v]\!] + \{\nabla v\} \cdot [\![u]\!] - \sigma [\![u]\!] \cdot [\![v]\!]) \,\mathrm{d}s = \langle f, v \rangle,$$

whose left-hand side is now symmetric with respect to the arguments u and v.

The above suggest the following numerical method:

$$\text{Find } u_h \in S_h^p \text{ such that } B(u_h, v_h) = \langle f, v_h \rangle, \quad \text{for all } v_h \in S_h^p, \tag{12}$$

where the bilinear form $B : S_h^p \times S_h^p \to \mathbb{R}$ is defined by

$$B(w,v) := \sum_{T\in\mathcal{T}} \int_T \nabla w \cdot \nabla v \,\mathrm{d}x - \int_\Gamma (\{\nabla w\}\cdot[v] + \{\nabla v\}\cdot[w] - \sigma[w]\cdot[v]) \,\mathrm{d}s. \tag{13}$$

This is the so-called *(symmetric) interior penalty discontinuous Galerkin method* for the Poisson problem.

Historically interior penalty methods were the first to appear in the literature [6, 3], but some of the ideas can be traced back to the treatment of non-homogeneous Dirichlet boundary conditions by penalties due to Nitsche [44]. Interior penalty dG methods are, perhaps, the most popular dG methods in the literature and in applications, so they will be our main focus in the present notes. In recent years, a number of other discontinuous Galerkin methods for second order elliptic problems have appeared in the literature; we refer to [4] and the references therein for a discussion on the unifying characteristics of these methods as well as on the particular advantages and disadvantages of each dG method.

4.1 Error Analysis of the DG Method

In the above derivation of the interior penalty dG method, we have been intentionally relaxed about the smoothness requirements of the arguments of the bilinear forms. The bilinear form (13) is well defined if the arguments w and v belong to the finite element space S_h^p.

However, it is well known from the theory of Sobolev spaces that functions in $L^2(\Omega)$ do not have a well-defined *trace* on $\partial\Omega$, that is, they are not uniquely defined up to boundary values. Therefore, $\{\nabla w\}$ and $\{\nabla v\}$ are not well defined on Γ_{int} in (13) when $w, v \in \mathcal{S}(= H_0^1(\Omega) + S_h^p)$. For the error analysis it is desirable that the bilinear form can be applied to the exact solution u. Fortunately, for (standard) a priori error analysis the exact solution u is assumed to admit at least H^2-regularity which, implies that all the terms in (13) can be taken to be well defined, so this issue does not pose any crucial restriction. For the derivation of a posteriori error bounds describe below, however, assuming the minimum possible regularity for u is essential for their applicability to the most general setting possible.

It is possible to overcome this hindrance by extending the bilinear form (13) from $S_h^p \times S_h^p$ to $\mathcal{S} \times \mathcal{S}$ in a non-trivial fashion. More specifically, we define

$$\tilde{B}(w,v) := \sum_{T\in\mathcal{T}} \int_T \nabla w \cdot \nabla v \,\mathrm{d}x - \int_\Gamma (\{\Pi\nabla w\}\cdot[v] + \{\Pi\nabla v\}\cdot[w] - \sigma[w]\cdot[v]) \,\mathrm{d}s,$$

where $\Pi : L^2(\Omega) \to S_h^p$ is the orthogonal L^2-projection with respect to the $\langle \cdot, \cdot, \rangle$- inner product. This way, the face integrals involving the terms $\{\Pi\nabla w\}$ and $\{\Pi\nabla v\}$ are well defined, as these terms are now traces of element-wise polynomial functions from the finite element space. Moreover, it is evident that

$$\tilde{B}(w,v) = B(w,v), \quad \text{if} \quad w, v \in S_h^p,$$

i.e., $\tilde{B}(\cdot,\cdot)$ is an extension of $B(\cdot,\cdot)$ to $\mathcal{S} \times \mathcal{S}$. However, $\tilde{B}(\cdot,\cdot)$ is *inconsistent* with respect to the Poisson problem, that is, it is not a weak form of (1). Indeed, suppose (1) admits a classical solution denoted by u_{cl}. Upon inserting u_{cl} into $\tilde{B}(\cdot,\cdot)$, and integrating by parts, we deduce

$$-\sum_{T\in\mathcal{T}} \int_T \Delta u_{cl} v \, \mathrm{d}x + \int_\Gamma \{\nabla u_{cl}\} \cdot [v] \, \mathrm{d}s - \int_\Gamma \{\Pi\nabla u_{cl}\} \cdot [v] \, \mathrm{d}s = \langle f, v \rangle,$$

for all $v \in \mathcal{S}$, noting that $[u_{cl}] = 0$ on Γ, which implies

$$\int_\Omega (f + \Delta u_{cl}) v \, \mathrm{d}x = \int_\Gamma \{\nabla u_{cl} - \Pi\nabla u_{cl}\} \cdot [v] \, \mathrm{d}s,$$

for all $v \in \mathcal{S}$; the right-hand side being a representation of the inconsistency. (If $\tilde{B}(\cdot,\cdot)$ was consistent, the right-hand side should have been equal to zero.) Nevertheless, as we shall see below, the inconsistency is of the same order as the convergence rate of the dG method and it is, therefore, a useful tool in the error analysis.

For the error analysis, we consider the following (natural) quantity:

$$|||w||| := \Big(\sum_{T\in\mathcal{T}} \int_T |\nabla w|^2 \, \mathrm{d}x + \int_\Gamma \sigma |[\![w]\!]|^2 \, \mathrm{d}s \Big)^{1/2},$$

for all $w \in \mathcal{S}$ and for $\sigma > 0$. Note that $|||\cdot|||$ is a norm in $\mathcal{S}$.

We begin the error analysis by assessing the coercivity and the continuity of the bilinear form.

Lemma 1. *Let constant* $c > 0$ *such that* $\operatorname{diam}(T)/\rho_T \leq c$ *for all* $T \in \mathcal{T}$, *where* ρ_T *is the radius of the incircle of* T. *Let also* $\sigma := C_\sigma p^2/h$ *with* $C_\sigma > 0$ *large enough and independent of* p, h *and of* $w, v \in \mathcal{S}$. *Then, we have*

$$\frac{1}{2} |||w|||^2 \leq \tilde{B}(w,w), \quad \textit{for all} \quad w \in \mathcal{S}, \tag{14}$$

and

$$\tilde{B}(w,v) \leq |||w||| \, |||v|||, \quad \textit{for all} \quad w, v \in \mathcal{S}. \tag{15}$$

Proof. We prove (14). For $w \in \mathcal{S}$, we have:

$$\tilde{B}(w,w) = |||w|||^2 - 2 \int_\Gamma \{\Pi\nabla w\} \cdot [w] \, \mathrm{d}s.$$

Now the last term on the right-hand side can be bounded from above as follows:

$$
\begin{aligned}
2\int_{\Gamma} \{\Pi\nabla w\}\cdot [w]\,\mathrm{d}s =& 2\int_{\Gamma} \left(\frac{\sigma}{2}\right)^{-1/2} \{\Pi\nabla w\}\cdot \left(\frac{\sigma}{2}\right)^{1/2} [w]\,\mathrm{d}s \\
\leq& 2\int_{\Gamma} \sigma^{-1}|\{\Pi\nabla w\}|^2\,\mathrm{d}s + \frac{1}{2}\int_{\Gamma}\sigma[w]^2\,\mathrm{d}s,
\end{aligned}
\tag{16}
$$

using in the last step an inequality of the form $2\alpha\beta \leq \alpha^2 + \beta^2$.

To bound from above the first term on the right-hand side of the last inequality, we make use of the *inverse inequality*:

$$
\|v\|_{\partial T}^2 \leq C_{\mathrm{inv}} p^2 |\partial T|/|T| \|v\|_T^2, \tag{17}
$$

for all $v \in \mathcal{P}_p(T)$, with $C_{\mathrm{inv}} > 0$ independent of p, h and v, with $|\partial T|$ and $|T|$ denoting the $(d-1)$- and d-dimensional volumes of ∂T and T, respectively. (We refer to Theorem 4.76 in [51] for a proof, when T is the reference element; the proof for a general element follows by a standard scaling argument.) To this end, we have

$$
2\int_{\Gamma} \sigma^{-1}|\{\Pi\nabla w\}|^2\,\mathrm{d}s \leq \sum_{T\in\mathcal{T}} \int_{\partial T} \sigma^{-1}|\Pi\nabla w|^2\,\mathrm{d}s, \tag{18}
$$

using an inequality of the form $(\alpha+\beta)^2 \leq 2\alpha^2 + 2\beta^2$. The right-hand side of (18) can be further bounded using (17), noting that $(\Pi\nabla w)|_T \in \mathcal{P}_p(T)$ for all $T \in \mathcal{T}$, giving

$$
\sum_{T\in\mathcal{T}} \int_{\partial T} \sigma^{-1}|\Pi\nabla w|^2\,\mathrm{d}s \leq \sum_{T\in\mathcal{T}} \frac{C_{\mathrm{inv}} p^2 |T|}{\sigma|\partial T|} \int_T |\Pi\nabla w|^2\,\mathrm{d}x. \tag{19}
$$

The orthogonal L^2-projection operator is stable in the L^2-norm, with $\|\Pi v\|_T \leq \|v\|_T$ for all $v \in L^2(T)$ which, in conjunction with (18) and (19), gives

$$
2\int_{\Gamma} \sigma^{-1}|\{\Pi\nabla w\}|^2\,\mathrm{d}s \leq \sum_{T\in\mathcal{T}} \frac{C_{\mathrm{inv}} p^2}{\sigma\rho_T} \int_T |\nabla w|^2\,\mathrm{d}x, \tag{20}
$$

noting that $\rho_T \leq |T|/|\partial T|$. Choosing $C_\sigma \geq 2c^2 C_{\mathrm{inv}}$, implies $C_{\mathrm{inv}} p^2 h/(\sigma\rho_T) \leq 1/2$, as $h \leq \mathrm{diam}(T) \leq c\rho_T$ and $\rho_T \leq c\rho_{T'}$ for all elements T' sharing a face with T (the latter is due to assumption $\mathrm{diam}(T)/\rho_T \leq c$ for all $T \in \mathcal{T}$). Hence, combining (20) with (16) already implies (14).

The proof (15) uses the Cauchy-Schwarz inequality along with the same tools as above and it is omitted for brevity. □

Remark 1. It can be seen from the proof that the coercivity and continuity constants in the previous result can be different, depending the choice of the penalty parameter σ.

A Priori Error Bounds

Since the bilinear form is now inconsistent, Galerkin orthogonality does not hold for dG (cf. (4) which, in turn complicates slightly the a priori error analysis. To this end, let $u_h \in S_h^p$ be the dG approximation to the exact solution u, arising from solving (12) and consider a $v_h \in S_h^p$. Then, we have

$$\begin{aligned}\frac{1}{2}|||v_h - u_h|||^2 \leq& \tilde{B}(v_h - u_h, v_h - u_h)\\ =& \tilde{B}(v_h - u, v_h - u_h) + \tilde{B}(u - u_h, v_h - u_h)\\ =& \tilde{B}(v_h - u, v_h - u_h) + \tilde{B}(u, v_h - u_h) - \langle f, v_h - u_h\rangle\end{aligned}$$

using coercivity (14), the linearity of the bilinear form and the definition of the dG method (12), respectively. Using the continuity (15) of the bilinear form, and diving by $|||v_h - u_h|||$, we arrive at

$$|||v_h - u_h||| \leq 2|||u - v_h||| + \sup_{w_h \in S_h^p} \frac{|\tilde{B}(u, w_h) - \langle f, w_h\rangle|}{|||w_h|||},$$

for all $v_h \in S_h^p$. Hence, we can conclude

$$|||v_h - u_h||| \leq 2 \inf_{v_h \in S_h^p} |||u - v_h||| + \sup_{w_h \in S_h^p} \frac{|\tilde{B}(u, w_h) - \langle f, w_h\rangle|}{|||w_h|||},$$

or

$$|||u - u_h||| \leq 3 \inf_{v_h \in S_h^p} |||u - v_h||| + \sup_{w_h \in S_h^p} \frac{|\tilde{B}(u, w_h) - \langle f, w_h\rangle|}{|||w_h|||}, \qquad (21)$$

using the triangle inequality. This result is a generalisation of Ceá's Lemma presented above for the case where the Galerkin orthogonality is not satisfied exactly; it is known in the literature as *Strang's Second Lemma* (see, e.g., [53, 12]). Indeed, if the bilinear form is consistent, then the last term on the right-hand side of (21) vanishes.

For the first term on the right-hand side of (21), we can use standard best approximation results (such as the Bramble-Hilbert Lemma, see, e.g., [9]) of the form

$$\inf_{v_h \in V_h^p} |||\nabla(u - v_h)||| \leq C h^{\min\{p, r-1\}} |u|_{H^r(\Omega)}, \qquad (22)$$

for $u \in H^r(\Omega) \cap H_0^1(\Omega)$, noting that the parameter σ in the dG-norm $|||\cdot|||$ scales like h^{-1} and using the standard trace estimate

$$\|w\|_{\partial T}^2 \leq C\big(\mathrm{diam}(T)^{-1}\|w\|_T^2 + \mathrm{diam}(T)\|\nabla w\|_T^2\big), \qquad (23)$$

on the term of $|||\cdot|||$ involving σ, $T \in \mathcal{T}$.

To bound the second term on the right-hand side (21), we begin by observing that

$$\tilde{B}(u, w_h) - \langle f, w_h \rangle = - \int_{\Gamma} \{\nabla u - \Pi \nabla u\} \cdot [w_h] \, \mathrm{d}s,$$

which implies

$$\sup_{w_h \in S_h^p} \frac{|\tilde{B}(u, w_h) - \langle f, w_h \rangle|}{|||w_h|||} \leq \Big(\int_{\Gamma} \sigma^{-1} |\{\nabla u - \Pi \nabla u\}|^2 \, \mathrm{d}s \Big)^{1/2}. \tag{24}$$

Letting $\eta := |\nabla u - \Pi \nabla u|$ for brevity and, working as before, the square of right-hand side of (24) can be bounded by

$$\int_{\Gamma} \sigma^{-1} |\{\eta\}|^2 \, \mathrm{d}s \leq \frac{1}{2} \sum_{T \in \mathcal{T}} \int_{\partial T} \sigma^{-1} \eta^2 \, \mathrm{d}s \leq C \sum_{T \in \mathcal{T}} \big(\|\eta\|_T^2 + \|h \nabla \eta\|_T^2 \big), \tag{25}$$

using the trace estimate (23) and the definition of σ. Now, standard best approximation results for the L^2-projection error (see, e.g., [51]) yield

$$\big(\|\eta\|_T^2 + \|h \nabla \eta\|_T^2 \big)^{1/2} \leq C h^{\min\{p, r-1\}} |u|_{H^r(\Omega)}, \tag{26}$$

for $u \in H^r(\Omega) \cap H_0^1(\Omega)$. Combining (26), (25), (24), and using the resulting bound together with (22) into (21), we arrive at the *a priori error bound*

$$|||u - u_h||| \leq C h^{\min\{p, r-1\}} |u|_{H^r(\Omega)},$$

for $u \in H^r(\Omega) \cap H_0^1(\Omega)$, for some $C > 0$ independent of u and of h.

We refer to [6, 3, 15, 50, 36, 4, 31, 28] and the references therein for discussion on a priori error analysis of interior penalty-type dG methods for elliptic problems.

A Posteriori Error Bounds

The above a priori bounds are relevant when we are interested in assessing the asymptotic error behaviour of the dG method. However, since they involve the unknown solution to the boundary-value problem u, they are not of relevance in practice. The derivation of computable bounds, usually referred to the finite element literature as *a posteriori estimates* is therefore, relevant to assess the accuracy of practical computations. Moreover, such bounds can be used to drive automatic mesh-adaptation procedures, usually termed *adaptive algorithms.* A posteriori bounds for dG methods for elliptic problems have been considered in [7, 40, 35, 11, 22, 23, 41]. Here, we shall only illustrate the main ideas in a simple setting.

We begin by decomposing the discontinuous finite element space S_h^p into the conforming finite element space $V_h^p \subset S_h^p$ and a *non-conforming* remainder part V_d, so as $S_h^p := V_h^p \oplus V_d$, where the uniqueness-of-the-decomposition property in the direct sum can be realised, once an inner product in S_h^p is selected. The approximation of functions in S_h^p by functions in the conforming finite element space V_h^p will play an important role in our derivation of the a posteriori bounds. This can be quantified by the following result, whose proof can be found in [40].

Lemma 2. *For a mesh $\mathcal{T}$, let constant $c > 0$ such that* $\text{diam}(T)/\rho_T \leq c$ *for all $T \in \mathcal{T}$, where ρ_T is the radius of the incircle of T. Then, for any function $v \in S_h^p$, there exists a function $v_c \in V_h^p$ such that*

$$\|\nabla(v - v_c)\| \leq C_1 \|\sqrt{\sigma}[\![v]\!]\|_{\Gamma}, \tag{27}$$

where the constant $C_1 > 0$ depends on c, p, but is independent of h, v, and v_c.

Using this lemma it is possible to derive an a posteriori bound for the dG method for the Poisson problem. This is the content of the following result.

Theorem 1. *Let u be the solution (2) and u_h its approximation by the dG method (12). Then, the following bound holds*

$$|\!|\!|u - u_h|\!|\!| \leq C\mathcal{E}(u_h, f, \mathcal{T}),$$

with

$$\mathcal{E}(u_h, f, \mathcal{T}) := \left(\|h(f + \Delta u)\|^2 + \|\sqrt{h}[\![\nabla u_h]\!]\|^2_{\Gamma_{\text{int}}} + \|\sqrt{\sigma}[\![u_h]\!]\|^2_{\Gamma}\right)^{1/2},$$

where $C > 0$ is independent of u_h, u, h and $\mathcal{T}$.

Proof. Let $u_h^c \in S_c$ the conforming part of u_h as in Lemma 2 and define

$$e := u - u_h = e_c + e_d, \quad \text{where} \quad e_c := u - u_h^c, \quad \text{and} \quad e_d := u_h^c - u_h,$$

yielding $e_c \in H_0^1(\Omega)$. Thus, we have $B(u, e_c) = \langle f, e_c\rangle$. Let $\Pi_0 : L^2(\Omega) \to \mathbb{R}$ denote the orthogonal L^2-projection onto the elementwise constant functions; then $\Pi_0 e_c \in S_h^p$ and we define $\eta := e_c - \Pi_0 e_c$. We then have, respectively,

$$\begin{aligned} B(e, e_c) =& B(u, e_c) - B(u_h, e_c) \\ =& \langle f, e_c\rangle - B(u_h, \eta) - B(u_h, \Pi_0 e_c) \\ =& \langle f, \eta\rangle - B(u_h, \eta), \end{aligned}$$

which, noting that $[\![e_c]\!] = 0$ on Γ, implies

$$\|\nabla e_c\|^2 = B(e_c, e_c) = \langle f, \eta\rangle - B(u_h, \eta) - B(e_d, e_c). \tag{28}$$

For the last term on the right-hand side of (28), we have

$$|B(e_d, e_c)| \leq \|\nabla e_d\|\|\nabla e_c\| + \frac{1}{2}\sum_{e\subset\Gamma}\sum_{T=T^+,T^-} \|\sqrt{h}(\Pi\nabla e_c)|_T\|_e \|h^{-1/2}[\![e_d]\!]\|_e, \tag{29}$$

where κ^+ and κ^- are the (generic) elements having e as common face. Using the inverse inequality (17) and the stability of the L^2-projection, we arrive at

$$|B(e_d, e_c)| \leq \|\nabla e_d\|\|\nabla e_c\| + C\|\nabla e_c\|\|\sqrt{\sigma}[\![e_d]\!]\|_{\Gamma}.$$

Finally, noting that $[\![e_d]\!] = [\![u_h]\!]$, and making use of (27) we conclude that

$$|B(e_d, e_c)| \leq C\|\nabla e_c\| \, \|\sqrt{\sigma}[\![u_h]\!]\|_\Gamma.$$

To bound the first two terms on the right-hand side of (28), we begin by an element-wise integration by parts yielding

$$\begin{aligned} \langle f, \eta \rangle - B(u_h, \eta) = \sum_{T \in \mathcal{T}} \int_T (f + \Delta u_h)\eta - \int_{\Gamma_{\text{int}}} \{\eta\}[\![\nabla u_h]\!] \, \mathrm{d}s \\ - \int_\Gamma \{\Pi \nabla \eta\} \cdot [\![u_h]\!] \, \mathrm{d}s - \int_\Gamma \sigma [\![u_h]\!] \cdot [\![\eta]\!] \, \mathrm{d}s. \end{aligned} \tag{30}$$

The first term on the right-hand side of (30) can be bounded as follows:

$$\begin{aligned} \Big| \sum_{T \in \mathcal{T}} \int_T (f + \Delta u_h)\, \eta \Big| \leq & \|h(f + \Delta u_h)\| \, \|h^{-1}\eta\| \\ \leq & C\|h(f + \Delta u_h)\| \, \|\nabla e_c\|, \end{aligned}$$

upon observing that $\|h^{-1}\eta\|_\kappa \leq C\|\nabla e_c\|_\kappa$.

For the second term on the right-hand side of (30), we use the trace estimate (23), the bound $\|h^{-1}\eta\|_\kappa \leq C\|\nabla e_c\|_\kappa$ and the observation that $\nabla \eta = \nabla e_c$, to deduce

$$\Big| \int_{\Gamma_{\text{int}}} \{\eta\}[\![\nabla u_h]\!] \, \mathrm{d}s \Big| \leq C\|\nabla e_c\| \, \|\sqrt{h}[\![\nabla u_h]\!]\|_{\Gamma_{\text{int}}}.$$

For the third term on the right-hand side of (30), we use $\nabla \eta = \nabla e_c$ and, similar to the derivation of (29), we obtain

$$\Big| \int_\Gamma \{\Pi \nabla \eta\} \cdot [\![u_h]\!] \Big| \leq C\|\nabla e_c\| \, \|\sqrt{\sigma}[\![u_h]\!]\|_\Gamma,$$

and finally, for the last term on the right-hand side of (30), we get

$$\Big| \int_\Gamma \sigma [\![\eta]\!] \cdot [\![u_h]\!] \Big| \leq C\|\nabla e_c\| \, \|\sqrt{\sigma}[\![u_h]\!]\|_\Gamma.$$

The result follows combining the above relations. □

5 DG Methods for First Order Hyperbolic Problems

The development of dG methods for elliptic problems, introduced above, is an interesting theoretical development and offers a number of advantages in particular cases, for instance, when using irregular meshes or, perhaps, "exotic" basis functions such as wavelets. However, the major argument for using dG methods lies with their ability to provide stable numerical methods for first order PDE problems, for which classical FEM is well known to perform poorly.

We consider the first order Cauchy problem

$$\mathcal{L}_0 u \equiv b \cdot \nabla u + cu = f \quad \text{in } \Omega, \tag{31}$$

$$u = g \quad \text{on } \partial_- \Omega, \tag{32}$$

where

$$\partial_- \Omega := \{x \in \partial\Omega : b(x) \cdot n(x) < 0\}$$

is the inflow part of the domain boundary $\partial\Omega$, $b := (b_1, \ldots, b_d) \in [C^1(\bar{\Omega})]^d$ and $g \in L^2(\partial_- \Omega)$.

We assume further that there exists a positive constant γ_0 such that

$$c(x) - \frac{1}{2}\nabla \cdot b(x) \geq \gamma_0 \quad \text{for almost every } x \in \Omega, \tag{33}$$

and we define $c_0 := (c - 1/2\nabla \cdot b)^{1/2}$.

Next, we consider a mesh $\mathcal{T}$ of the domain Ω as above, and we define

$$\partial_- T := \{x \in \partial T : b(x) \cdot n(x) < 0\}, \quad \partial_+ T := \{x \in \partial T : b(x) \cdot n(x) > 0\},$$

for each element T; we call these the *inflow* and *outflow* parts of ∂T respectively. For $T \in \mathcal{T}$, and a (possibly discontinuous) element-wise smooth function v, we consider the *upwind jump* across the inflow boundary $\partial_- T$, by

$$\lfloor v \rfloor(x) := \lim_{t \to 0+} \big(u(x + tb) - u(x - tb)\big),$$

for almost all $x \in \partial_- T$, when $\partial_- T \subset \Gamma_{\text{int}}$, and by $\lfloor v \rfloor(x) := v(x)$ for almost all $x \in \partial_- T$, when $\partial_- T \subset \partial_- \Omega$.

We require some more notation to describe the method. Let $u \in H^1(\Omega, \mathcal{T})$. Then, for every element $T \in \mathcal{T}$, we denote by u_T^+ the trace of u on $\partial\kappa$ taken from within the element T (interior trace). We also define the exterior trace u_T^- of $u \in H^1(\Omega, \mathcal{T})$ for almost all $x \in \partial_- T \backslash \Gamma$ to be the interior trace $u_{T'}^+$ of u on the element(s) T' that share the edges contained in $\partial_- T \backslash \Gamma$ of the boundary of element T. Then, the *jump* of u across $\partial_- T \backslash \Gamma$ is defined by

$$\lfloor u \rfloor_T := u_T^+ - u_T^-.$$

We note that this definition of jump is not the same as the one in the pure diffusion case discussed in the previous section; here the sign of the jump depends on the direction of the flow, whereas in the pure diffusion case it only depends on the element-numbering. Since they may genuinely differ up to a sign, we have used different notation for the jumps in the two cases. Again, we note that the subscripts will be suppressed when no confusion is likely to occur.

We shall now describe the construction of the discontinuous Galerkin weak formulation for the problem (31), (32), by imposing "weakly" the value of the solution on an outflow boundary of an element as an inflow boundary for the neighbouring downstream elements, we solve small local problems, until we have found the solution over the complete domain Ω.

We first construct a local weak formulation on every element T that is attached to the inflow boundary of the domain.

We define the space $\mathcal{S}_{\mathrm{adv}} := G_b + S_h^p$, for $p \geq 0$ (note that $p = 0$ is allowed in the dG discretization of first order hyperbolic problems), where

$$G_b := \{w \in L^2(\Omega) : b \cdot \nabla w \in L^2(\Omega)\},$$

is the graph space of the PDE (31). Multiplying with a test function $v \in \mathcal{S}_{\mathrm{adv}}$ and integrating over T we obtain

$$\int_T (\mathcal{L}_0 u) v \,\mathrm{d}x = \int_T f\, v \,\mathrm{d}x. \tag{34}$$

Now we impose the boundary conditions for the local problem. Since $\partial_- T \cap \Gamma_- \neq \emptyset$ we have $u^+ = g$ on $\partial_- T \cap \Gamma_-$. Therefore, after multiplication by $(b \cdot n)v^+$ and integration over $\partial_- T \cap \Gamma_-$, we get

$$\int_{\partial_- T \cap \Gamma_-} (b \cdot n) u^+ v^+ \,\mathrm{d}s = \int_{\partial_- T \cap \Gamma_-} (b \cdot n) g v^+ \,\mathrm{d}s. \tag{35}$$

Upon subtracting (35) from (34) we have

$$\int_T (\mathcal{L}_0 u) v \,\mathrm{d}x - \int_{\partial_- T \cap \Gamma_-} (b \cdot n) u^+ v^+ \,\mathrm{d}s = \int_T f\, v \,\mathrm{d}x - \int_{\partial_- T \cap \Gamma_-} (b \cdot n) g v^+ \,\mathrm{d}s. \tag{36}$$

We shall now deal with the remaining parts of the inflow boundary of the element T. The key idea in the discontinuous Galerkin method is to impose the boundary conditions "weakly", i.e., via integral identities. Therefore, we set as local boundary conditions for the element T on $\partial_- T \backslash \Gamma_-$, the exterior trace of the function u, and we impose them in the same way as the actual inflow boundary part:

$$\int_{\partial_- T \backslash \Gamma_-} (b \cdot n) u^+ v^+ \,\mathrm{d}s = \int_{\partial_- T \backslash \Gamma_-} (b \cdot n) u^- v^+ \,\mathrm{d}s, \tag{37}$$

which is equivalent to

$$\int_{\partial_- T \backslash \Gamma_-} (b \cdot n) \lfloor u \rfloor v^+ \,\mathrm{d}s = 0. \tag{38}$$

In order to justify the validity of (37) we have to resort to the classical theory of characteristics for hyperbolic equations. It is known that the solution of a first-order linear hyperbolic boundary-value problem can only exhibit jump discontinuities across characteristics. Thus the normal flux of the solution $bu \cdot n$ is a continuous function across the element faces $e \subset \Gamma_{\mathrm{int}}$ if $(b \cdot n)|_e \neq 0$, as in that case the element face does not lie on a characteristic. If $(b\cdot n)|_e = 0$, which is the case when e lies on a characteristic, then we have $bu \cdot n = (b \cdot n)u = 0$ on e. Hence in any case we have continuity of the normal flux and therefore (38) and thus (37) hold for all $T \in \mathcal{T}$.

Now, subtracting (37) from (36), we obtain

$$\int_T (\mathcal{L}_0 u)v \,\mathrm{d}x - \int_{\partial_- T\cap\Gamma_-} (b\cdot n)u^+v^+ \,\mathrm{d}s - \int_{\partial_- T\backslash\Gamma_-} (b\cdot n)\lfloor u\rfloor v^+ \,\mathrm{d}s$$
$$= \int_T fv \,\mathrm{d}x - \int_{\partial_- T\cap\Gamma_-} (b\cdot n)gv^+ \,\mathrm{d}s$$

for all $T \in \mathcal{T}$ such that $\partial_- T\backslash\Gamma_- \neq \emptyset$.

Arguing in the same way as above, we obtain the local weak formulation for the elements whose boundaries do not share any points with the inflow boundary Γ_- of the computational domain; in this case though the second terms on the left-hand side and the right-hand side of (39) do not appear:

$$\int_T (\mathcal{L}_0 u)v \,\mathrm{d}x - \int_{\partial_- T\backslash\Gamma_-} (b\cdot n)\lfloor u\rfloor v^+ \,\mathrm{d}s = \int_T fv \,\mathrm{d}x,$$

for all $T \in \mathcal{T}$ such that $\partial_- T \cap \Gamma_- = \emptyset$. Adding up all these and setting

$$B_{\mathrm{adv}}(u,v) := \sum_{T\in\mathcal{T}} \int_T (\mathcal{L}_0 u)v \,\mathrm{d}x - \sum_{T\in\mathcal{T}} \int_{\partial_- T\cap\Gamma_-} (b\cdot n)u^+v^+ \,\mathrm{d}s$$
$$- \sum_{T\in\mathcal{T}} \int_{\partial_- T\backslash\Gamma_-} (b\cdot n)\lfloor u\rfloor v^+ \,\mathrm{d}s,$$
$$l_{\mathrm{adv}}(v) := \sum_{T\in\mathcal{T}} \int_T fv \,\mathrm{d}x - \sum_{T\in\mathcal{T}} \int_{\partial_- T\cap\Gamma_-} (b\cdot n)gv^+ \,\mathrm{d}s$$

we can write the weak form for the problem (31):

$$\text{Find } u \in G_b \text{ such that } B_{\mathrm{adv}}(u,v) = l_{\mathrm{adv}}(v) \quad \forall v \in \mathcal{S}_{\mathrm{adv}}.$$

The discontinuous Galerkin method for the problem (31) then reads:

$$\text{Find } u_h \in S_h^p \text{ such that } B_{\mathrm{adv}}(u_h,v_h) = l_{\mathrm{adv}}(v_h) \quad \forall v_h \in S_h^p.$$

5.1 Error Analysis of the DG Method

We define the *energy norm*, denoted again by $|||\cdot|||$, (without causing, hopefully, any confusion) by

$$|||w|||_{\mathrm{adv}} := \Big(\sum_{T\in\mathcal{T}} \|c_0 w\|_\Omega^2 + \frac{1}{2}\|b_n[\![w]\!]\|_\Gamma^2\Big)^{1/2},$$

where $b_n := \sqrt{|b \cdot \mathbf{n}|}$, with n on ∂T denoting the outward normal to ∂T and σ as above. The choice of the above energy norm is related to the coercivity of the bilinear form $B_{\mathrm{adv}}(\cdot,\cdot)$.

The definition and properties of the dG method for the advection problem may become clearer, by studying the symmetric and the skew-symmetric parts of the bilinear form $B_{\mathrm{adv}}(\cdot,\cdot)$. Indeed, it is possible to rewrite the numerical fluxes as described in the following result.

Lemma 3. *The following identity holds:*

$$-\sum_{T\in\mathcal{T}}\left(\int_{\partial_- T\cap\Gamma_-}(b\cdot n)u^+v^+\,\mathrm{d}s+\int_{\partial_- T\setminus\Gamma_-}(b\cdot n)\lfloor u\rfloor v^+\,\mathrm{d}s\right)$$
$$=\int_{\Gamma}\left(\frac{1}{2}|b\cdot n|[\![u]\!]\cdot[\![v]\!]-[\![u]\!]\cdot\{bv\}\right)\mathrm{d}s+\frac{1}{2}\int_{\partial\Omega}(b\cdot n)u^+v^+\,\mathrm{d}s.$$

Proof. On each elemental inflow boundary, we have $-(b\cdot n)=|b\cdot n|$. Thus, on each $\partial_- T\setminus\Gamma_-$, we have

$$\begin{aligned}-(b\cdot n)\lfloor u\rfloor v^+ &= |b\cdot n|\lfloor u\rfloor v^+\\ &=\frac{1}{2}|b\cdot n|\lfloor u\rfloor\lfloor v\rfloor+|b\cdot n|\lfloor u\rfloor\{v\}\\ &=\frac{1}{2}|b\cdot n|[\![u]\!]\cdot[\![v]\!]-(b\cdot n)\lfloor u\rfloor\{v\}\\ &=\frac{1}{2}|b\cdot n|[\![u]\!]\cdot[\![v]\!]-[\![u]\!]\cdot\{bv\}.\end{aligned}$$

Hence,

$$-\sum_{T\in\mathcal{T}}\int_{\partial_- T\setminus\Gamma_-}(b\cdot n)\lfloor u\rfloor v^+\,\mathrm{d}s=\int_{\Gamma_{\mathrm{int}}}\left(\frac{1}{2}|b\cdot n|[\![u]\!]\cdot[\![v]\!]-[\![u]\!]\cdot\{bv\}\right)\mathrm{d}s. \quad (39)$$

Recalling the definitions of $[\![\cdot]\!]$ and $\{\cdot\}$ on the boundary $\partial\Omega$, along with the identities $-(b\cdot n)=|b\cdot n|$ and $(b\cdot n)=|b\cdot n|$ on the inflow and outflow parts of the boundary, respectively, it is immediate that

$$\begin{aligned}&\int_{\partial\Omega}\left(\frac{1}{2}|b\cdot n|[\![u]\!]\cdot[\![v]\!]-[\![u]\!]\cdot\{bv\}\right)\mathrm{d}s+\frac{1}{2}\int_{\partial\Omega}(b\cdot n)u^+v^+\,\mathrm{d}s\\ &=-\sum_{T\in\mathcal{T}}\int_{\partial_- T\cap\Gamma_-}(b\cdot n)u^+v^+\,\mathrm{d}s.\end{aligned} \quad (40)$$

By summing (39) and (40), the result follows. □

The above observation shows that the dG method for the advection problem contains a symmetric part on both the face terms and the elemental terms of the bilinear form [18, 10].

Motivated by identity (39), we decompose $B_{\mathrm{adv}}(\cdot,\cdot)$ into symmetric and skew-symmetric components.

Lemma 4. *The bilinear form can be decomposed into symmetric and skew-symmetric parts:*

$$B_{\mathrm{adv}}(w,v) = B_{\mathrm{adv}}^{\mathrm{symm}}(u,v) + B_{\mathrm{adv}}^{\mathrm{skew}}(u,v)$$

for all $u, v \in \mathcal{S}_{\mathrm{adv}}$, *where*

$$B_{\mathrm{adv}}^{\mathrm{symm}}(u,v) := \sum_{T\in\mathcal{T}} \int_T c_0^2\, u\, v \,\mathrm{d}x + \frac{1}{2}\int_\Gamma |b\cdot n| [\![u]\!] \cdot [\![v]\!] \,\mathrm{d}s \tag{41}$$

and

$$\begin{aligned} B_{\mathrm{adv}}^{\mathrm{skew}}(u,v) := &\frac{1}{2}\sum_{T\in\mathcal{T}} \int_T \big((b\cdot\nabla u)\, v - (b\cdot\nabla v)\, u\big)\,\mathrm{d}x \\ &+\frac{1}{2}\int_{\Gamma_{\mathrm{int}}} \big([\![v]\!]\cdot\{bu\} - [\![u]\!]\cdot\{bv\}\big)\,\mathrm{d}s. \end{aligned}$$

Proof. By adding and subtracting $1/2\sum_{T\in\mathcal{T}}\int_T \nabla\cdot buv\,\mathrm{d}x$ to the bilinear form, a straightforward calculation yields

$$B_{\mathrm{adv}}(u,v) = B_{\mathrm{adv}}^{\mathrm{symm}}(u,v) + \sum_{T\in\mathcal{T}} \int_T \Big((b\cdot\nabla u)\, v + \frac{\nabla\cdot b}{2}\, u\, v\Big)\,\mathrm{d}x - \int_\Gamma [\![u]\!]\cdot\{bv\}\,\mathrm{d}s, \tag{42}$$

with $B_{\mathrm{adv}}^{\mathrm{symm}}(u,v)$ as defined in (41). Integration by parts of the second term in the first integral on the right-hand side of (42) yields

$$\begin{aligned} \sum_{T\in\mathcal{T}}\int_T \frac{\nabla\cdot b}{2}\, u\, v\,\mathrm{d}x = &-\frac{1}{2}\sum_{T\in\mathcal{T}}\int_T \big((b\cdot\nabla u)\, v + (b\cdot\nabla v)\, u\big)\,\mathrm{d}x \\ &+\frac{1}{2}\sum_{T\in\mathcal{T}}\int_{\partial T} (b\cdot n) u^+ v^+ \,\mathrm{d}s. \end{aligned}$$

The result follows by making use of the (standard) identity (see, e.g., [4])

$$\sum_{T\in\mathcal{T}}\int_{\partial T} (b\cdot n) u^+ v^+\,\mathrm{d}s = \int_\Gamma [\![u]\!]\cdot\{bv\}\,\mathrm{d}s + \int_{\Gamma_{\mathrm{int}}} \{u\}[\![bv]\!]\,\mathrm{d}s \tag{43}$$

and by observing that $\{u\}[\![bv]\!] = \{bu\}\cdot[\![v]\!]$. □

Remark 2. We observe the coercivity of the bilinear form:

$$B_{\mathrm{adv}}(w,w) = |||w|||_{\mathrm{adv}}^2, \tag{44}$$

for all $w \in \mathcal{S}_{\mathrm{adv}}$, as $B_{\mathrm{adv}}^{\mathrm{symm}}(w,w) = |||w|||_{\mathrm{adv}}^2$ and $B_{\mathrm{adv}}^{\mathrm{skew}}(w,w) = 0$.

To prove a priori error bounds for the dG method, we begin by observing the Galerkin orthogonality property

$$B_{\text{adv}}(u - u_h, v_h) = 0, \tag{45}$$

for all $v_h \in \mathcal{S}_{\text{adv}}$, coming from subtracting the dG method from the weak form of the problem, tested again functions from the finite element space.

For simplicity of presentation, we shall assume in the sequel that

$$b \cdot \nabla v_h \in S_h^p. \tag{46}$$

Results for more general wind b are available, e.g., in [36, 27].

Using (44) and (45), we get the identity

$$|||v_h - u_h|||_{\text{adv}}^2 = B_{\text{adv}}(v_h - u_h, v_h - u_h) = -B_{\text{adv}}(u - v_h, v_h - u_h), \tag{47}$$

for all $v_h \in s_h^p$.

The next step is to bound the bilinear form from above by a multiple of $|||v_h - u_h|||_{\text{adv}}$. To this end, we work as follows. Integrating by parts the first term in the integrand of the first term on the right-hand side of (42) and using the standard identity (43), we come to

$$\begin{aligned} B_{\text{adv}}(u - v_h, v_h - u_h) =& B_{\text{adv}}^{\text{symm}}(u - v_h, v_h - u_h) \\ &- \sum_{T \in \mathcal{T}} \int_T (b \cdot \nabla(v_h - u_h))\,(u - v_h)\,\mathrm{d}x \\ &+ \int_\Gamma [\![v_h - u_h]\!] \cdot \{b(u - v_h)\}\,\mathrm{d}s. \end{aligned} \tag{48}$$

Setting $v_h = \Pi u$, where, as above, $\Pi : L^2(\Omega) \to S - h^p$ is the orthogonal L^2-projection operator onto the finite element space, we observe that the second term on the right-hand side of (48) vanishes in view of (46). The Cauchy-Schwarz inequality then yields

$$B_{\text{adv}}(u - \Pi u, \Pi u - u_h) \leq 2|||\Pi u - u_h|||_{\text{adv}} \big(|||u - \Pi u|||_{\text{adv}}^2 + \|\{u - \Pi u\}\|_\Gamma^2\big)^{1/2},$$

which can be used on (47) to deduce

$$|||\Pi u - u_h|||_{\text{adv}} \leq 2\big(|||u - \Pi u|||_{\text{adv}}^2 + \|\{u - \Pi u\}\|_\Gamma^2\big)^{1/2},$$

which, using triangle inequality and the approximation properties of the L^2-projection (see, e.g., [36] for details), yields the a priori error bound

$$|||u - u_h|||_{\text{adv}} \leq C h^{\min\{p+1,r\}-1/2} |u|_{H^r(\Omega)},$$

for $p \geq 0$ and $r \geq 1$.

6 Problems with Non-Negative Characteristic Form

Having considered the dG method for self-adjoint elliptic and first order hyperbolic problems respectively, we are now in position to combine the ideas presented above and present a dG method for a wide class of linear PDE problems.

Let Ω be a bounded open (curvilinear) polygonal domain in $\mathbb{R}^d$, and let $\partial\Omega$ signify the union of its $(d-1)$-dimensional open edges, which are assumed to be sufficiently smooth (in a sense defined rigorously later). We consider the convection-diffusion-reaction equation

$$\mathcal{L}u \equiv -\nabla \cdot (\bar{\mathbf{a}}\nabla u) + b \cdot \nabla u + cu = f \quad \text{in } \Omega, \tag{49}$$

where $f \in L^2(\Omega)$, $c \in L^\infty(\Omega)$, b is a vector function whose entries are Lipschitz continuous real-valued functions on $\bar{\Omega}$, and $\bar{\mathbf{a}}$ is the *symmetric* **diffusion tensor** whose entries are bounded, piecewise continuous real-valued functions defined on $\bar{\Omega}$, with

$$\zeta^T \bar{\mathbf{a}}(x)\zeta \geq 0 \quad \forall \zeta \in \mathbb{R}^d, \quad x \in \bar{\Omega}.$$

Under this hypothesis, (49) is termed a ***partial differential equation with a*** *nonnegative characteristic form.* By n we denote the unit outward normal vector to $\partial\Omega$. We define

$$\Gamma_0 = \left\{x \in \partial\Omega : n(x)^T \bar{\mathbf{a}}(x) n(x) > 0\right\},$$

$$\Gamma_- = \{x \in \partial\Omega \backslash \Gamma_0 : b(x) \cdot n(x) < 0\}, \quad \Gamma_+ = \{x \in \partial\Omega \backslash \Gamma_0 : b(x) \cdot n(x) \geq 0\}.$$

The sets Γ_- and Γ_+ are referred to as the *inflow* and *outflow* boundary, respectively. We can also see that $\partial\Omega = \Gamma_0 \cup \Gamma_- \cup \Gamma_+$. If Γ_0 has a positive $(d-1)$-dimensional Hausdorff measure, we also decompose Γ_0 into two parts Γ_{D} and Γ_{N}, and we impose Dirichlet and Neumann boundary conditions, respectively, via

$$\begin{aligned} u &= g_{\mathrm{D}} \text{ on } \Gamma_{\mathrm{D}} \cup \Gamma_-, \\ (\bar{\mathbf{a}}\nabla u) \cdot n &= g_{\mathrm{N}} \text{ on } \Gamma_{\mathrm{N}}, \end{aligned} \tag{50}$$

where we adopt the (physically reasonable) hypothesis that $b \cdot n \geq 0$ on Γ_{N}, whenever the latter is nonempty.

For a discussion on the physical models that are described by the above family of boundary-value problems, we refer to [36] and the references therein. The existence and uniqueness of solutions (in various settings) has been considered in [45, 25, 26, 37], under the standard assumption (33).

Then the *interior penalty dG method* for the problem (49), (50) is defined as follows:

$$\text{Find } u_h \in S_h^p \text{ such that } B(u_h, v_h) = l(v_h) \quad \forall v_h \in S_h^p,$$

where

$$\begin{aligned}B(w,v) := &\sum_{T\in\mathcal{T}} \int_T (\bar{\mathrm{a}}\nabla w\cdot\nabla v + (b\cdot\nabla w)\, v + c\, w\, v)\,\mathrm{d}x \\ &- \sum_{T\in\mathcal{T}} \int_{\partial_- T\cap(\Gamma_-\cup\Gamma_{\mathrm{D}})} (b\cdot n) w^+ v^+\,\mathrm{d}s - \sum_{T\in\mathcal{T}} \int_{\partial_- T\setminus\partial\Omega} (b\cdot n)\lfloor w\rfloor v^+\,\mathrm{d}s \\ &+ \int_{\Gamma_{\mathrm{D}}\cup\Gamma_{\mathrm{int}}} (\theta\{\bar{\mathrm{a}}\nabla v\}\cdot[w] - \{\bar{\mathrm{a}}\nabla w\}\cdot[v] + \sigma[w]\cdot[v])\,\mathrm{d}s\end{aligned}$$

and

$$\begin{aligned}l(v) : = &\sum_{T\in\mathcal{T}} \int_T fv\,\mathrm{d}x - \sum_{T\in\mathcal{T}} \int_{\partial_- T\cap(\Gamma_-\cup\Gamma_{\mathrm{D}})} (b\cdot n) g_{\mathrm{D}} v^+\,\mathrm{d}s \\ &+ \int_{\Gamma_{\mathrm{D}}} (\theta\bar{\mathrm{a}}\nabla v\cdot n + \sigma v) g_{\mathrm{D}}\,\mathrm{d}s + \int_{\Gamma_{\mathrm{N}}} g_{\mathrm{N}} v\,\mathrm{d}s\end{aligned}$$

for $\theta \in \{-1, 1\}$, with the function σ defined by

$$\sigma|_e := C_\sigma \left\{\frac{\mathrm{a}p^2}{h}\right\},$$

where $\mathrm{a} : \Omega \to \mathbb{R}$, with $\mathrm{a}|_T = \|(|\sqrt{\bar{\mathrm{a}}}|_2)^2\|_{L^\infty(T)}$, $T \in \mathcal{T}$, with $|\cdot|_2$ denoting the matrix-2-norm, and C_σ is a positive constant. We refer to the dG method with $\theta = -1$ as the *symmetric interior penalty dG method*, whereas $\theta = 1$ yields the *nonsymmetric interior penalty dG method*. This terminology stems from the fact that when $b \equiv \mathbf{0}$, the bilinear form $B(\cdot,\cdot)$ is symmetric if and only if $\theta = -1$.

Various types of error analysis for the variants of interior penalty DGFEMs can be found, e.g., in [6, 3, 15, 49, 36, 4, 27, 31, 29, 28, 21, 20, 38], along with an extensive discussion on the properties of this family of methods.

7 Numerical Examples

7.1 Example 1

We consider the first IAHR/CEGB problem (devised by workers at the CEGB for an IAHR workshop in 1981 as a benchmark steady-state convection-diffusion problem). For

$$b = \left(2y(1-x^2), -2x(1-y^2)\right)$$

and $0 \le \epsilon \ll 1$, we consider the convection-diffusion equation

$$-\epsilon\Delta u + b\cdot\nabla u = 0 \quad \text{for} \quad (x_1, x_2) \in (-1, 1)\times(0, 1),$$

subject to Dirichlet boundary conditions

$$u(-1,x_2) = u(x_1,1) = u(1,x_2) = 1 - \tanh(\alpha), \quad -1 \leq x_1 \leq 1,\ 0 \leq x_2 \leq 1,$$

on the tangential boundaries, with $\alpha > 0$ parameter, and inlet boundary condition

$$u(x_1,0) = 1 + \tanh\Big(\alpha(2x+1)\Big), \quad -1 \leq x_1 \leq 0. \tag{51}$$

Finally, a homogeneous Neumann boundary condition is imposed at the outlet $0 < x_1 \leq 1$, $x_2 = 0$.

We remark that this choice of convective velocity field b does not satisfy assumption (33). On the other hand b is incompressible, that is $\nabla \cdot b = 0$, and, therefore, $c_0 = 0$.

The inlet profile (51) involves the presence of a steep interior layer centred at $(-1/2, 0)$, whose steepness depends on the value of the parameter α. This layer travels clockwise circularly due to the convection and exits at the outlet.

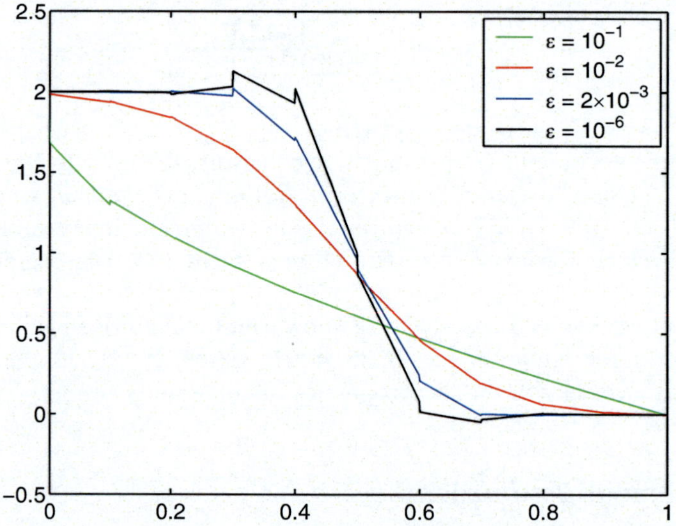

Fig. 3. Example 1. Outlet profiles for different values of ϵ.

Following MacKenzie & Morton [43] (cf. also Smith & Hutton [52]) we have chosen to work with $\alpha = 10$ on a uniform mesh of 20×10 elements, and for $\epsilon = 10^{-6}, 2 \times 10^{-3}, 10^{-2}, 10^{-1}$, respectively.

In Figure 3 the profiles of the outlet boundary $0 < x_1 \leq 1$, $x_2 = 0$ are plotted for different values of ϵ, and for $p = 1$. Note that the vertical line segments in the profiles correspond to the discontinuities across the element interfaces. To address the question of accuracy of the computation, in Figure 4 we compare the profile for $\epsilon = 10^{-6}$ (drawn in black in Figure 3) on the 20×10 mesh with the corresponding profile on a much finer mesh containing 80×40

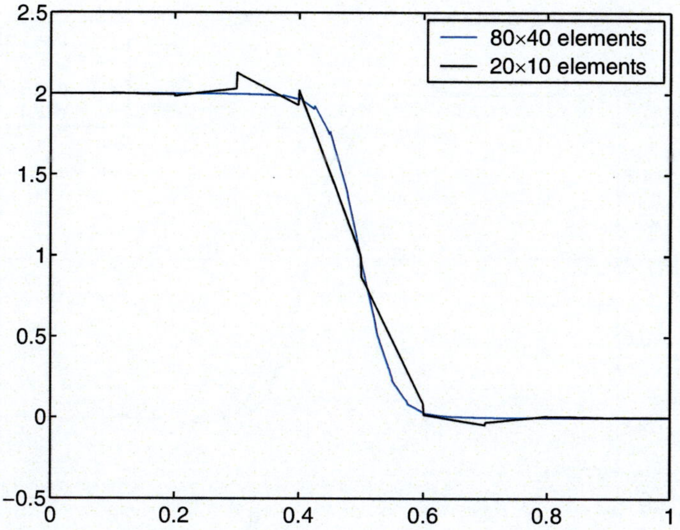

Fig. 4. Example 1. Outlet profile for $\epsilon = 10^{-6}$ when 20×10 elements and 80×40 elements are used.

elements. Also, in Figure 5 we present the computed outlet profiles when we use of uniform polynomial degrees $p = 1, \ldots, 4$ on the 20×10-mesh. Note that the quality of the approximation is better for $p = 4$ on the 20×10-mesh (DOF$=5000$), than the computed outlet profile for $p = 1$ on the 80×40 mesh (DOF$=12800$).

Finally, in Figure 6 we present the computed solutions on the 20×10-mesh for the different values of ϵ. We note that the quality of the approximations is remarkably good considering the computationally demanding features of the solutions.

7.2 Example 2

We consider the following equation on $\Omega = (-1, 1)^2$

$$\begin{aligned} -x_1^2 u_{x_2 x_2} + u_{x_1} + u &= 0, \quad \text{for } -1 \le x_1 \le 1,\ x_2 > 0, \\ u_{x_1} + u &= 0, \quad \text{for } -1 \le x_1 \le 1,\ x_2 \le 0, \end{aligned}$$

whose analytical solution is

$$u(x_1, x_2) = \begin{cases} \sin\left(\frac{1}{2}\pi(1 + x_2)\right) \exp\left(-\left(x_1 + \frac{\pi^2}{4}\frac{x_1^3}{3}\right)\right), & \text{if } x_1 \in [-1, 1], x_2 > 0; \\ \sin\left(\frac{1}{2}\pi(1 + x_2)\right) \exp(-x_1), & \text{if } x_1 \in [-1, 1], x_2 \le 0, \end{cases}$$

along with an appropriate Dirichlet boundary condition. This problem is of changing-type, as there exists a second order term for $x_2 > 0$, which is no

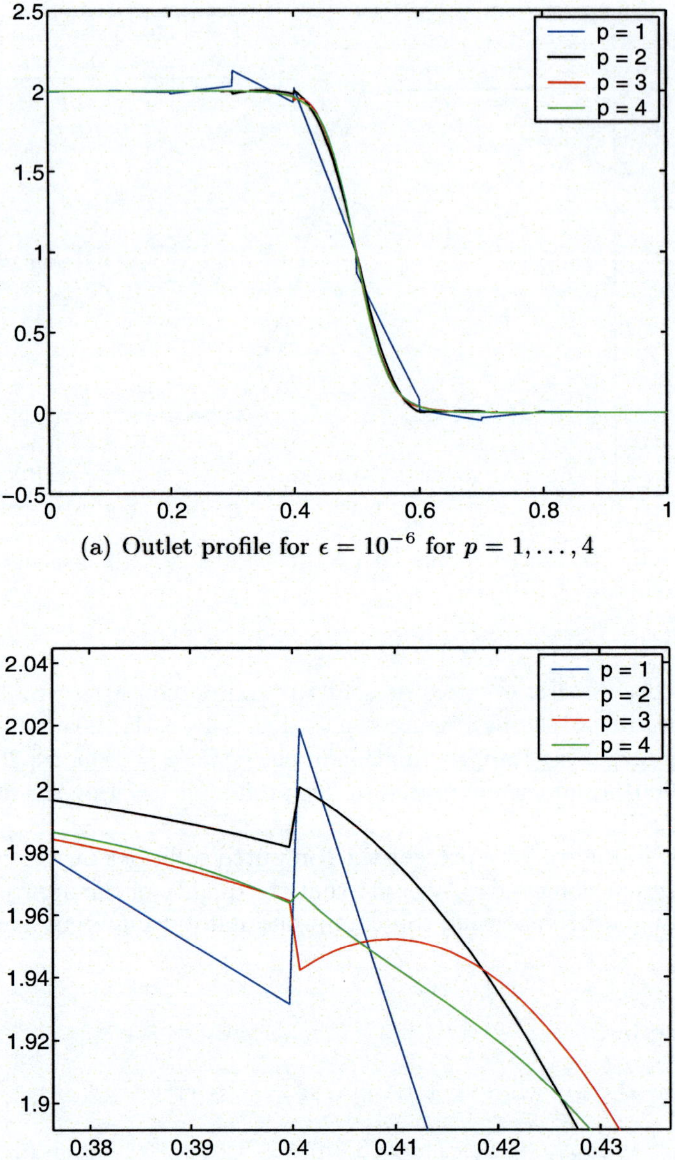

(a) Outlet profile for $\epsilon = 10^{-6}$ for $p = 1, \ldots, 4$

(b) Detail of (a)

Fig. 5. Example 1. Outlet profile for $\epsilon = 10^{-6}$ on the 20×10 mesh for $p = 1, \ldots, 4$.

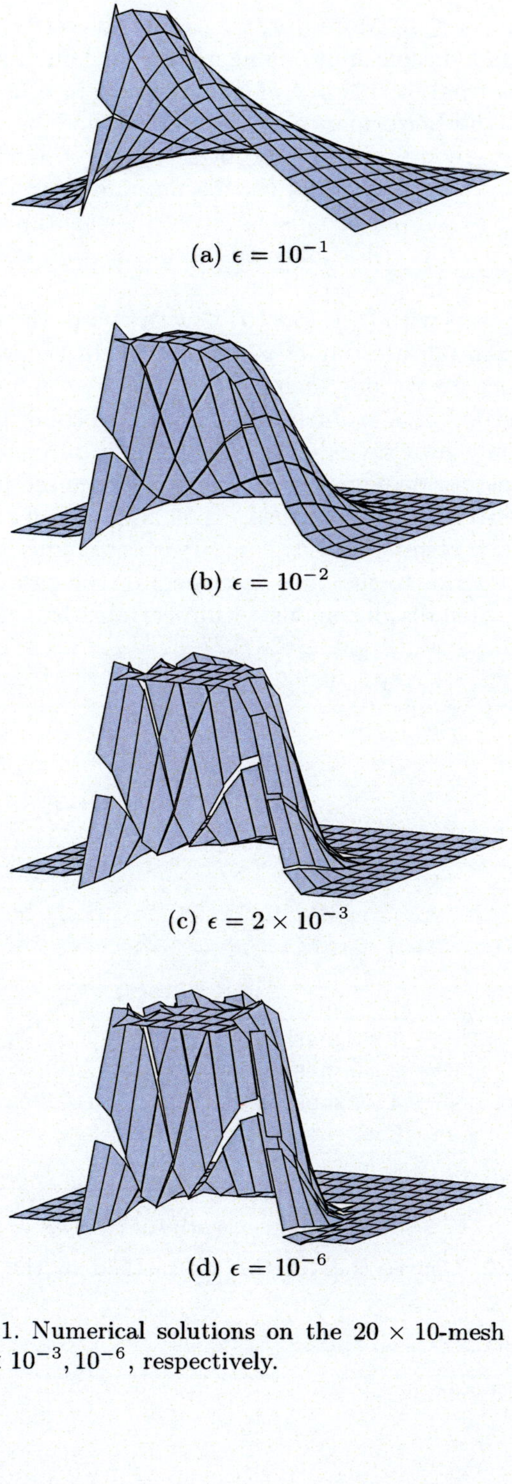

(a) $\epsilon = 10^{-1}$

(b) $\epsilon = 10^{-2}$

(c) $\epsilon = 2 \times 10^{-3}$

(d) $\epsilon = 10^{-6}$

Fig. 6. Example 1. Numerical solutions on the 20×10-mesh for $p = 1$ and for $\epsilon = 10^{-1}, 10^{-2}, 2 \times 10^{-3}, 10^{-6}$, respectively.

longer present for $x_2 \leq 0$. Moreover, we can easily verify that its analytical solution u exhibits a discontinuity along $x_2 = 0$, although the derivative of u, in the direction normal to this line of discontinuity in u, is continuous across $x_2 = 0$. We test the performance of the dG method by employing various meshes. We have to modify the method by setting $\sigma_e = 0$ for all element edges $e \subset (-1,1) \times \{0\}$, where σ_e denotes the discontinuity-penalisation parameter; this is done in order to avoid penalising physical discontinuities. Note that the diffusive flux $(\bar{a}\nabla u) \cdot n$ is still continuous across $x_2 = 0$, and thus the method still applies.

When subdivisions with $(-1,1) \times \{0\} \subset \bar{\Gamma}$ are used, the method appears to converge at exponential rates under p-enrichment. In Figure 7, we can see the convergence history for various such meshes. The reason for this excellent behaviour of the method, in a problem where standard conforming finite element methods would only provide us with low algebraic rates of convergence, lies in the fact that merely element-wise regularity is required for dG methods, as opposed to global regularity hypothesis that is needed for conforming methods to produce such results. If $(-1,1) \times \{0\}$ is not a subset of $\bar{\Gamma}$, the method produces results inferior to the ones described for the case $(-1,1) \times \{0\} \subset \bar{\Gamma}$, as the solution is then discontinuous within certain elements.

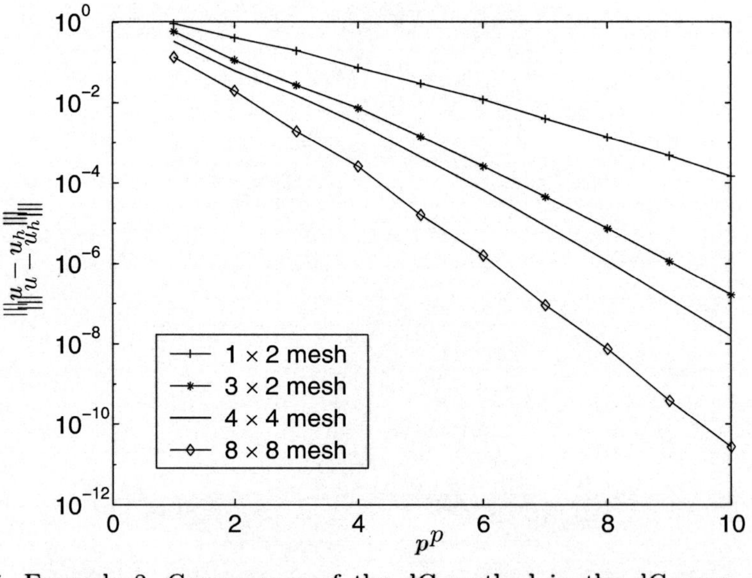

Fig. 7. Example 2. Convergence of the dG method in the dG-norm under p-enrichment.

8 Solving the Linear System

FEM and dG methods lead to large linear systems of the form $AU = F$, where usually the condition number $\kappa(A)$ of the matrix A increases as $h \to 0$; for the case of second order PDE problems we normally have $\kappa(A) = O(h^{-d})$. This is particularly inconvenient in the context of iterative methods for solving the linear system. Therefore, the construction of preconditioning strategies for the resulting linear system is of particular importance. Here we follow [30], where scalable solvers for linear systems arising from dG methods have been considered.

The classical preconditioning approach consists of designing a matrix P, called the *preconditioner*, such that the matrix $P^{-1}A$ is "well" conditioned compared to A (i.e., $\kappa(P^{-1}A) << \kappa(A)$) while at the same time P is cheaply inverted numerically. These two requirements are competing, so that the construction of efficient preconditioners a challenge. Once such a preconditioner P is known, one can solve recursively the linear systems $Py = F$ and $P^{-1}Ax = y$ efficiently to find x.

In [30] it is proposed to use a preconditioned GMRES iterative solver (see, e.g., [33]) with preconditioner

$$A_s := \frac{A + A^T}{2},$$

i.e., the symmetric part of the stiffness matrix A. This choice has a number of implications as we shall now see.

From an implementation point of view, employing GMRES with system matrix

$$A_s^{-\frac{1}{2}} A A_s^{-\frac{1}{2}},$$

is equivalent to running GMRES in the A_s-inner product and using A_s as a left-preconditioner.

It is not hard to see that

$$A_s^{-\frac{1}{2}} A A_s^{-\frac{1}{2}} = I + S,$$

where S is a skew-symmetric matrix. It is known that applying the GMRES algorithm to a matrix of the form $I + S$, where S is skew symmetric, is a 3-term recurrence, i.e., there is no need to compute using the whole Krylov subspace [2], but only the last two Krylov subspace vectors. Hence using A_s as a preconditioner within a GMRES algorithm leads to significant computational and storage savings.

Moreover, it is possible to show that the resulting preconditioned GMRES algorithm is *scalable* with respect to the size of the matrix, i.e., the number of GMRES iterations does not increase as the meshsize $h \to 0$ and/or as the polynomial degree $p \to \infty$. This property is shown theoretically in [30], but here we shall illustrate it via a numerical example.

To this end, we consider the convection-diffusion problem

$$-\epsilon\Delta u + u_x + u_y = f \quad \text{for } (x,y) \in (0,1)^2,$$

subject to Dirichlet boundary conditions, and right-hand side f, such that the analytical solution is given by

$$u(x,y) = x + y(1-x) + \frac{e^{-\frac{1}{\epsilon}} - e^{-\frac{(1-x)(1-y)}{\epsilon}}}{1 - e^{-\frac{1}{\epsilon}}}.$$

The solution exhibits boundary layer behaviour along $x = 1$ and $y = 1$, and the layers become steeper as $\epsilon \to 0$. We solve the problem for a range of ϵ using the dG method for a range of uniform meshsizes h and polynomial degrees p. The results are presented Table 1. As predicted by the theory in [30], the number of iterations is independent of discretization parameters.

For comparison purposes, we included the corresponding GMRES runs for the choice of a black-box preconditioner such as ILU. The results are presented in Table 2. We can see that, while the number of iterations is low for some

Table 1. GMRES iterations for DGFEM discretization of the convection-diffusion problem with constant wind $\mathbf{b}^T = (1,1)$ and with preconditioner A_s.

p	n	$\epsilon = 0.5$	$\epsilon = 0.1$	$\epsilon = 0.05$	$\epsilon = 0.01$
	2,500	7	15	22	77
1	10,000	7	15	22	80
	40,000	7	14	22	80
	5,625	7	14	22	80
2	22,500	6	14	22	80
	90,000	6	14	21	78
	10,000	6	14	22	79
3	40,000	6	14	22	78
	160,000	6	13	21	78

Table 2. GMRES iterations for dG discretization of the convection-diffusion problem with ILU(10^{-2}) preconditioning.

p	n	$\epsilon = 0.5$	$\epsilon = 0.1$	$\epsilon = 0.01$
	2,500	12	13	7
1	10,000	36	40	29
	40,000	124	117	69
	5,625	18	17	12
2	22,500	61	59	60
	90,000	235	231	137
	10,000	39	29	23
3	40,000	112	114	100
	160,000	> 300	> 300	> 300

values of the parameters, the overall convergence behavior is quite undesirable, with iteration counts growing with both discretization parameters. Thus, while the number of iterations appears to be decreasing with ϵ, it is exactly for this range that the discretization parameters have to be increased in order to resolve layers. The resulting convergence behavior becomes rapidly too costly to implement in practice. We note here that the ILU preconditioner is implemented with a standard *full* GMRES routine, which means that the storage increases with every iteration.

9 Concluding Remarks

These notes aim at giving a gentle introduction to discontinuous Galerkin methods used for the numerical solution of linear PDE problems of mixed type. The material is presented in a simple fashion in an effort to maximise accessibility. Indeed, this note is far from being exhaustive in any of the topics presented and, indeed, it is *not* meant to be a survey of the ever-growing subject of discontinuous Galerkin methods. For more material on dG methods we refer to the volumes [15, 34, 48] and the references therein.

References

1. R.A. Adams and J.J.F. Fournier: *Sobolev Spaces.* Pure and Applied Mathematics, vol. 140, Elsevier/Academic Press, Amsterdam, 2nd edition, 2003.
2. M. Arioli, D. Loghin, and A.J. Wathen: Stopping criteria for iterations in finite element methods. Numer. Math. **99**, 2005, 381–410.
3. D.N. Arnold: An interior penalty finite element method with discontinuous elements. SIAM J. Numer. Anal. **19**, 1982, 742–760.
4. D.N. Arnold, F. Brezzi, B. Cockburn, and L.D. Marini: Unified analysis of discontinuous Galerkin methods for elliptic problems. SIAM J. Numer. Anal. **39**, 2001, 1749–1779.
5. I. Babuška: The finite element method with penalty. Math. Comp. **27**, 1973, 221–228.
6. G.A. Baker: Finite element methods for elliptic equations using nonconforming elements. Math. Comp. **31**, 1977, 45–59.
7. R. Becker, P. Hansbo, and M.G. Larson: Energy norm a posteriori error estimation for discontinuous Galerkin methods. Comput. Methods Appl. Mech. Engrg. **192**, 2003, 723–733.
8. K.S. Bey and T. Oden: *hp*-version discontinuous Galerkin methods for hyperbolic conservation laws. Comput. Methods Appl. Mech. Engrg. **133**, 1996, 259–286.
9. S.C. Brenner and L.R. Scott: *The Mathematical Theory of Finite Element Methods.* Texts in Applied Mathematics, vol. 15, Springer, New York, 3rd edition, 2008.
10. F. Brezzi, L.D. Marini, and E. Süli: Discontinuous Galerkin methods for first-order hyperbolic problems. Math. Models Methods Appl. Sci. **14**, 2004, 1893–1903.

11. C. Carstensen, T. Gudi, and M. Jensen: A unifying theory of a posteriori error control for discontinuous Galerkin FEM. Numer. Math. **112**, 2009, 363–379.
12. P.G. Ciarlet: *The Finite Element Method for Elliptic Problems.* Studies in Mathematics and its Applications, vol. 4, North-Holland Publishing Co., Amsterdam, 1978.
13. B. Cockburn: Discontinuous Galerkin methods for convection-dominated problems. In: *High-order Methods for Computational Physics*, Springer, Berlin, 1999, 69–224.
14. B. Cockburn, S. Hou, and C.-W. Shu: The Runge-Kutta local projection discontinuous Galerkin finite element method for conservation laws IV: the multidimensional case. Math. Comp. **54**, 1990, 545–581.
15. B. Cockburn, G.E. Karniadakis, and C.-W. Shu (eds.): *Discontinuous Galerkin Methods.* Theory, computation and applications. Papers from the 1st International Symposium held in Newport, RI, May 24–26, 1999. Springer-Verlag, Berlin, 2000.
16. B. Cockburn, S.Y. Lin, and C.-W. Shu: TVB Runge-Kutta local projection discontinuous Galerkin finite element method for conservation laws III: onedimensional systems. J. Comput. Phys. **84**, 1989, 90–113.
17. B. Cockburn and C.-W. Shu: TVB Runge-Kutta local projection discontinuous Galerkin finite element method for conservation laws II: general framework. Math. Comp. **52**, 1989, 411–435.
18. B. Cockburn and C.-W. Shu: The local discontinuous Galerkin method for timedependent convection-diffusion systems. SIAM J. Numer. Anal. **35**, 1998, 2440–2463.
19. B. Cockburn and C.-W. Shu: The Runge-Kutta discontinuous Galerkin method for conservation laws V: multidimensional systems. J. Comput. Phys. **141**, 1998, 199–224.
20. A. Ern and J.-L. Guermond: Discontinuous Galerkin methods for Friedrichs' systems I: general theory. SIAM J. Numer. Anal. **44**, 2006, 753–778.
21. A. Ern and J.-L. Guermond: Discontinuous Galerkin methods for Friedrichs' systems II: second-order elliptic PDEs. SIAM J. Numer. Anal. **44**, 2006, 2363–2388.
22. A. Ern and A.F. Stephansen: A posteriori energy-norm error estimates for advection-diffusion equations approximated by weighted interior penalty methods. J. Comp. Math. **26**, 2008, 488–510.
23. S.A.F. Ern, A. and P. Zunino: A discontinuous Galerkin method with weighted averages for advectiondiffusion equations with locally small and anisotropic diffusivity. IMA J. Numer. Anal. **29**, 2009, 235–256.
24. R.S. Falk and G.R. Richter: Local error estimates for a finite element method for hyperbolic and convection-diffusion equations. SIAM J. Numer. Anal. **29**, 1992, 730–754.
25. G. Fichera: Sulle equazioni differenziali lineari ellittico-paraboliche del secondo ordine. Atti Accad. Naz. Lincei. Mem. Cl. Sci. Fis. Mat. Nat. Sez. I. **5**(8), 1956, 1–30.
26. G. Fichera: On a unified theory of boundary value problems for elliptic-parabolic equations of second order. In: *Boundary Problems in Differential Equations.* Univ. of Wisconsin Press, Madison, 1960, 97–120.
27. E.H. Georgoulis: *Discontinuous Galerkin Methods on Shape-Regular and Anisotropic Meshes.* D.Phil. Thesis, University of Oxford, 2003.

28. E.H. Georgoulis, E. Hall, and J.M. Melenk: On the suboptimality of the p-version interior penalty discontinuous Galerkin method. J. Sci. Comput. **42**, 2010, 54–67.
29. E.H. Georgoulis and A. Lasis: A note on the design of hp-version interior penalty discontinuous Galerkin finite element methods for degenerate problems. IMA J. Numer. Anal. **26**, 2006, 381–390.
30. E.H. Georgoulis and D. Loghin: Norm preconditioners for discontinuous Galerkin hp-finite element methods. SIAM J. Sci. Comput. **30**, 2008, 2447–2465.
31. E.H. Georgoulis and E. Süli: Optimal error estimates for the hp-version interior penalty discontinuous Galerkin finite element method. IMA J. Numer. Anal. **25**, 2005, 205–220.
32. D. Gilbarg and N.S. Trudinger: *Elliptic Partial Differential Equations of Second Order*. Grundlehren der Mathematischen Wissenschaften, vol. 224, Springer-Verlag, Berlin, 2nd edition, 1983.
33. G.H. Golub and C.F. Van Loan: *Matrix Computations*. Johns Hopkins Studies in the Mathematical Sciences, Johns Hopkins University Press, Baltimore, MD, 3rd edition, 1996.
34. J.S. Hesthaven and T. Warburton: *Nodal Discontinuous Galerkin Methods*. Algorithms, analysis, and applications. Texts in Applied Mathematics, vol. 54, Springer, New York, 2008.
35. P. Houston, D. Schötzau, and T.P. Wihler: Energy norm a posteriori error estimation of hp-adaptive discontinuous Galerkin methods for elliptic problems. Math. Models Methods Appl. Sci. **17**, 2007, 33–62.
36. P. Houston, C. Schwab, and E. Süli: Discontinuous hp-finite element methods for advection-diffusion-reaction problems. SIAM J. Numer. Anal. **39**, 2002, 2133–2163.
37. P. Houston and E. Süli: Stabilised hp-finite element approximation of partial differential equations with nonnegative characteristic form. Computing **66**, 2001, 99–119.
38. M. Jensen: *Discontinuous Galerkin Methods for Friedrichs Systems*. D.Phil. Thesis, University of Oxford, 2005.
39. C. Johnson and J. Pitkäranta: An analysis of the discontinuous Galerkin method for a scalar hyperbolic equation. Math. Comp. **46**, 1986, 1–26.
40. O.A. Karakashian and F. Pascal: A posteriori error estimates for a discontinuous Galerkin approximation of second-order elliptic problems. SIAM J. Numer. Anal. **41**, 2003, 2374–2399.
41. O.A. Karakashian and F. Pascal: Convergence of adaptive discontinuous Galerkin approximations of second-order elliptic problems. SIAM J. Numer. Anal. **45**, 2007, 641–665.
42. P. Lesaint and P.-A. Raviart: On a finite element method for solving the neutron transport equation. In: *Mathematical Aspects of Finite Elements in Partial Differential Equations*, Math. Res. Center, Univ. of Wisconsin-Madison, Academic Press, New York, 1974, 89–123.
43. J.A. Mackenzie and K.W. Morton: Finite volume solutions of convection-diffusion test problems. Math. Comp. **60**, 1993, 189–220.
44. J. Nitsche: Über ein Variationsprinzip zur Lösung von Dirichlet-Problemen bei Verwendung von Teilräumen, die keinen Randbedingungen unterworfen sind. Abh. Math. Sem. Uni. Hamburg **36**, 1971, 9–15.

45. O.A. Oleĭnik and E.V. Radkevič: *Second Order Equations with Nonnegative Characteristic Form*. Translated from Russian by Paul C. Fife. Plenum Press, New York, 1973.
46. W.H. Reed and T.R. Hill: *Triangular Mesh Methods for the Neutron Transport Equation*. Technical Report LA-UR-73-479, Los Alamos Scientific Laboratory, 1973.
47. M. Renardy and R.C. Rogers: *An Introduction to Partial Differential Equations*. Texts in Applied Mathematics, vol. 13, Springer, New York, 1993.
48. B. Rivière: *Discontinuous Galerkin Methods for Solving Elliptic and Parabolic Equations*. Frontiers in Applied Mathematics. Theory and implementation, vol. 35, Society for Industrial and Applied Mathematics (SIAM), Philadelphia, PA, 2008.
49. B. Rivière, M.F. Wheeler, and V. Girault: Improved energy estimates for interior penalty, constrained and discontinuous Galerkin methods for elliptic problems I. Comput. Geosci. **3**, 1999, 337–360.
50. B. Rivière, M.F. Wheeler, and V. Girault: A priori error estimates for finite element methods based on discontinuous approximation spaces for elliptic problems. SIAM J. Numer. Anal. **39**, 2001, 902–931.
51. C. Schwab: *p- and hp-Finite Element Methods: Theory and Applications in Solid and Fluid Mechanics*. Oxford University Press, Numerical mathematics and scientific computation, 1998.
52. R.M. Smith and A.G. Hutton: The numerical treatment of convention – a performance/comparison of current methods. Numer. Heat Transfer **5**, 1982, 439–461.
53. G. Strang and G.J. Fix: *An Analysis of the Finite Element Method*. Prentice-Hall Series in Automatic Computation. Prentice-Hall Inc., Englewood Cliffs, N.J., 1973.
54. M.F. Wheeler: An elliptic collocation-finite element method with interior penalties. SIAM J. Numer. Anal. **15**, 1978, 152–161.

A Numerical Analyst's View of the Lattice Boltzmann Method

Jeremy Levesley, Alexander N. Gorban, and David Packwood

Department of Mathematics, University of Leicester, LE1 7RH, UK

Summary. The purpose of this paper is to raise the profile of the Lattice Boltzmann method (LBM) as a computational method for solving fluid flow problems. We put forward the point of view that the method need not be seen as a discretisation of the Boltzmann equation, and also propose an alternative route from microscopic to macroscopic dynamics, traditionally taken via the Chapman-Enskog procedure. In that process the microscopic description is decomposed into processes at different time scales, parametrised with the Knudsen number. In our exposition we use the time step as a parameter for expanding the solution. This makes the treatment here more amenable to numerical analysts. We explain a method by which one may ameliorate the inevitable instabilities arising when trying to solve a convection-dominated problem, entropic filtering.

1 Introduction

The most commonly used model for high Reynold's number flow is the Navier-Stokes equations. The two dimensional version of this equation is given below:

E.H. Georgoulis, A. Iske, J. Levesley (eds.), *Approximation Algorithms for Complex Systems*, Springer Proceedings in Mathematics 3, DOI: 10.1007/978-3-642-16876-5_6,
© Springer-Verlag Berlin Heidelberg 2011

$$\frac{\partial \rho}{\partial t} = -\nabla \cdot (\rho \mathbf{u}),$$

$$\frac{\partial}{\partial t}(\rho u_1) = -\sum_{j=1}^{2} \frac{\partial}{\partial x_j}(\rho u_1 u_j) - \frac{\partial P}{\partial x_1} + \mu \left(\frac{\partial}{\partial x_1} P \left(\frac{\partial u_1}{\partial x_1} - \frac{\partial u_2}{\partial x_2} \right) + \frac{\partial}{\partial x_2} P \left(\frac{\partial u_2}{\partial x_1} + \frac{\partial u_1}{\partial x_2} \right) \right),$$

$$\frac{\partial}{\partial t}(\rho u_2) = -\sum_{j=1}^{2} \frac{\partial}{\partial x_j}(\rho u_2 u_j) - \frac{\partial P}{\partial x_2} + \mu \left(\frac{\partial}{\partial x_2} P \left(\frac{\partial u_2}{\partial x_2} - \frac{\partial u_1}{\partial x_1} \right) + \frac{\partial}{\partial x_1} P \left(\frac{\partial u_1}{\partial x_2} + \frac{\partial u_2}{\partial x_1} \right) \right),$$

$$\frac{\partial E}{\partial t} = -\sum_{i=1}^{2} \frac{\partial}{\partial x_i} \{u_i (E + P)\} + \tau \sum_{i=1}^{2} \frac{\partial}{\partial x_i} \left(P \frac{\partial}{\partial x_i} \frac{P}{\rho} \right).$$

where ρ, $\mathbf{u} = (u_1, u_2)$, P and E are density, velocity, pressure and energy respectively. These equations model the conservation of mass, momentum and energy. The number μ is the coefficient of viscosity, and as this number tends to zero we recover the Euler equations for inviscid flow.

The standard approach is to discretise this equation, using either finite differences, finite elements, or finite volumes. The Godunov theorem [11] tells us that we should expect oscillation in solutions near to evolving discontinuities in solutions of the differential equations. Standard methods for dealing with such oscillations are slope limiters, artificial viscosity, and more recently ENO, WENO, and ADER schemes [17, 23, 24].

So we have a philosophical issue to consider. High order numerical methods for PDEs provide excellent solutions to a problem where the solution is smooth, and here model errors (Navier-Stokes' is a model) are probably much higher than numerical errors. Where the solution is not smooth it may be that Navier-Stokes' is a poor model, in which case we might need to ask why we would try to get very accurate solutions. Perhaps the best rationale for numerical discretisation of PDEs is in understanding qualitative behaviour of the fluid.

In this paper we discuss the Lattice Boltzmann method for simulating fluid flow. The usual justification for this method is via the finite difference discretisation of the Boltzmann equation, which governs the motion of probability distributions in phase space. Via the Chapman-Enskog procedure, one can show that the macroscopic variables obtained via integration of this discretisation reproduce the Navier-Stokes' equations up to the viscous term, involving second order derivatives. For a very nice tutorial discussing the LBM see [18]. The term in third order derivatives obtained via this approach (leading to the Burnett equations) are well-known to be unstable [9].

We will demonstrate that for low viscosity a finite difference discretisation of the Boltzmann equation does not approximate the equation except with unfeasibly small time step. Thus we adopt a different view point. We will consider the LBM as a fluid flow model in its own right, and examine the macroscopic dynamics we obtain as a function of the step size in our method. We will show that, under certain conditions (the nice conditions), the LBM approximates Navier-Stokes well. When conditions are extreme, as we have shown in other publications [6, 7, 8], the perspective we have on the LBM allows for relatively non-invasive control of artificial diffusion. Of course, we cannot be sure we have reproduced the real dynamics of the fluid, but we do have a good rationale for the targeted control of oscillation. Our aim is to put the LBM on a firm footing in the arsenal of techniques for simulating fluid flow, even though, there is essentially no numerical analysis, because the method is the model.

2 The Boltzmann Equation

Let $f = f(\mathbf{x}, \mathbf{v}, t)$ be the one-particle distribution function, i.e., the probability of finding a particle in a volume dV around a point $(\mathbf{x}, \mathbf{v})$, at a time t, in phase space is $f(\mathbf{x}, \mathbf{v}, t)dV$. Then, the Boltzmann kinetic transport equation is the following time evolution equation for f,

$$\frac{df}{dt} = \frac{\partial f}{\partial t} + \mathbf{v} \cdot \nabla f = \frac{\partial f}{\partial t} + \nabla \cdot (\mathbf{v}f) = C(f). \tag{1}$$

Here df/dt denotes the *material derivative* and the *collision integral*, C, describes the interactions of the populations f, at sites $\mathbf{x}$ for different values of $\mathbf{v}$. We have also used $\nabla \cdot (\mathbf{v}f) = \mathbf{v} \cdot \nabla f$ since the spacial derivatives are independent of $\mathbf{v}$.

Equation (1) describes the microscopic dynamics of our model. We will wish to recover the macroscopic dynamics, the fluid density, momentum density and energy density. We do this by integrating the distribution function:

$$\begin{aligned}
\rho(\mathbf{x}, t) &= \int f(\mathbf{x}, \mathbf{v}, t)\, d\mathbf{v} \\
\rho u_i(\mathbf{x}, t) &= \int v_i f(\mathbf{x}, \mathbf{v}, t)\, d\mathbf{v}, \quad i = 1, 2, \\
E(\mathbf{x}, t) &= \frac{1}{2} \int \mathbf{v}^2 f(\mathbf{x}, \mathbf{v}, t)\, d\mathbf{v}.
\end{aligned}$$

Such functionals of the distribution are called *moments*. The pressure P is given by

$$E = P + \frac{1}{2}\rho \mathbf{u}^2.$$

Let m be a mapping which takes us from microscopic variables f to the vector of macroscopic variables M. It is clear that this is a linear. In 2 dimensions the vector M has 4 components, and in d-dimensions, $d+2$ components.

There are an infinite number of distribution functions which give rise to any particular macroscopic configuration M. Given a concave entropy functional $S(f)$, for any fixed M there will be a unique f which is the solution of the optimisation

$$Qf = \text{argmax}\{S(f) : m(f) = M\}.$$

We call Qf the quasiequilibrium as it is not a global equilibrium. Let us call the set of quasiequilibria the *quasiequilibrium manifold* QE.

If the entropy is the Gibbs entropy

$$S(f) = \int f \log f d\mathbf{v}$$

the quasi-equilibrium is the Maxwellian distribution, which in two dimensions is

$$Qf(\mathbf{v},\mathbf{x}) = \frac{\rho^2}{2\pi P} \exp\left(-\frac{\rho}{P}(\mathbf{v}-\mathbf{u})^2\right),$$

which we note is independent of $\mathbf{x}$. Also, for each set of macroscopic variables M, we have a unique $Qf_M \in$ QE; see Figure 2.

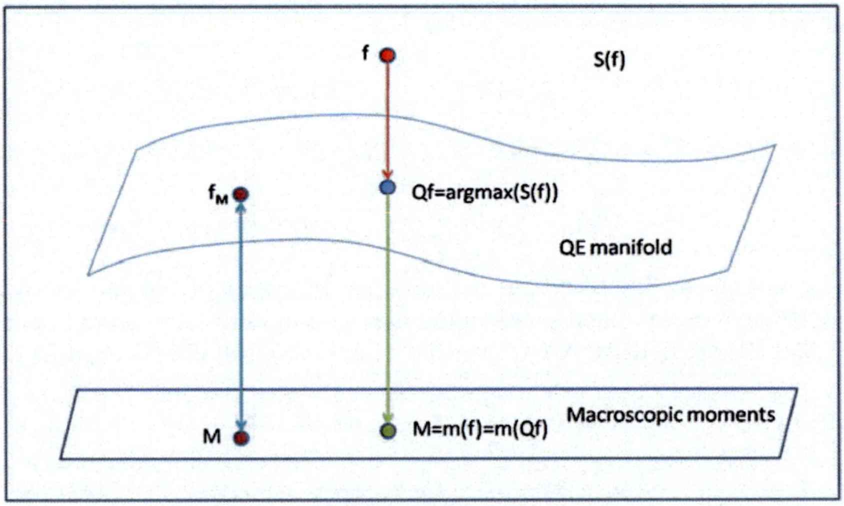

Fig. 1. The quasiequilibrium manifold.

Of course, in general we cannot compute the above integrals to find the above macroscopic variables. Therefore we need to a numerical integration technique which approximates the integrals well, and at least preserves low order degree polynomials, in other words, the macroscopic variables of interest. Since $M(f) = M(Qf)$ then an integration rule which evaluates

$$\int g(\mathbf{v}) Qf(\mathbf{v},\mathbf{x}) d\mathbf{v} = \int g(\mathbf{v}) \frac{\rho^2}{2\pi P} \exp\left(-\frac{\rho}{P}(\mathbf{v}-\mathbf{u})^2\right) d\mathbf{v}$$

when g is a low degree polynomials will preserve the conservation of the macroscopic variables M. The obvious candidate for this is Gauss-Hermite type integration formulae. If we do this we get an integration formula

$$\int g(\mathbf{v})f(\mathbf{v},\mathbf{x})d\mathbf{v} \approx \sum_i W_i g(\mathbf{v}_i)f(\mathbf{v}_i).$$

If we write $f_i(\mathbf{x}) = f(\mathbf{x}, \mathbf{v}_i)$ then we can view the lattice Boltzmann equation (see Section 3 below) as a quadrature approximation in the velocity variable to the Boltzmann equation (1). For a complete treatment of this point of view see Shan and He [27].

In this article we will be interested in collision integrals of the form

$$C(f) = -\sigma_{\mathbf{v}}(f - Qf),$$

with $\sigma_{\mathbf{v}} \in \mathbb{R}$ (later we will discuss appropriate ranges of values of this parameter). We immediately remark that these collisions do not result in changes in the macroscopic variables as f and Qf have the same macroscopic variables by definition of the quasiequilibrium. In the case that $\sigma_{\mathbf{v}} = 1/\gamma$ we have the much used Bhatnagar-Gross-Crook (BGK) collision [4] (we will see below that this corresponds to a rescaling of fast nonequilibrium variables). If γ is small then the derivative is high and the time to equilibrium is long. Thus γ is a measure of the viscosity in the system, and we will quantify this more precisely later. In fact this collision is the linear part of a more general collision integral expanded about the quasi-equilibrium:

$$Cf = Qf + [D_f C]|_{Qf}(f - Qf) + \cdots,$$

where D_f is the Frechet derivative. In what remains we will assume that $\sigma_{\mathbf{v}} = 1/\gamma$.

For the following argument we will need to assume that

Assumption 1 *A) The distributions f are all differentiable in space. The distributions and their gradients remains bounded through the motion:*

$$\max\{f(\mathbf{x},\mathbf{v}), |\nabla_{\mathbf{x}} f(\mathbf{x},\mathbf{v})|\} \le F, \mathbf{x} \in \mathbb{R}^d.$$

B) The nonlinear operators Q are bounded, i.e. for some real $C > 0$

$$|(Qf)| \le C\|f\|_\infty.$$

C) The nonlinear operators Q are differentiable as functions of the macroscopic moments M:

$$|\nabla_M (Qf)| \le C\|f\|_\infty.$$

Since the quasiequilibrium distributions are parametrised by the macroscopic moments, they depend on time through the macroscopic moments:

$$\frac{\partial(Qf)}{\partial t} = \nabla_M(Qf)\cdot\frac{\partial M}{\partial t}. \tag{2}$$

Now, since m is linear, we have

$$\begin{aligned}\frac{\partial M}{\partial t} &= m\left(\frac{\partial f}{\partial t}\right)\\ &= m\left(-\nabla\cdot(\mathbf{v}f) - \frac{1}{\gamma}(f-Qf)\right)\\ &= \frac{1}{\gamma}m\left(-\gamma\nabla\cdot(\mathbf{v}f)-(f-Qf)\right).\end{aligned}$$

Since m is bounded as an operator from $L_\infty(\mathbb{R}^{2d}) = \{f : \|f\|_\infty < \infty\}$ to $L_\infty(\mathbb{R}^{d+2})$, we have

$$\left|\frac{\partial M}{\partial t}\right| \le \frac{C}{\gamma}\max\{\|\nabla f\|_\infty, \|f\|_\infty, \|Qf\|_\infty\}.$$

Substituting into (2), and using the boundedness of $\nabla_M(Qf)$ we have

$$\left|\frac{\partial}{\partial t}(Qf)\right| \le \frac{CF}{\tau}.$$

Hence, if we perform the material derivative of the Boltzmann equation we obtain

$$\begin{aligned}\left|\frac{d^2 f}{dt^2}\right| &= -\frac{1}{\gamma}\left(\frac{df}{dt} - \frac{d}{dt}Qf\right)\\ &\le \frac{CF}{\gamma^2},\end{aligned}$$

for some constant C. More generally

$$\left|\frac{d^k f}{dt^k}\right| = \frac{CF}{\gamma^k}.$$

In order to develop a numerical method we first consider the time discretisation of the Boltzmann equation. A simple Euler scheme with time interval $[0,\tau]$ for instance would give us:

$$\begin{aligned}f(\mathbf{x}+\mathbf{v}\tau,\mathbf{v},\tau) &= f(\mathbf{x},\mathbf{v},0) + \int_0^\tau \frac{df}{dt}(\mathbf{x},\mathbf{v},t)dt\\ &\approx f(\mathbf{x},\mathbf{v},0) + \tau\frac{df}{dt}(\mathbf{x},\mathbf{v},0),\end{aligned}$$

with error term

$$\begin{aligned}E(\tau) &\le \tau^2\max_{t\in[0,\tau]}\left|\frac{d^2 f}{dt^2}\right|\\ &\le \frac{CF\tau^2}{\gamma^2}.\end{aligned}$$

Remark 1. 1. It has become orthodox to view the LBM as a discretisation of the Boltzmann equation. However, we see that the error does not go to zero in the above finite difference approximation unless $\tau \ll \gamma$. This is fine if we are approximating a flow of high viscosity, but for high Reynolds number flow, where γ may be very small, the notion of using a time step which is smaller is computationally unfeasible. We develop below an alternative point of view.

2. We should remark that Qf does not directly depend on time, but only on the macroscopic moments, and the velocity. It depends on position via the macroscopic moments at that position. The macroscopic moments depend on populations moving with all velocities, so that $Qf(\mathbf{x}, \mathbf{v}) = Q(f(\mathbf{x}, \mathbf{u}) : \mathbf{u} \in \mathbb{R}^d)$. We will use this in Section 7 when we discuss stability.

3 The Lattice Boltzmann Method

In the Lattice Boltzmann method we have only a finite number of populations $f_1, \cdots, f_N$, with f_i moving with velocity $\mathbf{v}_i$. We can think of this as a discretisation of the velocity part of phase space, but we need not. The computational domain is a τ-scaled grid, where τ is the time step of the method:

$$X = \left\{ \tau \sum_{i=1}^{N} z_i \mathbf{v}_i : z_i \in \mathbb{Z} \right\};$$

see Figure 2, where we have 7 velocities, one from a point to each of its nearest neighbours, plus the zero velocity. Thus, in one time step (which we think of as of length 1), the populations f_i move from $\mathbf{x} \in X$ to $\mathbf{x} + \tau \mathbf{v}_i \in X$. Hence, our dynamics happen on the grid. We should not think of X as the discretisation of computational space. It is a set of reference points at which we know our populations.

As indicated in the introduction we compute the macroscopic moments at the kth timestep with

$$\rho(\mathbf{x}, k\tau) = \sum_{i=1}^{N} W_i f_i(\mathbf{x}, k\tau),$$

$$(\rho \mathbf{u})(\mathbf{x}, k\tau) = \sum_{i=1}^{N} W_i \mathbf{v}_i f_i(\mathbf{x}, k\tau),$$

$$E(\mathbf{x}, k\tau) = \frac{1}{2} \sum_{i=1}^{N} W_i \mathbf{v}_i^2 f_i(\mathbf{x}, k\tau).$$

Let us call this mapping $M(\mathbf{x}, k\tau) = m(f_1(\mathbf{x}, k\tau), \cdots, f_N(\mathbf{x}, k\tau))$.

Given that we know our populations at time step k we compute the the populations at time step $k + 1$ via the following set of rules. For each $i = 1, 2, \cdots, N$,

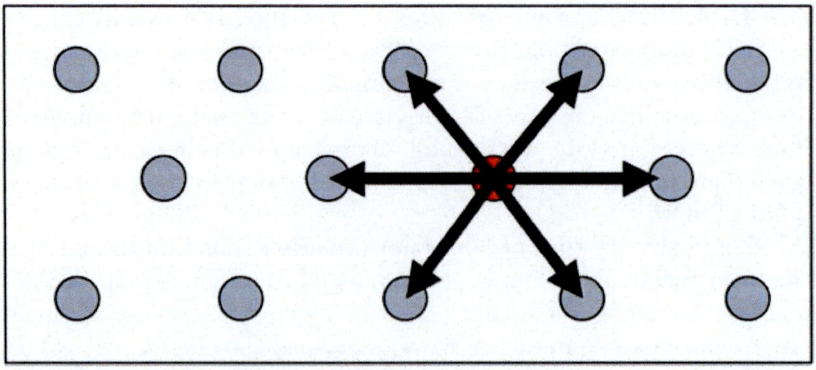

Fig. 2. The computational grid.

a) compute intermediate populations

$$f_i^{\text{int}}(\mathbf{x},(k+1)\tau) = f_i(\mathbf{x} - \tau\mathbf{v}_i, k\tau);$$

b) compute the macroscopic moments

$$M(\mathbf{x},(k+1)\tau) := m(f_1^{\text{int}}(\mathbf{x},(k+1)\tau), \cdots, f_N^{\text{int}}(\mathbf{x},(k+1)\tau));$$

c) compute $Qf_i^{\text{int}}(\mathbf{x},(k+1)\tau)$ which depends on $M(\mathbf{x},(k+1)\tau)$;
d) compute the new populations

$$f_i(\mathbf{x},(k+1)\tau) = f_i^{\text{int}}(\mathbf{x},(k+1)\tau) - \beta(f_i^{\text{int}}(\mathbf{x},(k+1)\tau) - Qf_i^{\text{int}}(\mathbf{x},(k+1)\tau)).$$

The computation of the intermediate population in (a) above is typically called the *free flight* or *streaming step*. The macroscopic variables are transported through the computational domain in this step. The steps (b)–(d) are the collision step, in which the macroscopic variables are redistributed between the different populations f_i arriving at the point $\mathbf{x}$. As mentioned above, this redistribution is done conserving the macroscopic variables, but possibly with an increase in entropy.

The choice of β in (iv) above is crucial. In particular, if $\beta = 1$, the collision returns the distribution to equilibrium. Such a step is called an Ehrenfests' step in honour of Tanya and Paul Ehrenfest [10], who introduced course graining, which results in entropy increase. Equilibration is an example of such a course graining.

4 The Chapman Enskog Procedure

The Chapman-Enskog procedure is the standard route via which the Boltzmann equation is linked to the Navier-Stokes equation. This discussion is based on that provided by Gorban [12], with the fast-slow variable point of view described in Jones [13]. We should make the point that the purpose of the procedure is not to produce an approximation to the Boltzmann equation, but rather to seek a manifold which is, in some sense, one to one correspondence with the macroscopic moments. If this is so then we have observable slow variables M (the macroscopic moments), with corresponding unique member $f_M \in \mathrm{QE}$. Substituting Qf into (1) we obtain the following:

$$\frac{\partial(f_M)}{\partial t} + \nabla \cdot \mathbf{v}(f_M) = 0.$$

If we now multiply this by $1, \mathbf{v}, \mathbf{v}^2$ and integrate we obtain the Euler equations (we will not spell out the details here which can be found in e.g. [18])

$$\frac{\partial \rho}{\partial t} + \nabla \cdot \rho \mathbf{u} = 0,$$

$$\rho \frac{\partial \mathbf{u}}{\partial t} + \rho(\mathbf{u} \cdot \nabla)\mathbf{u} = -\rho(\mathbf{u} \cdot \nabla)\mathbf{u} - \nabla P,$$

$$\frac{\partial E}{\partial t} = -\nabla \cdot (\mathbf{u}(P + E)).$$

Following [13] we split the variables into fast and slow. The dynamics of the fast variables appear via $f^1 = f - Qf$, since $Qf \in \mathrm{QE}$ is the unique distribution with these macroscopic moments. Therefore, the Boltzmann equation

$$\frac{\partial f}{\partial t} + \nabla \cdot (\mathbf{v} f) = -\frac{1}{\gamma} f^1. \tag{3}$$

can be viewed as saying that the rate of change of f is proportional to nonequilibrium part of the distribution, which has a natural time scale γt, if t is the timescale of the slow variables.

We can write

$$f = Qf + \gamma f^1, \tag{4}$$

Since $M(f) = M(Qf)$, it is clear that

$$M(f^1) = 0.$$

In many expositions the parameter in the expansion γ is identified with the Knudsen number

$$\gamma = \lambda / L$$

where λ is the mean free path (the average distance between collisions) and L is a length scale in the problem (the size of an obstacle for instance).

If we substitute (4) into (1) we get

$$\frac{\partial(Qf + \gamma f^1)}{\partial t} + \nabla \cdot (\mathbf{v}(Qf + \gamma f^1)) = -f^1.$$

If we now equate the terms of order 0 in γ we see that

$$f^1 = -\left(\frac{\partial(Qf)}{\partial t} + \nabla \cdot (\mathbf{v}(Qf))\right). \tag{5}$$

If we substitute this into (3) and integrate we obtain the Navier-Stokes equations as given at the start of the paper, with viscosity γ.

Of course, we could introduce longer asymptotic expansions, and derive expressions for higher order corrections to the Navier-Stokes equations. If we do this we obtain the Burnett and super Burnett equations, which are known to be unphysical (see e.g. [18, Page 31]). In the next section we use a different philosophy for our expansion. We think of the LBM as a computational model, and that the notion of smallness should be the time step, not any physical parameter. This produces similar, but not the same results.

5 The Time Discretisation Expansion

In this section we will describe a different procedure by which we obtain the Navier-Stokes equations. We will assume that we are not performing any discretisation in space. A full description of the modified Navier-Stokes equations one obtains from the fully discrete process is given in [21]. We have described this procedure in a previous paper [5], but we repeat some of it here for completeness. The idea is to study the rate of change of the macroscopic variables induced by the sequence of free flight by time step τ, followed by equilibration, and then repeated. Since such a collision is called an Ehrenfests' step, we call this dynamic chain an Ehrenfests' chain: see Figure 3. We seek the form of $\frac{\partial M}{\partial t}$ which leads to these values of the macroscopic variables at each of the times $0, \tau, 2\tau, \cdots$.

Let us recall that the LBM has a free flight phase followed by a collision phase.

$$\frac{\partial f}{\partial t} = -\mathbf{v} \cdot \nabla f,$$

with solution

$$\Theta_t(f_0)(\mathbf{x}, \mathbf{v}, t) = f_0(\mathbf{x} - \mathbf{v}t, \mathbf{v}, t).$$

For the moment, let us suppose that the collision phase returns the process to the appropriate quasiequilibrium. Thus, if we start at $f_0 \in \mathrm{QE}$, the next point in our iteration is $f_1 = Q(\Theta_\tau(f_0)) \in \mathrm{QE}$. If we iterate we get a sequence f_i, $i = 0, 1, \cdots$. We wish to determine the macroscopic dynamics which passes through the points $m(f_i)$, $i = 0, 1, \ldots,$. This will depend on the parameter τ, so we get an equation of the form

$$\frac{\partial M}{\partial t} = F(M, \tau).$$

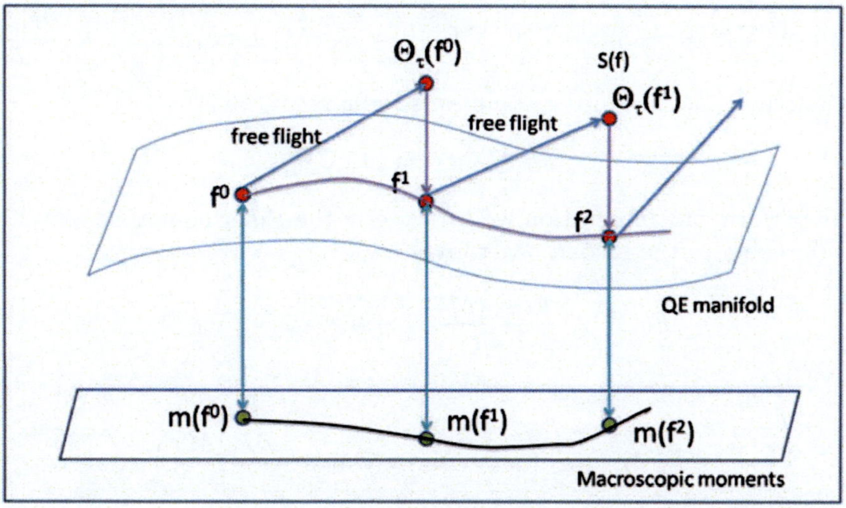

Fig. 3. The Ehrenfests' chain.

We will expand this for small τ in a series $F(M,\tau) = F_0(M) + \tau F_1(M) + \mathcal{O}(\tau^2)$ and match terms in powers of τ to determine F_0 and F_1. In other words we wish to have

$$m(\Theta_\tau(f_0)) = M(\tau)$$

to second order in τ.

The second order expansion in time for the dynamics of the distribution f is, to order τ^2,

$$\begin{aligned}\Theta_\tau(f_0) &= \Theta_0(f_0) + \tau \left.\frac{\partial \Theta_t}{\partial t}\right|_{t=0} + \frac{\tau^2}{2} \left.\frac{\partial^2 \Theta_t}{\partial t^2}\right|_{t=0} \\ &= f_0 - \tau \mathbf{v} \cdot \nabla f_0 + \frac{\tau^2}{2} \mathbf{v} \cdot \nabla(\mathbf{v} \cdot \nabla f_0).\end{aligned}$$

Thus, to second order,

$$\begin{aligned}&m(\Theta_\tau(f^*)) \\ &= m(\Theta_0(f_0)) - \tau \frac{\partial}{\partial t} m\left(\mathbf{v} \cdot \nabla f_0\right) + \frac{\tau^2}{2} m(\mathbf{v} \cdot \nabla(\mathbf{v} \cdot \nabla f_0)).\end{aligned}$$

Similarly, to second order,

$$\begin{aligned}M(\tau) &= M(0) + \tau \left.\frac{\partial M}{\partial t}\right|_{t=0} + \frac{\tau^2}{2} \left.\frac{\partial^2 M}{\partial t^2}\right|_{t=0} \\ &= M(0) + \tau(F_0(M) + \tau F_1(M)) + \frac{\tau^2}{2} \frac{\partial F_0(M)}{\partial t}.\end{aligned}$$

Since $M(0) = m(\Theta_0(f^*))$, we have

$$-\tau m\left(\mathbf{v}\cdot\nabla f_0\right)+\frac{\tau^2}{2}m(\mathbf{v}\cdot\nabla(\mathbf{v}\cdot\nabla f_0))=\tau(F_0(M)+\tau F_1(M))+\frac{\tau^2}{2}\frac{\partial F_0(M)}{\partial t}.$$

Matching the first order conditions we have

$$F_0(M)=-m\left(\mathbf{v}\cdot\nabla f_0\right).$$

If we perform the integration we get exactly the Euler equations as before.

Matching to the second order gives

$$F_1(M)+\frac{1}{2}\frac{\partial F_0(M)}{\partial t}=\frac{1}{2}m(\mathbf{v}\cdot\nabla(\mathbf{v}\cdot\nabla f_0)),$$

and rearranging we get

$$F_1(M)=\frac{1}{2}\left(m(\mathbf{v}\cdot\nabla(\mathbf{v}\cdot\nabla f_0))-\frac{\partial F_0(M)}{\partial t}\right).$$

These are an integrated version of (5).

Hence, to second order, the macroscopic equations are

$$\frac{\partial}{\partial t}m(f_0)=-m\left(\mathbf{v}\cdot\nabla f_0\right)+\frac{\tau}{2}\left(m(\mathbf{v}\cdot\nabla(\mathbf{v}\cdot\nabla f_0))-\frac{\partial F_0(M)}{\partial t}\right).$$

Integration of these gives the Navier-Stokes equations with coefficient of viscosity $\tau/2$; see [5].

In order to perform these asymptotic expansions we require that higher order derivatives of the distribution function behave well. This line of enquiry is developed further in [21].

6 Decoupling Time Step and Viscosity

There is of course a difficulty in simulating a Navier-Stokes flow where viscosity is given, with a numerical scheme in which the viscosity is directly proportional to the time step, as the free flight and equilibration scheme (the Ehrenfests' step) detailed above. We can write this scheme in the form

$$f_i(\mathbf{x},(k+1)\tau)=f_i(\mathbf{x}-\tau\mathbf{v}_i,k\tau)-(f_i(\mathbf{x}-\tau\mathbf{v}_i,k\tau)-Qf_i(\mathbf{x}-\tau\mathbf{v}_i,k\tau)).$$

Thus, after free flight dynamics we move along the vector from free flight to equilibrium. With the BGK collision ([4] and Section 2 above) we move some part of the way along this direction. This suggests then a more general numerical simulation process

$$f_i(\mathbf{x},(k+1)\tau)=f_i(\mathbf{x}-\tau\mathbf{v}_i,k\tau)-\beta(f_i(\mathbf{x}-\tau\mathbf{v}_i,k\tau)-Qf_i(\mathbf{x}-\tau\mathbf{v}_i,k\tau)),$$

where the β may be chosen to satisfy a physically relevant condition. A choice of $\beta=1$ gives the Ehrenfests' step, whilst $\beta=2$ gives the so-called LBGK method [25]. In this latter case we are reflecting in the quasi-equilibrium to give us a microscopic description with the same macroscopic variable, and, to order τ^2, the same entropy.

The revolution in LBMs achieved by Succi and coworkers [25], was the overrelaxation step, with $\beta > 1$. Here we pass through the quasi-equilibrium so that the next phase of free flight takes us back through the quasi-equilibrium manifold. One variant of this is the so-called *entropic* LBM (ELBM) in which β is chosen to that $S(f_i(\mathbf{x} + \tau\mathbf{v}_i, (k+1)\tau)) = S(f_i(\mathbf{x}, k\tau))$.

Both LBGK and ELBM uncouple the viscosity parameter from the time step. There are a number of other ways in which one can achieve the same goal; see [7]. We describe one procedure here, giving an intuitive justification, using the idea that the quasiequilibrium manifold is flat (see Figure 4). If the manifold is not flat we only introduce $\mathcal{O}(\tau^2)$ errors in the following argument. Free flight of time τ, followed by an Ehrenfests' step adds viscosity of $\tau/2$.

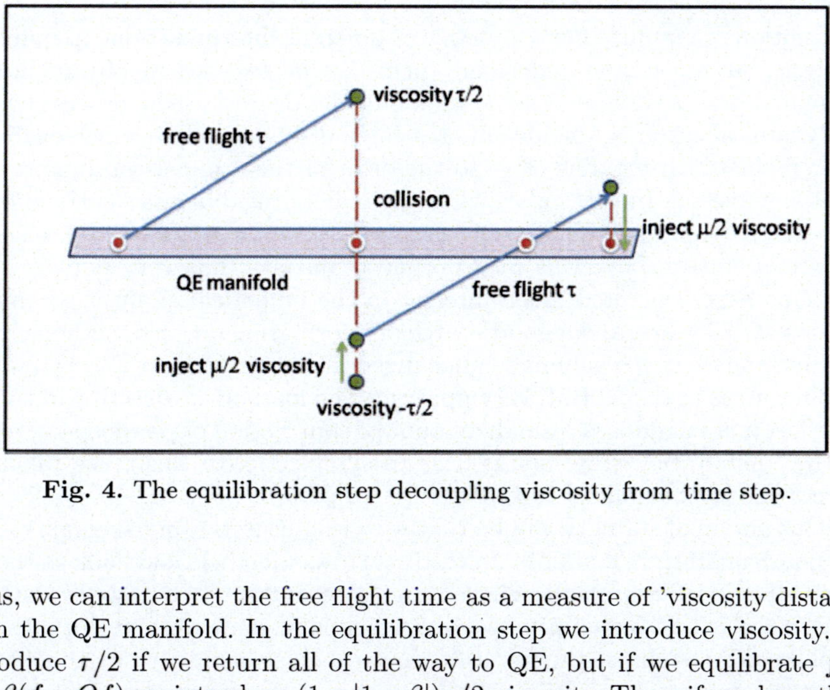

Fig. 4. The equilibration step decoupling viscosity from time step.

Thus, we can interpret the free flight time as a measure of 'viscosity distance' from the QE manifold. In the equilibration step we introduce viscosity. We introduce $\tau/2$ if we return all of the way to QE, but if we equilibrate $f \to f - \beta(f - Qf)$ we introduce $(1 - |1 - \beta|)\tau/2$ viscosity. Thus, if we do nothing ($\beta = 0$), or reflect in the QE manifold ($\beta = 2$) we do not introduce any viscosity. On the other hand, an Ehrenfests' step ($\beta = 1$) introduces $\beta/2$ viscosity.

Suppose when we perform our equilibration process we set $\beta = 2 - \mu/\tau$. Then we introduce $\mu/2$ viscosity, and we are a free flight distance $\mu/2 - \tau/2$ from the QE manifold (this is negative). Thus, free flight by time τ moves us a distance $\tau/2$ towards the manifold, i.e. to a distance $\mu/2$. An Ehrenfests' step will now introduce viscosity of $\mu/2$. Hence we have added viscosity of μ (to $\mathcal{O}(\tau^2)$).

Unfortunately, as we will see in Section 8 below, there are instabilities (see the first paragraph of the next section to see what we mean by this) in the simulation with LBGK and ELBM. This is because the free flight dynamics sometimes takes us too far from the quasi-equilibrium manifold. In this case we apply a single Ehrenfests' step and return to the equilibrium manifold. As you will see, this stabilises the method beautifully. In order to retain an order τ^2 method we can only apply Ehrenfest stabilising at a bounded number of sites. Thus we fix a tolerance δ which measures the distance from the equilibrium manifold, and then we choose the k (a fixed number) most distant points and return these to equilibrium.

7 Stability

The notion of stability has a number of different interpretations in numerical analysis, but a common underlying theme is that two nearby objects at some stage in a process do not become unboundedly distant as the process evolves. In dynamical systems, the idea of creating a discrete scheme which evolves on a manifold which remains close to the orbit of the original continuous equations is universal. In particular, when the underlying dynamics is *Hamiltonian* (conserves volume in phase space) then *symplectic methods*, those which are also conservative, have this nice property, and are thus very popular. Such methods have generated much interest in the numerical analysis community in the past 20 years, and a good source of information on such methods is [22].

Since phase space volume is unchanged in free-flight, and also by entropic involution, as in the ELBM, is symplectic. The method we describe in the previous section introduces a small amount of contraction of phase space volume in the equilibration step, but is near to symplectic. In this sense we should expect good behaviour of the LBM.

Our notion of stability will be that we wish our iteration to remain close to the quasiequilibrium manifold. In the first subsection below we look at the case when all quantities are well-behaved. In the second subsection we examine the situation when an instability evolves (which is in some sense inevitable), and what we do to stabilise the iteration. We recognise instability by the distance the iteration gets from QE, so in some sense, the extent to which the fast variables are dominating the dynamics.

In [21] the stability of the method under perturbation by high frequency signal is examined. This is the standard method of stability analysis for finite difference methods. Pictures which give different stability regions are presented there.

7.1 The Well-Behaved Case

Let us consider one time step. From the description of the LBM algorithm from Section 3 we have

$$f_i(\mathbf{x},(k+1)\tau) = f_i^{\text{int}}(\mathbf{x},(k+1)\tau) - \beta(f_i^{\text{int}}(\mathbf{x},(k+1)\tau) - Qf_i^{\text{int}}(\mathbf{x},(k+1)\tau).$$

Let us write Q_i as the operator which equilibrates the ith population.

Let us assume that the operators Q_i and distributions f_i satisfy the conditions of Assumptions 1, and additionally,

D) The nonlinear operators Q_i are differentiable as functions of the distributions $f_1, \cdots, f_N$:

$$|\nabla Q_i(f_1, \cdots, f_N)| \leq C \max_{i=1,\cdots,N} \|f_i\|_\infty.$$

Assumption D above just ensures that the equilibrated population depends in a smooth way on the populations which give rise to it. Thus we are saying that the QE manifold does not bend around too much.

We now wish to compare

$$d_i(\mathbf{x},k\tau) = |f_i(\mathbf{x},k\tau) - Q_i(f_1(\mathbf{x},k\tau), \cdots, f_N(\mathbf{x},k\tau))|$$

with $d_i(\mathbf{x},(k+1)\tau)$. We have

$$\begin{aligned}
&d_i(\mathbf{x},(k+1)\tau)\\
&= |f_i^{\text{int}}(\mathbf{x},(k+1)\tau)\\
&\qquad -\beta(f_i^{\text{int}}(\mathbf{x},(k+1)\tau) - Q_i(f_1(\mathbf{x}-\tau\mathbf{v}_1,k\tau), \cdots, f_N(\mathbf{x}-\tau\mathbf{v}_N,k\tau))|\\
&= |f_i(\mathbf{x}-\mathbf{v}_i,k\tau)\\
&\qquad -\beta(f_i(\mathbf{x}-\tau\mathbf{v}_i,k\tau) - Q_i(f_1(\mathbf{x}-\tau\mathbf{v}_1,k\tau), \cdots, f_N(\mathbf{x}-\tau\mathbf{v}_N,k\tau))\\
&\qquad\qquad -Q_i(f_1(\mathbf{x}-\tau\mathbf{v}_1,k\tau), \cdots, f_N(\mathbf{x}-\tau\mathbf{v}_N,k\tau))|\\
&= |(1-\beta)(f_i(\mathbf{x}-\tau\mathbf{v}_i,k\tau) - Q_i(f_1(\mathbf{x}-\tau\mathbf{v}_1,k\tau), \cdots, f_N(\mathbf{x}-\tau\mathbf{v}_N,k\tau))|\\
&\leq |(1-\beta)(f_i(\mathbf{x}-\tau\mathbf{v}_i,k\tau) - Q_i(f_1(\mathbf{x}-\tau\mathbf{v}_i,k\tau), \cdots, f_N(\mathbf{x}-\tau\mathbf{v}_i,k\tau))|\\
&\qquad +|(1-\beta)(Q_i(f_1(\mathbf{x}-\tau\mathbf{v}_i,k\tau), \cdots, f_N(\mathbf{x}-\tau\mathbf{v}_i,k\tau))\\
&\qquad\qquad -Q_i(f_1(\mathbf{x}-\tau\mathbf{v}_1,k\tau), \cdots, f_N(\mathbf{x}-\tau\mathbf{v}_N,k\tau))|\\
&= (1-\beta)d_i(\mathbf{x}-\tau\mathbf{v}_i,k\tau)\\
&\qquad +|(1-\beta)(Q_i(f_1(\mathbf{x}-\tau\mathbf{v}_i,k\tau), \cdots, f_N(\mathbf{x}-\tau\mathbf{v}_i,k\tau))\\
&\qquad\qquad -Q_i(f_1(\mathbf{x}-\tau\mathbf{v}_1,k\tau), \cdots, f_N(\mathbf{x}-\tau\mathbf{v}_N,k\tau))| \qquad (6)
\end{aligned}$$

Let us now bound the second term above. To do this we write

$$\Delta_j = f_j(\mathbf{x}-\tau\mathbf{v}_i) - f_j(\mathbf{x}-\tau\mathbf{v}_j).$$

Then,

$$
\begin{aligned}
|\Delta_j| &\leq \tau |\mathbf{v}_i - \mathbf{v}_j| \|\nabla f_j\|_\infty \\
&\leq 2VF\tau,
\end{aligned}
$$

using Assumption A above, where $V = \max_{i=1,\cdots,N}\{|\mathbf{v}_i|\}$. We define the vector

$$\mathbf{D} = (\Delta_1, \cdots, \Delta_N)^T.$$

Then,

$$
\begin{aligned}
&|Q_i(f_1(\mathbf{x} - \tau\mathbf{v}_i, k\tau), \cdots, f_N(\mathbf{x} - \tau\mathbf{v}_i, k\tau)) \\
&\quad -Q_i(f_1(\mathbf{x} - \tau\mathbf{v}_1, k\tau), \cdots, f_N(\mathbf{x} - \tau\mathbf{v}_N, k\tau))| \\
&\leq \max_{f_1,\cdots,f_N} |\nabla Q_i(f_1, \cdots, f_N)| \|\mathbf{D}\|_2 \\
&\leq CN \max_{j=1,\cdots,N} |\Delta_j| \\
&\leq CN(2VF\tau) = 2CNVF\tau
\end{aligned}
$$

using (7).

Substituting into (6) we see that

$$d_i(\mathbf{x}, k+1) \leq (1-\beta)(d_i(\mathbf{x} - \tau\mathbf{v}_i, k) + 2KNVF\tau).$$

Hence, for small enough τ, given the assumptions above, the motion remains close to the quasiequilibrium manifold. These assumptions apply when the motion is well-behaved.

7.2 The Difficult Case

Of course, the challenge is to simulate fluid flow as shocks emerge. At this stage, the theory described does not apply because we get unbounded derivatives in the derivatives of the populations, and Navier-Stokes equations and LBM become different models for fluid flow. In [3] the Karlin-Succi involution is used for computation and it is reported that this exact implementation of the entropy balance (guaranteeing the fulfilment of the 2nd law of thermodynamics) significantly regularizes the post-shock dispersive oscillations. This regularization is very surprising, because, as described here, the entropic lattice BGK model gives a second-order approximation to the Navier–Stokes equation and due to the Godunov theorem [11], second-order methods have to be non monotonic. Moreover, Lax [15] and Levermore with Liu [16] demonstrated that these *dispersive oscillations* are unavoidable in classical numerical methods. Schemes with precise control of entropy production, studied by Tadmor with Zhong [26], also demonstrated post-shock oscillations. In [20] the author tests the hypothesis that artificial viscosity is introduced using imprecise numerical solvers for the problem $S(f+\beta(f-Qf)) = S(f)$ which is required for entropic involution. Our model problem is the shock tube which we describe below. In Figure 5 we present results for the LBM with the BGK collision (LBGK), with the parameter γ chosen to give the stated viscosity ν, with polynomial and exact equilibria, and the entropic LBM (ELBM), with high precision solution of the last equation. We are supposed to see an improvement in the stability in the right hand pictures. We see this is not so,

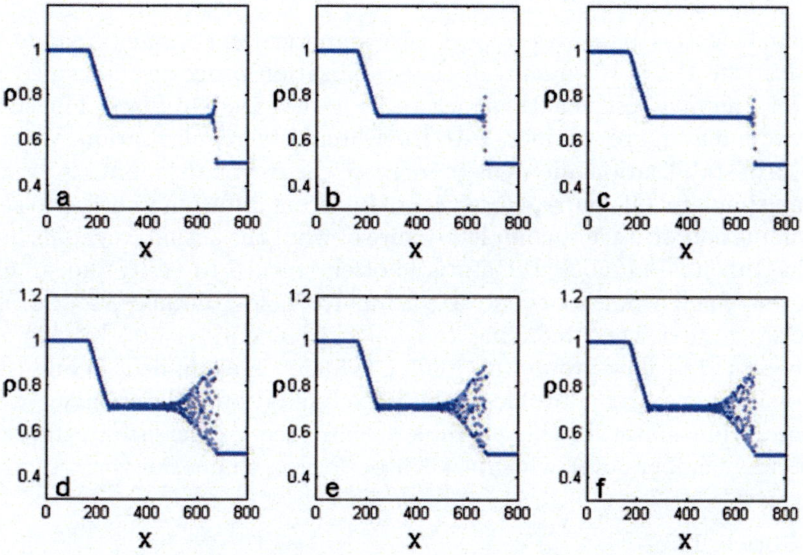

Fig. 5. Density profile of the simulation of the shock tube problem following 400 time steps using **(a)** LBGK with polynomial quasi-equilibria [$\mu = (1/3) \cdot 10^{-1}$]; **(b)** LBGK with entropic quasi-equilibria [$\mu = (1/3) \cdot 10^{-1}$]; **(c)** ELBM [$\mu = (1/3) \cdot 10^{-1}$]; **(d)** LBGK with polynomial quasi-equilibria [$\mu = 10^{-9}$]; **(e)** LBGK with entropic quasi-equilibria [$\mu = 10^{-9}$]; **(f)** ELBM [$\mu = 10^{-9}$].

suggesting that entropic preservation is not enough for stability.

We see evolving instabilities in numerical simulation when the free flight phase of the LBM carried the distribution too far away from the quasiequilibrium manifold, i.e.

$$\Delta_i S(\mathbf{x}) = |S(f_i(\mathbf{x})) - S(Qf_i(\mathbf{x}))| > \delta,$$

for some tolerance δ. It turns out that this happens only very locally to the instability, so that one can locally modify the microscopic entropy in a very precise way.

In previous work we have considered different ways of dealing with too much non-equilibrium entropy. Ehrenfests' stabilisation is simply the return of the population to equilibrium, as described in Section 3. In Section 8 we will see how this works for the lid driven cavity.

A more sophisticated and less extreme approach is to stabilise the LBGK method using median filtering at a single point. We will describe this approach for the shock tube simulation on the next section, and in Figure 6 one can see how well the simulation is stabilised.

8 Numerical Experiments

To conclude this paper we report two numerical experiments conducted to demonstrate the performance of the stabilisation processes described in this article. The first test is a 1D shock tube, with a median filter. The second is flow around a square cylinder, with the Ehrenfests' regularisation. We see that the use of stabilisation allows us to increase the Reynolds' numbers over which we can simulate. Of course, since we are injecting diffusion in order to stabilise the simulation we are not completely sure of what the actual Reynolds' number is. The Strouhal number, the statistic often tracked to verify the simulation, does not change much beyond a certain Reynolds' number, so we need to exercise caution when inferring too much from our results. In this current volume, in [14], flow around a circular cylinder is examined. While this is a physically less extreme problem, the implementation of boundary conditions is more problematic. In this example we also see an increase in the range of Reynolds' numbers over which we can perform a simulation.

8.1 Shock Tube

A standard experiment for the testing of LBMs is the one-dimensional shock tube problem. The lattice velocities used are $\mathbf{v} = (-1, 0, 1)$ therefore space shifts of the velocities give lattice sites separated by the unit distance. We call the populations associated with these velocities f_-, f_0, f_+ respectively. 800 lattice sites are used and are initialised with the density distribution

$$\rho(x) = \begin{cases} 1, & 1 \leq x \leq 400, \\ 0.5, & 401 \leq x \leq 800. \end{cases}$$

Initially all velocities are set to zero. The polynomial equilibria mentioned prior to Figure 5 are given in [25]:

$$\begin{aligned} f_-^* &= \frac{\rho}{6}\left(1 - 3u + 3u^2\right), \\ f_0^* &= \frac{2\rho}{3}\left(1 - \frac{3u^2}{2}\right), \\ f_+^* &= \frac{\rho}{6}\left(1 + 3u + 3u^2\right). \end{aligned}$$

The entropic quasi-equilibria also used by the ELBM are available explicitly as the maximum of the entropy function

$$S(\mathbf{f}) = -f_- \log(f_-) - f_0 \log(f_0/4) - f_+ \log(f_+) :$$

$$\begin{aligned} f_-^* &= \frac{\rho}{6}(-3u - 1 + 2\sqrt{1+3u^2}), \\ f_0^* &= \frac{2\rho}{3}(2 - \sqrt{1+3u^2}), \\ f_+^* &= \frac{\rho}{6}(3u - 1 + 2\sqrt{1+3u^2}). \end{aligned}$$

The governing equations for the simulation are

$$\begin{aligned} f_-(x, t+\tau) &= f_-(x+1, t) - \alpha\beta(f_-^*(x+1, t) - f_-(x+1, t)), \\ f_0(x, t+\tau) &= f_0(x, t) + \alpha\beta(f_0^*(x, t) - f_0(x, t)), \\ f_+(x, t+\tau) &= f_+(x-1, t) + \alpha\beta(f_+^*(x-1, t) - f_+(x-1, t)). \end{aligned}$$

Here the parameter α is used for entropy control in order to perform a stable simulation. For instance, in ELBM, α is chosen so that

$$S(f(x, t+\tau)) = S(f(x-1, t) + \alpha(f_+^*(x-1, t) - f_+(x-1, t))).$$

Then $\beta = \tau/(\tau + 2\mu)$ controls the viscosity introduced into the model so that we simulate Navier-Stokes's equations with parameter μ.

As we intimated above, if the non-equilibrium entropy is too high at a single site x, i.e.

$$\Delta_i S(x) = |S(f_i(x)) - S(Qf_i(x))| > \delta,$$

we filter the populations in the following way. Instead of being updated using the standard BGK over-relaxation this single site is updated as follows:

$$\begin{aligned} f_-(x, t+1) &= f_-^*(x+1, t) + \sqrt{\frac{\Delta S_{med}}{\Delta S_x}}(f_-(x+1, t) - f_-^*(x+1, t)), \\ f_0(x, t+1) &= f_0^*(x, t) + \sqrt{\frac{\Delta S_{med}}{\Delta S_x}}(f_0(x, t) - f_0^*(x, t)), \\ f_+(x, t+1) &= f_+^*(x-1, t) + \sqrt{\frac{\Delta S_{med}}{\Delta S_x}}(f_+(x-1, t) - f_+^*(x-1, t)), \end{aligned}$$

where

$$\Delta_i S_{\text{med}} = \text{median}\,\{S(f_j(\mathbf{x} - \tau\mathbf{v}_j)) - S(Qf_j(\mathbf{x} - \tau\mathbf{v}_j)) : j = -, 0, +\}.$$

More generally, we might find the median value of ΔS over the set of nodes which have free flight ending up at the node in question:

$$\Delta_i S_{\text{med}} = \text{median}\,\{S(f_i(\mathbf{x} - \tau\mathbf{v}_j)) - S(Qf_i(\mathbf{x} - \tau\mathbf{v}_j)) : j = 1, \cdots, N\}$$

(recall that 0 is one of the velocities).

We observe that median filtering applied at only one point has stabilised the simulation very effectively. Of course, for very small viscosity the noise behind the shock is more pronounced than for higher viscosity.

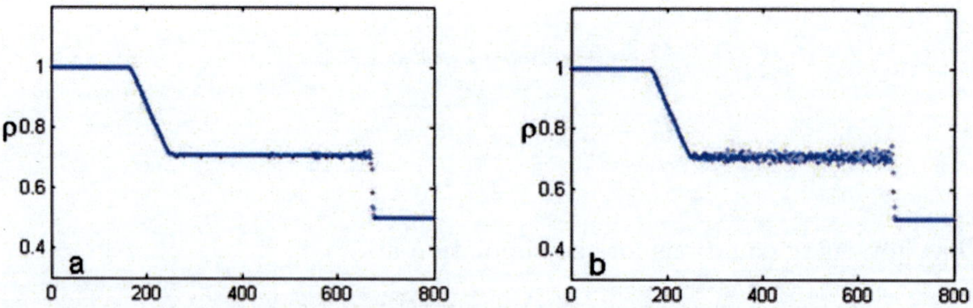

Fig. 6. Density profile of the simulation of the shock tube problem following 400 time steps using (a) LBGK with entropic quasi-equilibria and one point median filtering [$\mu = (1/3) \cdot 10^{-1}$]; (b) LBGK with entropic quasi-equilibria and one point median filtering [$\mu = 10^{-9}$].

8.2 Flow around a Square Cylinder

Our second test is the 2D unsteady flow around a square-cylinder. We use a uniform 9-speed square lattice with discrete velocities

$$v_i = \begin{cases} 0, & i = 0, \\ \left(\cos\left((i-1)\frac{\pi}{2}\right), \sin\left((i-1)\frac{\pi}{2}\right)\right), & i = 1,2,3,4, \\ \sqrt{2}\left(\cos\left((i-5)\frac{\pi}{2}+\frac{\pi}{4}\right), \sin\left((i-5)\frac{\pi}{2}+\frac{\pi}{4}\right)\right), & i = 5,6,7,8. \end{cases}$$

The numbering $f_0, f_1, \dots, f_8$ are for the static, east-, north-, west-, south-, northeast-, northwest-, southwest- and southeast-moving populations, respectively. The quasiequilibrium states, $Q_i f$, can be uniquely determined by maximising an entropy functional

$$S(f) = -\sum_i f_i \log\left(\frac{f_i}{W_i}\right),$$

subject to the constraints of conservation of mass and momentum:

$$Q_i f_i = nW_i \prod_{j=1}^{2} \left(2 - \sqrt{1+3u_j^2}\right)\left(\frac{2u_j + \sqrt{1+3u_j^2}}{1-u_j}\right)^{v_{i,j}}$$

Here, the *lattice weights*, W_i, are given lattice-specific constants: $W_0 = 4/9$, $W_{1,2,3,4} = 1/9$ and $W_{5,6,7,8} = 1/36$. The macroscopic variables are given by the expressions

$$\rho := \sum_i f_i, \quad (u_1, u_2) := \frac{1}{\rho}\sum_i v_i f_i.$$

We consider a square-cylinder of side length L, is emersed in a constant flow in a rectangular channel of length $30L$ and height $25L$. The cylinder is place on the centre line in the y-direction. The centre of the cylinder is placed at a distance $10.5L$ from the inlet. The free-stream velocity is fixed at $(u_\infty, v_\infty) = (0.05, 0)$ (in lattice units) for all simulations.

On the north and south channel walls a free-slip boundary condition is imposed (see, e.g., [25]). At the inlet, the inward pointing velocities are replaced with their quasiequilibrium values corresponding to the free-stream velocity. At the outlet, the inward pointing velocities are replaced with their associated quasiequilibrium values corresponding to the velocity and density of the penultimate row of the lattice.

As a test of the Ehrenfests' regularisation, a series of simulations were performed for a range of Reynolds numbers

$$Re = \frac{Lu_\infty}{\nu}.$$

We perform an Ehrenfests' step at, at most, $L/2$ sites, where the nonequilibrium entropy $\Delta_i(S) > 10^{-3}$.

We are interested in computing the Strouhal–Reynolds relationship. The Strouhal number S is a dimensionless measure of the vortex shedding frequency in the wake of one side of the cylinder:

$$S = \frac{L\omega}{u_\infty},$$

where ω is the shedding frequency. For the precise details of how to compute the shedding frequency see [7].

The computed Strouhal–Reynolds relationship using the Ehrenfests' regularisation of LBGK is shown in Figure 7. The simulation compares well with Okajima's data from wind tunnel and water tank experiment [19]. The simulation reported here extends previous LBM studies of this problem e.g. [2] which have been able to quantitatively capture the relationship up to $Re = \mathcal{O}(1000)$. Figure 7 also shows the ELBM simulation results from [2]. Furthermore, the computational domain was fixed for all the present computations, with the smallest value of the kinematic viscosity attained being $\nu = 5 \times 10^{-5}$ at $Re = 20000$. It is worth mentioning that, for this characteristic length, LBGK exhibits numerical divergence at around $Re = 1000$. We estimate that, for the present set up, the computational domain would require at least $\mathcal{O}(10^7)$ lattice sites for the kinematic viscosity to be large enough for LBGK to converge at $Re = 20000$. This is compared with $\mathcal{O}(10^5)$ sites for the present simulation.

9 Conclusions

The purpose of this paper is to try to establish the lattice Boltzmann method as a computational model for fluid flow. When the flow is nice in some sense

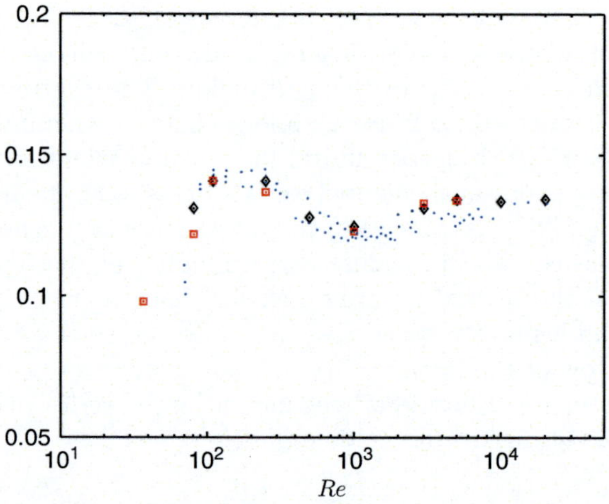

Fig. 7. Variation of Strouhal number as a function of Reynolds. Dots are Okajima's experimental data [19] (the data has been digitally extracted from the original paper). Diamonds are the Ehrenfests' regularisation of LBGK and the squares are the ELBM simulation from [2].

LBM is close to the Navier-Stokes' model, but when the flow is unpleasant in the sense that there is large production of nonequilibrium entropy, then these two models will diverge. We would not like to guess at which is the best model in these circumstances. What the LBM allows us to do is to introduce artificial viscosity into the model in a very precise way and computationally efficient way, unlike stabilisation methods used by finite element and finite volume practitioners. Our numerical experiments have allowed us to simulate fluid flow up to high Reynolds' number. Of course, we do not know what the effective Reynolds' number of the flow really is once we have introduced significant amounts of artificial viscosity, but we are pleased that we can produce stable simulation and observe reasonable statistics in regimes where we believe other methods struggle to work at all. We look forward to the LMB being adopted by numerical analysis along side the more traditional methods, and suggest that it is a very promising method for the simulation of fluid flow in three dimensions at high Reynolds' number.

References

1. S. Ansumali and I.V. Karlin: Kinetic boundary conditions in the lattice Boltzmann method. Phys. Rev. E **66**(2), 2002, 026311.
2. S. Ansumali, S.S. Chikatamarla, C.E. Frouzakis, and K. Boulouchos: Entropic lattice Boltzmann simulation of the flow past square-cylinder. Int. J. Mod. Phys. C **15**, 2004, 435–445.
3. S. Ansumali and I.V. Karlin: Stabilization of the lattice Boltzmann method by the H theorem: a numerical test. Phys. Rev. E **62**(6), 2000, 7999–8003.

4. P.L. Bhatnagar, E.P. Gross, and M. Krook: A model for collision processes in gases I. Small amplitude processes in charged and neutral one-component systems. Phys. Rev. **94**(3), 1954, 511–525.
5. R.A. Brownlee, A.N. Gorban, and J. Levesley: Stable simulation of fluid flow with high Reynolds number using Ehrenfests' steps. Numerical Algorithms **45**, 2007, 389–408.
6. R.A. Brownlee, A.N. Gorban, and J. Levesley: Stability and stabilization of the lattice Boltzmann method. Phys. Rev. E **75**, 2007, 036711.
7. R.A. Brownlee, A.N. Gorban, and J. Levesley: Stabilisation of the lattice Boltzmann method using the Ehrenfests' coarse-graining. Phys. Rev. E **74**, 2006, 037703.
8. R.A. Brownlee, A.N. Gorban, and J. Levesley: Nonequilibrium entropy limiters in lattice Boltzmann methods. Physica A **387**(2-3), 2008, 385–406.
9. D. Burnett: The distribution of velocities and mean motion in a slight nonuniform gas. Proc. London Math. Soc. **39**, 1935, 385–430.
10. P. Ehrenfest and T. Ehrenfest: *The Conceptual Foundation of the Statistical Approach in Mechanics.* Dover, New York, 1990.
11. S.K. Godunov: A difference scheme for numerical solution of discontinuous solution of hydrodynamic equations. Math. Sbornik **47**, 1959, 271–306.
12. A.N. Gorban: Basic types of coarse-graining. In *Model Reduction and Coarse-Graining Approaches for Multiscale Phenomena*, A.N. Gorban, N. Kazantzis, I.G. Kevrekidis, H.-C. Öttinger, and C. Theodoropoulos (eds.), Springer, Berlin, 2006, 117–176.
13. C.K.R.T. Jones: *Geometric Singular Perturbation Theory.* Lecture Notes in Mathematics **1609**, Springer, Berlin, 1995.
14. T.S. Khan and J. Levesley: Stabilising lattice Boltzmann simulation of flow past a circular cylinder with Ehrenfests' limiter. Submitted for publication.
15. P.D. Lax: On dispersive difference schemes. Phys. D **18**, 1986, 250–254.
16. C.D. Levermore and J.-G. Liu: Oscillations arising in numerical experiments. Physica D **99**, 1996, 191-216.
17. X.D. Liu, S.J. Osher, and T. Chan: Weighted essentially non-oscillatory schemes. J. Comput. Physics **115**, 1994, 200–212.
18. R.R. Nourgaliev, T.N. Dinh, T.G. Theofanous, and D. Joseph: The lattice Boltzmann method: theoretical interpretation, numerics and implications. Intern. J. Multiphase Flows **29**, 2003, 117–169.
19. A. Okajima: Strouhal numbers of square cylinders. Journal of Fluid Mechanics **123**, 1982, 379–398.
20. D.J. Packwood: Entropy balance and dispersive oscillations in lattice Boltzmann methods. Phys. Rev. E **80**, 067701.
21. D.J. Packwood, J. Levesley, and A.G. Gorban: Time step expansions and ther invariant manifold approach to lattice Boltzmann models. Submitted for publication.
22. J.M. Sanz-Serna: Symplectic integrators for Hamiltonian problems. Acta Numerica **1**, 1992, 243–286.
23. T. Schwartzkopff, C.D. Munz, and E.F. Toro: ADER: a high-order approach for hyperbolic systems in 2d. J. Sci. Computing **17**, 2002, 231–240.
24. C. Shu and S.J. Osher: ENO and WENO shock capturing schemes II. J. Comp. Phys. **83**, 1989, 32–78.
25. S. Succi: *The Lattice Boltzmann Equation for Fluid Dynamics and Beyond.* Oxford University Press, Oxford, 2001.

26. E. Tadmor, W. Zhong: Entropy stable approximations of Navier-Stokes equations with no artificial numerical viscosity. J. of Hyperbolic DEs **3**, 2006, 529–559.
27. X. Shan and X. He: Discretization of the velocity space in the solution of the Boltzmann equation. Phys. Rev. Lett. **80**, 1998, 65–68.

Approximating Probability Measures on Manifolds via Radial Basis Functions

Jeremy Levesley[1] and Xingping Sun[2]

[1] Department of Mathematics, University of Leicester, LE1 7RH, UK
[2] Department of Mathematics, Missouri State Univ., Springfield, MO 65897, USA

Summary. Approximating a given probability measure by a sequence of normalized counting measures is an interesting problem and has broad applications in many areas of mathematics and engineering. If the target measure is the uniform distribution on a manifold then such approximation gives rise to the theory of uniform distribution of point sets and the corresponding discrepancy estimates. If the target measure is the equilibrium measure on a manifold, then such approximation leads to the minimization of certain energy functionals, which have applications in discretization of manifolds, best possible site selection for polynomial interpolation and Monte Carlo method, among others. Traditionally, polynomials are the major tool in this arena, as have been demonstrated in the celebrated Weyl's criterion, Erdős-Turán inequalities. Recently, the novel approach of employing radial basis functions (RBFs) has been successful, especially in higher dimensional manifolds. In its general methodology, RBFs provide an efficient vehicle that allows a certain type of linear translation operators to act in various function spaces, including reproducing kernel Hilbert spaces (RKHS) associated with RBFs. This approach is crucial in the establishment of the LeVeque type inequalities that are capable of giving discrepancy estimates for some minimal energy configurations. We provide an overview of the recent developments outlined above. In the final section we show that many results on the sphere can be generalised to other compact homogeneous manifolds. We also propose a few research topics for future investigation in this area.

1 Introduction

Let M be a d-dimensional manifold embedded in $\mathbb{R}^m$ ($m \geq d$). Let a probability measure ν be given on M. Let $N \in \mathbb{N} \setminus \{1\}$. We are interested in finding a set of N distinct points $x_1, \ldots, x_N$ in M such that the normalized counting measure

$$\sigma_N := \frac{1}{N} \sum_{j=1}^{N} \delta_{x_j} \tag{1}$$

approximates ν well according to a given criterion which will be specified later. Here δ_{x_j} denotes the unit mass at the point x_j. Note that we have eliminated

E.H. Georgoulis, A. Iske, J. Levesley (eds.), *Approximation Algorithms for Complex Systems*, Springer Proceedings in Mathematics 3, DOI: 10.1007/978-3-642-16876-5_7,
© Springer-Verlag Berlin Heidelberg 2011

the interesting (but trivial) case $N = 1$, in which the unit mass δ_{x_1} at the *center of gravity* x_1 (may or may not be in M) of the measure ν is often the undisputable choice.

Lets first consider the simple example in which we are approximating the uniform distribution μ on the interval $[0, 1)$. Note that the density function of μ is the constant function: $t \mapsto 1, \;\; t \in [0, 1)$. Let a triangular array

$$\{x_{N,1}, \ldots, x_{N,N}\}_{N=2}^{\infty}$$

be given. We say that the set $\{x_{N,1}, \ldots, x_{N,N}\}$ is uniformly distributed in $[0, 1)$ (as $n \to \infty$) if for each fixed $0 < x < 1$, we have

$$\lim_{N\to\infty} \frac{1}{N} \#\{x_{N,j} : j = 1, \ldots, N, x_{N,j} \in [0, x)\} = x.$$

We remark that the above limit is equivalent to the following:

$$\lim_{N\to\infty} \frac{1}{N} \sum_{j=1}^{N} \chi_{[0,x)}(x_{N,j}) = \int_0^1 \chi_{[0,x)}(t)dt = x,$$

where $\chi_{[0,x)}$ is the indicator function of the interval $[0, x)$. The collection of the indicator functions $\{\chi_{[0,x)} : x \in [0, 1)\}$ plays an important role here. They provide a *testing ground* for the approximation of the uniform distribution by a sequence of normalized counting measures. In his study of uniform distribution of points, Weyl [47] used trigonometrical polynomials to approximate these indicator functions, and obtained the celebrated Weyl's criterion that asserts that the set $\{x_{N,1}, \ldots, x_{N,N}\}$ is uniformly distributed in $[0, 1)$ (as $n \to \infty$) if and only if for each integer k, $k \neq 0$, we have

$$\lim_{N\to\infty} \frac{1}{N} \sum_{j=1}^{N} e^{2\pi i k x_{N,j}} = 0.$$

Weyl's criterion can be considered as a qualitative characterization of uniform distribution of point sets. To measure uniform distribution of point sets in a quantitative way, one needs the notion of "discrepancy". There are many different (but similar) ways of defining discrepancy. For the time being, we use the so called "star" discrepancy $D^*(N)(\sigma_1, \sigma_2)$ between the two probability measures σ_1, σ_2 on the interval $[0, 1)$ defined by

$$D^*(\sigma_1, \sigma_2) := \sup_{x\in[0,1)} \left| \int_0^1 \chi_{[0,x)}(t)(d\sigma_1(t) - d\sigma_2(t)) \right|.$$

The star discrepancy $D^*(N)(\sigma_N, \mu)$ will be simply denoted by $D^*(N)$.

Erdős and Turán [11] refined Weyl's trigonometrical polynomial approximation scheme and proved the following theorem that has since been called the Erdős-Turán Inequality:

Theorem 1. *For each $x \in [0,1)$. There exist trigonometrical polynomials T^- and T^+ of degree at most K such that*

$$T^-(t) \le \chi_{[0,x)}(t) \le T^+(t), \quad t \in [0,1]; \quad \int_0^1 T^\pm(t)dt = x + O(K^{-1}). \quad (2)$$

Therefore the following inequality holds true:

$$D^*(N) \ll \frac{1}{K} + \frac{1}{N}\sum_{k=1}^{K}\frac{1}{k}\left|\sum_{j=1}^{N} e^{2\pi i k x_j}\right|. \quad (3)$$

Here we have employed Vinogradov's $\ll$ notation, more precisely, that $f \ll g$ is equivalent to $f = O(g)$. Making a connection to the "large sieve method" in number theory, Beurling, Selberg, Vaaler (see [24]) obtained the optimal majorizing and minorizing trigonometrical polynomials for the function $\chi_{[0,x)}$, and found a sharp constant in the Erdős-Turán Inequality. An episode of beautiful classical analysis and its broad applications notwithstanding, the above development shows that trigonometrical polynomials can be used to "conquer" all the indicator functions, and therefore can be used as test functions for the approximation of measures.

With the choice $x_{N,j} = j/N$ ($j = 0, 1, \ldots, (N-1)/N$), the normalized counting measure σ_N as in the form of Equation 1 provides an excellent approximation to the uniform distribution μ on $[0,1)$. In fact, a simple application of the Erdős and Turán Inequality shows that

$$D^*(N) \ll N^{-1}.$$

The classical Koskma [17] Theorem then asserts that for every continuous function f on $[0,1]$, we have

$$\left|\frac{1}{N}\sum_{j=1}^{N} f(x_{N,j}) - \int_0^1 f(t)dt\right| \le \omega_f(N^{-1}),$$

where ω_f denotes the *modulus of continuity* of f.

The focus of the above discussion is how to select N distinct points so that $D^*(N)$ is minimized, and the conclusion is that equally spaced points give the best (or near best) result. Consider now a different problem in which one wants to select N points from $[-1,1]$ as sites for polynomial interpolation. If one makes the choice of the N equally spaced points:

$$x_{N,j} = -1 + \frac{2j}{N}, \quad j = 0, 1, \ldots, N-1,$$

then Rouge [33] showed that the Lebesgue constant (the norm in $C([-1,1])$ of the polynomial interpolation operator) grows exponentially with N. Thus, they should be avoided. To seek good polynomial interpolation sites on which the Lebesgue constant is relatively small, we choose to take a "lifting" approach and work on the unit circle. Assume that we are given $(N+1)$ points in $[-1,1]$:

$$-1 \le x_1 < x_2 < \cdots < x_N \le 1.$$

Consider the mapping

$$x = \cos t \tag{4}$$

of the interval $[-1,1]$ on to the upper semi-circle parameterized by the angular variable t, $0 \le t \le \pi$. The map transforms a function $F(x)$ defined on the interval $-1 \le x \le 1$ into the function

$$f(t) := F(\cos t),$$

and the points $x_0, x_1, \dots, x_N$ into points $t_0, t_1, \dots, t_N$, in which

$$x_j = \cos t_j \quad (0 \le j \le N).$$

The polynomial $P_N(x)$ interpolating $F(x)$ at the points $x_0, x_1, \dots, x_N$ becomes $P_N(\cos t)$ that interpolates $f(t)$ at the points $t_0, t_1, \dots, t_N$.

Conversely, let $f(t)$ be a function defined on the upper semi-circle $0 \le t \le \pi$. Suppose that $0 \le t_0 < t_1 < \cdots < t_N \le \pi$ and that $C_N(t)$ is a cosine polynomial, that is, an element of the linear span of the $(N+1)$ functions

$$1, \cos t, \dots, \cos Nt,$$

that interpolates $f(t)$ at the points $t_0, t_1, t_2, \dots, t_N$. Observe that $\cos kt$ can be written as a polynomial of degree k in $\cos t$. Therefore the transformation as in Equation 4 carries $C_N(t)$ into a polynomial $P_N(x)$ that interpolates $F(x)$ at the points $x_0, x_1, x_2, \dots, x_N$.

Hence, interpolating at the points $t_0, t_1, t_2, \dots, t_N$ by a cosine polynomial is equivalent to interpolating at the points $x_0, x_1, x_2, \dots, x_N$ by a polynomial.

Let

$$t_j^{(N)} := \frac{2\pi j}{2N+1}, \quad j = 0, 1, \dots, N.$$

If no confusion is likely to occur, we will simply denote $t_j^{(N)}$ by t_j. Consider the Dirichlet kernel D_N, i.e.,

$$D_N(t) := \sum_{k=-N}^{N} e^{ikt} = \frac{\sin(N+\frac{1}{2})t}{\sin\frac{t}{2}}.$$

It is easy to see that

$$\frac{1}{2N+1} D_N(t_j) = \delta_{0,j}.$$

Thus $\frac{1}{2N+1} D_N(t-t_j)$ is a cosine polynomial of order N, and is the fundamental Lagrange function for t_j with which we can write the interpolating cosine polynomial $I_N(f,t)$ in the following form:

$$I_N(f,t) := \frac{1}{2N+1}\sum_{j=0}^{N} f(t_j)D_N(t-t_j).$$

This is a linear operator sending a continuous function $f(t)$ on the upper semi-circle to a cosine polynomial of degree N or less. The operator norm of $I_N(f,t)$ can be estimated as follows.

$$\begin{aligned}
&\max_{0\le t\le\pi} |I_N(f,t)| \\
&\le \frac{1}{2N+1}\max_{0\le t\le\pi} |f(t)| \sum_{j=0}^{N} |D_N(t-t_j)| \\
&\le C \max_{0\le t\le\pi} |f(t)| \log N,
\end{aligned}$$

where C is a positive constant independent of f and N. This shows that if we choose

$$x_j = \cos\frac{2\pi j}{2N+1}, \quad j = 0,1,\ldots,N,$$

as sites for interpolation, then the corresponding Lebesgue constant is of the order $\log N$, which is the optimal order. Note that with the x_j chosen as above, the normalized counting measure

$$\frac{1}{N}\sum_{j=0}^{N}\delta_{x_j},$$

provides an excellent approximation to the "arcsine" distribution with the density function

$$\left(\pi\sqrt{1-x^2}\right)^{-1}.$$

In both examples, the central question is how to select a large number of points so that the resulted normalized counting measure approximates a certain continuous probability measure well. In the first example, the target measure is the uniform distribution on the interval $[0,1]$. In the second example, the target measure is the arcsine distribution on the interval $[-1,1]$. These measures are some special equilibrium measures; see [20].

Turning our attention to higher dimensional manifolds, we immediately find ourself confronting a much more daunting problem: effectively selecting a large number of points has become much less tractable. We also find that it is no longer efficient to use polynomials as test functions for the simple reason that the dimension of the polynomial space needed grows too fast with the increase of the dimension of the manifold. This difficulty calls for the use of radial basis functions (RBF's). The $d-$dimensional sphere $\mathbb{S}^d$ embedded in $\mathbb{R}^{d+1}$ is considered by many to be the canonical choice of $d-$dimensional manifolds. Being important for its own sake, $\mathbb{S}^d$ can also be used as a springboard to derive results on other manifolds, and we deal with a class of these in the final section of the paper. In the second example above, we first carried out the interpolation on the unit circle, and then used the cosine transform to obtain interpolation result on the interval $[-1, 1]$. As is well known, the equilibrium measure on S^1 is the uniform probability measure. Furthermore, the equilibrium measure on the interval $[-1, 1]$ can be obtained from the uniform probability measure on S^1 by the transform as in Equation 4. Further study shows that many useful equilibrium measures are the results of "transforming" (in various ways) the uniform probability measures on $\mathbb{S}^d$ onto the underlying manifolds. If we can approximate the uniform probability measures on $\mathbb{S}^d$, then the approximation of many other useful equilibrium measures is just a "change of variable" away.

The current paper is arranged as follows. In Section 2, we describe a new approach that features radial basis functions (RBFs) as test functions. The key idea is to use a certain type of "translation" operators in various function spaces to study uniform distribution of points in $\mathbb{S}^d$. In particular, we find that the translation operators work very well in the native space $\mathcal{N}_\phi$ of a strictly positive definite (SPD) function ϕ; see [7, 32, 48]. In Section 3, we give a brief account of what the traditional approach (using polynomials) has achieved. This is showcased by the Erdős-Turán Inequality on $\mathbb{S}^d$ established by Grabner [13], and Li and Vaaler [23], respectively. In Section 4, we present the fruition of the new approach outlined in Section 2, which will culminate in the establishment of LeVeque type inequality on $\mathbb{S}^d (d \geq 2)$. We will also show the advantage of LeVeque type inequality in determining the discrepancy of certain normalized counting measures obtained by minimizing some discrete energy functionals. In Section 5, we demonstrate, in the native space setting, how equilibrium measures can be efficiently approximated by normalized counting measures supported on minimal energy configuration and its application to quadrature rules. In Section 6 we generalize some of the results from the earlier sections on more abstract manifolds, the compact homogeneous spaces. We have not yet specialized these results to, for instance, the projective spaces, in which case we could get more quantitative estimates, but this is one of the directions of our future research.

2 The New Approach

To make the presentation more accessible, we need to start with a brief introduction of Fourier analysis on spheres. There are many standard references in the mathematical literature addressing Fourier analysis on $\mathbb{S}^d$ that are familiar to analysts, applied mathematicians, and theoretical statisticians. Here we recommend [25] and [37]. We will let $L^2(\mathbb{S}^d)$ be the Hilbert space equipped with the inner product

$$\langle f, g \rangle := \int_{\mathbb{S}^d} f(x)\overline{g(x)} d\sigma(x),$$

where σ is the uniform probability measure on $\mathbb{S}^d$. The $Y_{\ell,m}$'s will be taken to be the usual orthonormal basis of spherical harmonics [25], which we may assume to be real. For ℓ fixed, these span the eigenspace of the Laplace-Beltrami operator on $\mathbb{S}^d$ corresponding to the eigenvalue $\lambda_\ell = \ell(\ell + d - 1)$. Here, $m = 1, \ldots, q_\ell$, where q_ℓ is the dimension of the eigenspace corresponding to λ_ℓ and is given by [25, p. 4]

$$q_\ell = \begin{cases} 1, & \ell = 0, \\ \dfrac{(2\ell + d - 1)\Gamma(\ell + d - 1)}{\Gamma(\ell + 1)\Gamma(d)}, & \ell \geq 1 . \end{cases}$$

Legendre Polynomials and the Addition Formula. Let P_ℓ^d denote the degree-ℓ Legendre polynomials in $(d + 1)$ dimensions, which is the notation used by Grabner in [13]; Müller [25] denotes them by $P_\ell(d + 1; x)$. The Legendre polynomials are related to the Gegenbauer polynomials via $P_\ell^d(x) = C_\ell^{\frac{d-1}{2}}(x)/C_\ell^{\frac{d-1}{2}}(1)$, and to Jacobi polynomials via

$$P_\ell^d(x) = \frac{P_\ell^{(\frac{d-2}{2},\frac{d-2}{2})}(x)}{P_\ell^{(\frac{d-2}{2},\frac{d-2}{2})}(1)} = \frac{\Gamma(\ell + d/2)}{\ell!\Gamma(d/2)} P_\ell^{(\frac{d-2}{2},\frac{d-2}{2})}(x)$$

In this notation, the addition formula for spherical harmonics is the following:

$$\sum_{m=1}^{q_\ell} Y_{\ell,m}(x) Y_{\ell,m}(y) = q_\ell P_\ell^d(x \cdot y).$$

On $\mathbb{S}^1$, we may use the angular variable u and adapt the following orthonormal system:

$$1, \sqrt{2}\cos u, \sqrt{2}\sin u, \sqrt{2}\cos 2u, \sqrt{2}\sin 2u, \ldots .$$

The addition formula on $\mathbb{S}^1$ is simply $\cos \ell(u - v) = \cos \ell u \cos \ell v + \sin \ell u \sin \ell v$. As an easy consequence of the Addition Formula, we have the following useful inequality:

$$\sum_{m=1}^{q_\ell} |Y_{\ell,m}(x) Y_{\ell,m}(y)| \leq \sum_{m=1}^{q_\ell} Y_{\ell,m}^2(x) = q_\ell.$$

Funk-Hecke Formula. Suppose $g(t) \in L^2[-1, 1]$. For each nonnegative integer ℓ, let

$$\tilde{g}(\ell) := \frac{\omega_{d-1}}{\omega_d} \int_{-1}^{1} g(t) P_\ell^d(t)(1 - t^2)^{\frac{d-2}{2}} dt.$$

Then for every spherical harmonic $Y_{\ell,m}$, we have

$$\int_{\mathbb{S}^d} g(x \cdot y) Y_{\ell,m}(y) d\sigma(y) = \tilde{g}(\ell) Y_{\ell,m}(x).$$

For a fixed $x \in \mathbb{S}^d$, and $0 < r < 2$, let $C(x, r) := \{y : |y - x| \leq r\}$, where $|y - x|$ denotes the Euclidean distance between x and y. We will call $C(x, r)$ a spherical cap centered at x and having radius r, or just a spherical cap when the center and radius are not important in the context.

Definition 1. *For each $N \geq 1$, let $\{x_{N,1}, \ldots, x_{N,N}\}$ be a set of N points in $\mathbb{S}^d$. The collection $\{x_{N,1}, \ldots, x_{N,N}\}_{N=1}^{\infty}$ is a triangular array. It is called "uniformly distributed in $\mathbb{S}^d$" if for each spherical cap $C(x,r)$, we have*

$$\lim_{N\to\infty} \frac{\#\{x_{N,j} : x_{N,j} \in C(x,r)\}}{N} = \sigma(C(x,r)).$$

We will also say, with a tint of ambiguity, that the points $x_{N,1}, \ldots, x_{N,N}$ are uniformly distributed in $\mathbb{S}^d$. The following theorem is known as the spherical version of Weyl's criterion; see [19].

Theorem 2. *Let $\{x_{N,1}, \ldots, x_{N,N}\}_{N=1}^{\infty}$ be a triangular array of points in $\mathbb{S}^d$. Then the following three statements are equivalent:*

1. *The points $x_{N,1}, \ldots, x_{N,N}$ are uniformly distributed in $\mathbb{S}^d$.*
2. *For each fixed integer $\ell \geq 1$, and each fixed m, $1 \leq m \leq q_\ell$, we have*

$$\lim_{N\to\infty} \frac{1}{N} \sum_{j=1}^{N} Y_{\ell,m}(x_{N,j}) = 0.$$

3. *For every continuous function f on $\mathbb{S}^d$, we have*

$$\lim_{N\to\infty} \frac{1}{N} \sum_{j=1}^{N} f(x_{N,j}) = \int_{\mathbb{S}^d} f(x) d\sigma(x).$$

Part 3 of the Weyl's criterion states that the points $x_{N,1}, \ldots, x_{N,N}$ are uniformly distributed in $\mathbb{S}^d$ if and only if the sequence of normalized counting measures σ_N converges to the measure σ in the weak star topology. Quantitative estimates of the weak star convergence are possible by restricting the measures to suitable subsets of functions. Let P_1 and P_2 be two probability measures on $\mathbb{S}^d$. The "spherical cap discrepancy", or simply discrepancy, between P_1 and P_2 is defined by

$$D(P_1, P_2) := \sup_{C(x,r)} |P_1(C(x,r)) - P_2(C(x,r))|,$$

where the supremum is taken over all spherical caps $C(x,r)$. The discrepancy between a probability measure P and σ will be referred to as the discrepancy of P. The discrepancy of a normalized counting measure will be denoted by $D(N)$. Literature abounds in discrepancy estimates. This is an important topic in analytic number theory [24] and in Monte Carlo and quasi-Monte Carlo methods [30]. Generally speaking, point sets with small discrepancy yield small errors in quasi-Monte Carlo integration [30, p. 21]. As a result, tremendous effort has been devoted to the searching for sequences that enjoy low discrepancy; see [9]. We will first summarize a general method that uses RBFs to study uniform distribution of points and discrepancy (Theorem 3). In Sections 4 and 5, we will illustrate how the method can be used to show that certain point sets generated in some deterministic ways (for example, the minimal energy points associated with an SPD function) have low discrepancy.

We can think of the discrepancy as an upper bound estimate for the convergence of measures σ_N to the measure σ in the uniform topology with respect to the set of all the indicator functions of spherical caps. These indicator functions can be regarded as certain "test functions". This observation leads us to view discrepancy from a new angle.

Let $CV(\mathbb{S}^d)$ denote the class of kernels of the form:

$$\phi(x \cdot y) := \sum_{\ell=0}^{\infty} \hat{\phi}(\ell) \sum_{m=1}^{q_\ell} Y_{\ell,m}(x) Y_{\ell,m}(y),$$

in which $x \cdot y$ denotes the Euclidean inner product of x and y, and the Fourier-Legendre coefficients $\hat{\phi}(\ell)$ are defined by[3]

$$\hat{\phi}(\ell) = \int_{\mathbb{S}^d} \phi(x \cdot y) Y_{\ell,m}(x) d\sigma(x),$$

and are required to satisfy:

$$|\hat{\phi}(\ell)| > 0, \quad \text{and} \quad \sum_{\ell=0}^{\infty} |\hat{\phi}(\ell)|\, q_\ell < \infty.$$

Each function in $CV(\mathbb{S}^d)$ is continuous on $\mathbb{S}^d \times \mathbb{S}^d$, and the rotational invariance has earned them the name "zonal kernels". Each $\phi \in CV(\mathbb{S}^d)$ generates the so called native space $\mathcal{N}_\phi$ of ϕ defined by

$$\mathcal{N}_\phi := \left\{ f \in L^2(\mathbb{S}^d) : \sum_{\ell=0}^{\infty} |\hat{\phi}(\ell)|^{-1} \sum_{m=1}^{q_\ell} \hat{f}_{\ell,m}^2 < \infty \right\}.$$

Here $\hat{f}_{\ell,m} = \displaystyle\int_{\mathbb{S}^d} f(x) Y_{\ell,m}(x) d\sigma(x)$. Note that $\mathcal{N}_\phi$ is a Reproducing Kernel Hilbert Space (RKHS) with inner product $\langle f, g \rangle_\phi$, defined by

$$\langle f, g \rangle_\phi = \sum_{\ell=0}^{\infty} |\hat{\phi}(\ell)|^{-1} \sum_{m=1}^{q_\ell} \hat{f}_{\ell,m} \hat{g}_{\ell,m}, \quad f, g \in \mathcal{N}_\phi$$

and the reproducing kernel (or Mercer kernel) being ϕ^*, defined by

$$\phi^*(x \cdot y) = \sum_{\ell=0}^{\infty} |\hat{\phi}(\ell)| \sum_{m=1}^{q_\ell} Y_{\ell,m}(x) Y_{\ell,m}(y), \quad (x, y) \in \mathbb{S}^d \times \mathbb{S}^d.$$

[3] By Funk-Hecke formula, the definition of $\hat{\phi}(\ell)$ does not depend on m and y.

For a given triangular array $\{x_{N,1},\ldots,x_{N,N}\}_{N=1}^{\infty}$ in $\mathbb{S}^d$ and a given $\phi \in CV(\mathbb{S}^d)$, define the sequence of functions $T_{\phi,N}$ by

$$T_{\phi,N}(x) := \frac{1}{N}\sum_{j=1}^{N}\phi(x\cdot x_{N,j}), \quad x \in \mathbb{S}^d.$$

Let $A_\phi := \int_{\mathbb{S}^d}\phi(x\cdot y)d\sigma(y)$. Note that, since σ is rotation invariant, A_ϕ is a constant not depending on x.

The following theorem shows how translation operators can be related to uniform distribution of point sets.

Theorem 3. *Let $\phi \in CV(\mathbb{S}^d)$. Then we have the following equivalent statements:*

1. *The triangular array $\{x_{N,1},\ldots,x_{N,N}\}_{N=1}^{\infty}$ are uniformly distributed in $\mathbb{S}^d$.*
2. *The following limit holds true: $\lim_{N\to\infty}\|T_{\phi,N}-A_\phi\|_{\mathcal{N}_\phi}=0$.*
3. *For every fixed p, $0<p\le\infty$, we have $\lim_{N\to\infty}\|T_{\phi,N}-A_\phi\|_p=0$.*
4. *The following limit holds true:*

$$\lim_{N\to\infty}\sum_{\ell=1}^{\infty}|\hat{\phi}(\ell)|^2\sum_{m=1}^{q_\ell}\left(\frac{1}{N}\sum_{j=1}^{N}Y_{\ell,m}(x_{N,j})\right)^2=0.$$

We first remind readers that the range for p in Part 3, $p>0$ is not a typo. As a matter of fact, we can use p in the range $0<p<1$ to study minimal energy configurations associated with the Riesz s-kernels; see Hardin and Saff [14, 15, 18] and the references therein. This provides an interesting and yet challenging future research project. Most parts of the above theorem were proved in [40]. A full proof can be carried out with a minor modification of the proof given in [40].

Note that Parseval identity implies that

$$\|T_{\phi,N}-A_\phi\|_2^2=\sum_{\ell=1}^{\infty}|\hat{\phi}(\ell)|^2\sum_{m=1}^{q_\ell}\left(\frac{1}{N}\sum_{j=1}^{N}Y_{\ell,m}(x_{N,j})\right)^2.$$

Therefore, Part 4 follows from Part 3 in a trivial way. Some parts of the theorem can be generalized in useful ways. For example, we can show that uniform distribution of points is equivalent to the pertinent sequence of linear operators defined on various Banach spaces converging weakly to zero.

Theorem 3 and some of its corollaries have given us an in-depth understanding of uniform distribution of points, and have provided us with new tools in dealing with discrepancy. Furthermore, Theorem 3 shows that various norms of the function $(T_{\phi,N}-A_\phi)$ can be used to quantify uniform distribution of points. The connection between Theorem 3 and (spherical) radial basis functions is self-evident; see [6, 26, 27, 29].

3 Erdős-Turán Type Inequalities

A natural question to ask is: Can we find a function $\phi \in CV(\mathbb{S}^d)$ so that there exist a $g \in \mathcal{N}_\phi$ or a sequence $g_n \in \mathcal{N}_\phi$, such that

$$\|T_{g,N} - A_g\|_\infty = D(N) \quad \text{or} \quad \lim_{n\to\infty} \|T_{g_n,N} - A_{g_n}\|_\infty = D(N)$$

for each fixed N?

On $\mathbb{S}^1$ (or $\mathbb{R}/\mathbb{Z}$), let $\varsigma(t)$ be the saw-tooth function defined by

$$\varsigma(t) = \begin{cases} t - \lfloor t \rfloor - 1/2, \; t \neq 0, \pm 1, \pm 2, \ldots, \\ \quad 0, \qquad\qquad t = 0, \pm 1, \pm 2, \ldots . \end{cases}$$

Montgomery [24] showed that the discrepancy $D(N)$ on $\mathbb{R}/\mathbb{Z}$ satisfies the following inequality (noting that $A_\varsigma = 0$):

$$\|T_{\varsigma,N}\|_\infty \leq D(N) \leq 2\|T_{\varsigma,N}\|_\infty.$$

On $\mathbb{S}^d$ ($d > 1$), similar (albeit less precise) inequalities were implicit in Li and Vaaler [23]. These inequalities serve as the main impetus for the proof of several optimal Erdős-Turán type inequalities; see [24, 41, 42]. Note that $\varsigma \notin CV(\mathbb{S}^1)$. Therefore, an optimal $\phi \in CV(\mathbb{S}^1)$ is selected, and some intricately designed extremal functions from $\mathcal{N}_\phi$ are used as "sieves". In obtaining the optimal Erdös-Turán inequality on $\mathbb{S}^1$, a pair of Selberg polynomials $T^\pm$ (built from Vaaler polynomials; see [24]) are used so that

$$T^-(t) \leq \varsigma(t) \leq T^+(t),$$

for all t, and

$$\int_{\mathbb{S}^1} \left[T^+(t) - T^-(t)\right] d\sigma(t)$$

is as small as possible.

Using a similar approach, Grabner [13], and Li and Vaaler [23] independently established the Erdös-Turán type inequality on $\mathbb{S}^d$. Here we cite the version given by Li and Vaaler.

$$D(N) \leq \frac{2\sqrt{\pi}\Gamma\left(\frac{d+1}{2}\right)}{\Gamma(\frac{d}{2})} K^{-1}$$
$$+ \sum_{\ell=1}^{K-1} \sqrt{(d+1)2^{d-1}}\Gamma\left(\frac{d+1}{2}\right)\left(\frac{\sqrt{2^{d-2}}}{\ell\sqrt{\pi}} + \frac{2}{K}\right) \sum_{m=1}^{q_\ell} \frac{1}{N}\left|\sum_{j=1}^{N} Y_{\ell,m}(x_{N,j})\right|,$$

for every positive integer K.

The above inequality is quite useful, especially on low dimensional spheres. Brauchart [5] used the above theorem to show that the minimal logarithmic energy points on spheres are uniformly distributed. However, the optimality of the above inequality is difficult to determine for $d > 1$. As dimensions of spheres increase, the dimensions of polynomial spaces grow very rapidly. Thus, the above theorem suffers from "the curse of dimensionality". In the next section, we will present an alternative: the LeVeque [21] type inequality on $\mathbb{S}^d$.

4 LeVeque Type Inequalities

In 1965, a little more than a decade after the advent of the Erdős-Turán inequality [11], LeVeque [21] established the inequality below on $\mathbb{S}^1$:

$$D^*(N) \leq \left[\frac{6}{\pi^2} \sum_{\ell=1}^{\infty} \ell^{-2} \left| \frac{1}{N} \sum_{j=1}^{N} e^{i\ell x_j} \right|^2 \right]^{1/3}. \tag{5}$$

As one reads along, one finds that this bound has in it an implicit RBF element. The bound is different in nature from the Erdős-Turán bound as in Inequality (3), and so is the method LeVeque employed to prove it. LeVeque [21] also elaborated the sharpness of his inequality as follows. Let $x_1 = x_2 = \cdots = x_N = 0$. Then it is easy to see that the star discrepancy $D^*(N)$ for the point set $\{x_1, x_2, \cdots, x_N\}$ is 1. Using Euler's formula:

$$\sum_{\ell=1}^{\infty} \ell^{-2} = \frac{\pi^2}{6},$$

we see that the right hand side of Inequality (5) is also 1. From this simple example, LeVeque concluded that the constant $\frac{6}{\pi^2}$ is best possible. To show that the exponent $1/3$ is also best possible, LeVeque constructed a uniformly distributed sequence $\{x_n\}_{n=1}^{\infty}$ for which the star discrepancy $D^*(N)$ satisfies, for any given $\epsilon > 0$,

$$D^*(N) > \left[\left(\frac{3}{2\pi^2} - \epsilon \right) \sum_{\ell=1}^{\infty} \ell^{-2} \left| \frac{1}{N} \sum_{j=1}^{N} e^{i\ell x_j} \right|^2 \right]^{1/3},$$

for infinitely many N.

Recently, Narcowich, Sun, Ward, and Wu [28] proved the following LeVeque-type inequality for $D(N)$ on $\mathbb{S}^d$.

Theorem 4. *Let $x_1, x_2, \cdots, x_N$ be N points in $\mathbb{S}^d$ (not necessarily distinct). Then the discrepancy $D(N)$ of the point set $\{x_1, x_2, \cdots, x_N\}$ satisfies the following estimate:*

$$D(N) \le A(d) \left[\sum_{\ell=1}^{\infty} \ell^{-(d+1)} \sum_{m=1}^{q_\ell} \left(\frac{1}{N} \sum_{j=1}^{N} Y_{\ell,m}(x_j) \right)^2 \right]^{\frac{1}{d+2}}, \tag{6}$$

where the constant $A(d)$ is given by

$$A(d) := c_1(d)\,(c_2(d))^{-\frac{1}{d+2}}\,(c_3(d))^{\frac{2}{d+2}},$$

and where

$$c_1(d) := \left[\left(\frac{2}{d}\right)^{\frac{2}{d+2}} + \left(\frac{2}{d}\right)^{-\frac{d}{d+2}} \right] \left(\frac{\Gamma(\frac{d+1}{2})}{\Gamma(\frac{d}{2})\sqrt{\pi}} \right)^{\frac{d-1}{d+2}},$$

$$c_2(d) := \inf_{0<h<\pi} h^{-d} \left(\int_0^h \sin^{d-1}\theta d\theta \right),$$

and

$$c_3(d) := \inf\{C\colon \sup_{0<r<\pi} \left| \int_{C(x,r)} P_\ell^d(x\cdot y)d\sigma(y) \right| \le C\ell^{-\frac{d+1}{2}} \text{ for all } \ell \ge 1\}.$$

The two constants $c_2(d)$, $c_3(d)$ expressed as above do not seem readily serviceable. Therefore, the authors of [28] devoted some efforts developing bounds for them. It is easy to see that $c_2(1) = 1$, and $c_2(d) > 1/d$ for $d = 2, 3$. A crude but also convenient lower bound for $c_2(d)$ was obtained in [28] by using the inequality

$$\sin\theta > (2/\pi)\theta, \quad 0 < \theta < \pi/2.$$

Let $h' := \min(h, \pi/2)$. Then for $0 < h < \pi$, we have $h' \ge (1/2)h$. It follows that

$$\begin{aligned} c_2(d) &\ge h^{-d} \int_0^{h'} \sin^{d-1}\theta d\theta \\ &\ge h^{-d} \int_0^{h'} ((2/\pi)\theta)^{d-1} d\theta \\ &= (2/\pi)^{d-1}\, h^{-d}\, (h')^d/d \\ &= (2d)^{-1}\, (\pi)^{-(d-1)}. \end{aligned}$$

A closed form of the constant $c_3(d)$ is known for some small d; see [13]. In particular, an easy calculation shows that $c_3(1) = 1/\pi$ for all $\ell \ge 1$. An upper bound estimate for $c_3(d)$ is also given in [28] as:

$$c_3(d) \le \frac{2^{d-\frac{3}{2}}\Gamma((d+1)/2)}{\sqrt{\pi}}.$$

For the $d = 1$ case, these constants can be easily calculated. We have $c_1(1) = 2^{2/3} + 2^{-1/3} = 3 \cdot 2^{-1/3}$, $c_2(1) = 1$, and $c_3(1) = 1/\pi$, hence

$$A(1) = 3 \cdot 2^{-1/3}(1/\pi)^{2/3}.$$

With the normalization dictated by the probability measure, the real $d = 1$ spherical harmonics $Y_{\ell,m}$ ($\ell \geq 1$) have the form

$$\sqrt{2}\sin(\ell\theta), \sqrt{2}\cos(\ell\theta).$$

Taking this into account in converting from the form in (6) to one with complex exponentials and using the value of $A(1)$ found above, the authors of [28] have the following estimate derived from Theorem 4:

$$D(N) \leq \left[\frac{27}{\pi^2} \sum_{\ell=1}^{\infty} \ell^{-2} \left| \frac{1}{N} \sum_{j=1}^{N} e^{i\ell x_j} \right|^2 \right]^{1/3}.$$

LeVeque's original inequality (5) is for $D^*(N)$. To compare it to the one above, which is for $D(N)$, one uses the simple inequality $D(N) \leq 2D^*(N)$ to get

$$D(N) \leq 2D^*(N) \leq \left[\frac{48}{\pi^2} \sum_{\ell=1}^{\infty} \ell^{-2} \left| \frac{1}{N} \sum_{j=1}^{N} e^{i\ell x_j} \right|^2 \right]^{1/3}.$$

This underscores that the constant obtained from Theorem 4 is comparable to LeVeque's for the case $d = 1$.

In the proof of Theorem 4, compactly supported "spherical basis functions" [29] are used as majorants and minorants for the indicator functions of spherical caps. The size of the supports of the spherical basis functions is first made as a free parameter. At a later stage of the proof, an optimization procedure on the parameter is applied to get the desired inequality. Readers may find it worth comparing the proof to those of the Erdős-Turán type inequalities (see [13, 23]) in which trigonometrical polynomials are employed as majorants and minorants of the indicator functions of spherical caps. In proving Inequality (5) on $\mathbb{S}^1$, LeVeque relied on the fact that the measures Q_N can be identified with a step function. However, this is no longer valid on $\mathbb{S}^d$ for $d > 1$.

The superiority of the Erdős-Turán type inequalities on $\mathbb{S}^1$ is evident in spite of the sharpness in various senses of Inequality (5) as demonstrated by LeVeque himself. Using the Erdős-Turán inequality on $\mathbb{S}^1$ and the Cauchy-Schwarz Inequality, Montgomery [24, p. 9] derived the LeVeque inequality (with a larger constant). Montgomery also discussed several sequences for which the Erdős-Turán inequality gives sharper discrepancy estimates than LeVeque's inequality. This may have explained why LeVeque's inequality has been in a sort of obscurity since its birth in 1965. Montgomery's derivation of LeVeque's inequality (using the Erdős-Turán inequality) naturally motivates one to perform a similar act on $\mathbb{S}^d$: deriving Inequality (6) using the Erdős-Turán type inequality already established on $\mathbb{S}^d$ $(d > 1)$ by Grabner [13], and Li and Vaaler [23]. However, such a maneuver does not seem to be capable of producing the desired result. In Section 5, we will demonstrate that Inequality (6) yields optimal discrepancy estimates for normalized counting measures associated with certain minimal energy configurations. The likelihood of obtaining comparable estimates by using the Erdős-Turán type inequality on $\mathbb{S}^d$ $(d > 1)$ does not seem promising.

Let $\{x_1, \ldots, x_N\}$ be a subset of $\mathbb{S}^1$. Su [38] proved the following lower bound for the discrepancy of the normalized counting measure supported on $\{x_1, \ldots, x_N\}$:

$$D(N) \geq \left[\frac{2}{\pi^2} \sum_{\ell=1}^{\infty} \ell^{-2} \left| \frac{1}{N} \sum_{j=1}^{N} e^{i\ell x_j} \right|^2 \right]^{1/2}. \tag{7}$$

Su [38] also showed that both the order and the constant are sharp. He applied this inequality in the study of random walks [38] and [39]. Making use of Stolarsky's invariance principle [36], the authors of [28] have proved the following:

Theorem 5. *Let $x_1, x_2, \cdots, x_N$ be N points (not necessarily distinct) on $\mathbb{S}^d$ $(d \geq 2)$. Then the discrepancy $D(N)$ of the point set $\{x_1, x_2, \cdots, x_N\}$ satisfies the following estimate:*

$$D(N) \geq \left[\frac{2^{d-3}\Gamma^2((d+1)/2)}{\pi} \sum_{\ell=1}^{\infty} \frac{\Gamma(\ell - 1/2)}{\Gamma(\ell + d + 1/2)} \sum_{m=1}^{q_\ell} \left(\frac{1}{N} \sum_{j=1}^{N} Y_{\ell,m}(x_j) \right)^2 \right]^{1/2}. \tag{8}$$

The formula $\Gamma(x+1) = x\Gamma(x)$ can be utilized to reduce the expansion coefficients as follows.

$$\begin{aligned} &\frac{\Gamma(\ell - 1/2)}{\Gamma(\ell + d + 1/2)} \\ &= \frac{1}{(\ell + d - 1/2)(\ell + d - 1 - 1/2)(\ell + d - 2 - 1/2) \cdots (\ell - 1/2)} \\ &\approx \ell^{-(d+1)}. \end{aligned} \tag{9}$$

In the case $d = 1$, the spherical harmonics of degree (≥ 1) are of the form:

$$\sqrt{2} \sin \ell u, \sqrt{2} \cos \ell u, \quad \ell \geq 1.$$

Therefore, Inequality (8) can be rewritten as:

$$D(N) \geq \left[\frac{1}{2\pi} \sum_{\ell=1}^{\infty} \frac{1}{\ell^2 - 1/4} \sum_{\ell=1}^{\infty} \left(\frac{1}{N} \sum_{j=1}^{N} e^{i\ell x_j} \right)^2 \right]^{1/2}.$$

Both the expansion coefficients and the constant are on par with those of Inequality (7).

The best constants in the LeVeque type inequalities carry important geometrical information, as has been shown by LeVeque [21] and Su [38] in the case $d = 1$. As a result, some efforts are warranted to pursue them for the cases $d > 1$. In particular, it may be interesting to determine whether or not the constants obtained in Theorems 4 and 5 are best possible. We caution that such an undertaking seems to be rather difficult.

5 Applications of LeVeque Type Inequalities

For a real number $\alpha > 0$, we define the nonnegative integer k_α by

$$k_\alpha := \lfloor \frac{\alpha + 2}{2} \rfloor,$$

and consider the function T_α defined for $x \in \mathbb{R}^{d+1} \setminus \{0\}$,

$$T_\alpha(x) = \begin{cases} (-1)^{k_\alpha} |x|^\alpha, & \alpha/2 \notin \mathbb{N}, \\ (-1)^{k_\alpha} |x|^\alpha \log |x|, & \alpha/2 \in \mathbb{N}. \end{cases} \tag{10}$$

The function T_α is an order k_α conditionally positive definite function; see [12]. The function has a simple distributional Fourier transform ([12]):

$$\xi \mapsto (-1)^{k_\alpha} 2^{\alpha+d+1} \pi^{\frac{d+1}{2}} \frac{\Gamma\left(\frac{\alpha+d+1}{2}\right)}{\Gamma\left(\frac{-\alpha}{2}\right)} |\xi|^{-(\alpha+d+1)}, \quad \xi \in \mathbb{R}^{d+1} \setminus \{0\}. \tag{11}$$

The usefulness of these functions (especially for α in the range $0 < \alpha < 2$) has been exhibited in many areas, including scattered data interpolation on spheres and other Riemannian manifolds [10], distance geometry and embedding theory [34], minimal energy and uniform distribution of points on spheres [36, 43, 44, 45]. In the current context, we use these functions to estimate the discrepancies of normalized counting measures on spheres. To proceed, we need to expand the kernels

$$(x, y) \mapsto T_\alpha(x - y), \quad (x, y) \in \mathbb{S}^d \times \mathbb{S}^d,$$

in spherical harmonics. We remark that such expansion formulas are already available in the literature. In fact, Pólya and Szegő [31] formulated the expansion for the cases $0 < \alpha < 2$ as early as in 1931 in their study of transfinite diameters. Baxter and Hubbert [1] developed expansions based on integrals involving Gegenbauer polynomials. In what follows, we assume that α is not a positive even integer. In [26], a simple method of using Equation (11), Proposition 3.1 in [26] (see also [6]), and Formula (2) in Watson [46, Section 13.41] yields that for $\ell \geq k_\alpha$,

$$\begin{aligned}\hat{T}_\alpha(\ell) &= (-1)^{k_\alpha} 2^{\alpha+d+1} \pi^{\frac{d+1}{2}} \omega_d^{-1} \frac{\Gamma\left(\frac{\alpha+d+1}{2}\right)}{\Gamma\left(\frac{-\alpha}{2}\right)} \int_0^\infty t^{-(d+\alpha)} J_\nu^2(t) dt \\ &= (-1)^{k_\alpha} 2^{\alpha+d+1} \pi^{\frac{d+1}{2}} \omega_d^{-1} \frac{\Gamma\left(\frac{\alpha+d+1}{2}\right)}{\Gamma\left(\frac{-\alpha}{2}\right)} \frac{\Gamma(d+\alpha)\Gamma(\ell-\alpha/2)}{2^{d+\alpha}\Gamma^2((d+\alpha+1)/2)\Gamma(\ell+d+\alpha/2)} \\ &= (-1)^{k_\alpha} \frac{\Gamma(d+\alpha)\Gamma((d+1)/2)\Gamma(\ell-\alpha/2)}{\Gamma\left(\frac{-\alpha}{2}\right)\Gamma((d+\alpha+1)/2)\Gamma(\ell+d+\alpha/2)},\end{aligned}$$

which is of the order $\ell^{-(d+\alpha)}$ as $\ell \to \infty$.

Let

$$c_{d,\alpha} := (-1)^{k_\alpha} \frac{\Gamma(d+\alpha)\Gamma((d+1)/2)}{\Gamma\left(\frac{-\alpha}{2}\right)\Gamma((d+\alpha+1)/2)} > 0.$$

Let $K_\alpha(x,y)$ denote the "truncated" kernel

$$K_\alpha(x,y) := c_{d,\alpha} \sum_{\ell=k_\alpha}^{\infty} \frac{\Gamma(\ell-\alpha/2)}{\Gamma(\ell+d+\alpha/2)} \sum_{m=1}^{q_\ell} Y_{\ell,m}(x) Y_{\ell,m}(y).$$

From the asymptotic relations $q_\ell \approx \ell^{d-1}$, and

$$\frac{\Gamma(\ell-\alpha/2)}{\Gamma(\ell+d+\alpha/2)} \approx \ell^{-(d+\alpha)},$$

we conclude that the above series converges uniformly for all $(x,y) \in \mathbb{S}^d \times \mathbb{S}^d$ for $\alpha > 0$. Hence $K_\alpha(x,y)$ is a continuous function on $\mathbb{S}^d \times \mathbb{S}^d$. Since all the expansion coefficients are nonnegative, it follows from Schoenberg's result [35] that $K_\alpha(x,y)$ is a positive definite function on $\mathbb{S}^d$. Of course, we can say that $K_\alpha(x,y)$ is an order zero conditionally positive definite function on $\mathbb{S}^d$.

For each fixed $x \in \mathbb{S}^d$, we use $T_{\alpha,x}$ to denote the function

$$y \mapsto T_\alpha(x-y), \quad y \in \mathbb{S}^d.$$

Consider the set of functions $E_\alpha := \{T_{\alpha,x} : x \in \mathbb{S}^d\}$. We define a bilinear form $\langle \cdot, \cdot \rangle$ on span(E_α) as follows. Firstly, for $x_1, x_2 \in \mathbb{S}^d$, we define

$$\langle T_{\alpha,x_1}, T_{\alpha,x_2} \rangle = K_\alpha(x_1, x_2).$$

A reminder is in order. There is a difference between the two kernels T_α and K_α. To be precise, the kernel K_α is a "truncated version" of the kernel T_α. Extending the above bilinear form linearly throughout span(E_α), the authors of [28] showed that the result is an inner product on span(E_α). One completes this inner product space to have a Hilbert space. Denote it by $\mathcal{N}_\alpha$. It is convenient to view the elements in this Hilbert space as equivalence classes. Two functions f and g are in the same equivalence class if and only if $f - g$ is a polynomial of degree $(k_\alpha - 1)$ or less. For $g \in \mathcal{N}_\alpha$, let $\|g\|_{\mathcal{N}_\alpha}$ denote the norm of g in $\mathcal{N}_\alpha$. Then we have that $\|g\|_{\mathcal{N}_\alpha} = 0$ if and only if g is a polynomial of degree $(k_\alpha - 1)$ or less. The following result is proved in [28].

Proposition 1. *For each $f \in \mathcal{N}_\alpha$ and each fixed $x \in \mathbb{S}^d$, we have*

$$f(x) - c_{d,\alpha} \sum_{\ell=0}^{k_\alpha - 1} \frac{\Gamma(\ell - \alpha/2)}{\Gamma(\ell + d + \alpha/2)} \sum_{m=1}^{q_\ell} \langle f, Y_{\ell,m} \rangle Y_{\ell,m}(x)$$
$$= \int_{\mathbb{S}^d} f(y) K_\alpha(x,y) d\sigma(y).$$

In other words, the above result asserts that the kernel $K_\alpha(x,y)$ can be mobilized to reproduce each function (up to a polynomial of degree k_α or less) in the Hilbert space. The "reproducing" structure is particularly effective for the case $0 < \alpha < 2$, in which we have: $T_\alpha(x - y) = -|x - y|^\alpha$, and

$$-|x - y|^\alpha + A_{d,\alpha} = c_{d,\alpha} \sum_{\ell=1}^{\infty} \frac{\Gamma(\ell - \alpha/2)}{\Gamma(\ell + d + \alpha/2)} \sum_{m=1}^{q_\ell} Y_{\ell,m}(x) Y_{\ell,m}(y)$$

where $A_{d,\alpha} := \int_{\mathbb{S}^d} |x - y|^\alpha d\sigma(y)$, which is independent of x due to the rotational invariance of the measure σ. Let $\Omega_N := \{x_1, \ldots, x_N\}$ be a set of N points in $\mathbb{S}^d$. Let

$$U_\alpha(x, \Omega_N) := \frac{1}{N} \sum_{j=1}^{N} T_\alpha(x - x_j) + A_{d,\alpha}$$
$$= c_{d,\alpha} \sum_{\ell=1}^{\infty} \frac{\Gamma(\ell - \alpha/2)}{\Gamma(\ell + d + \alpha/2)} \sum_{m=1}^{q_\ell} \left[\frac{1}{N} \sum_{j=1}^{N} Y_{\ell,m}(x_j) \right] Y_{\ell,m}(x),$$

and let

$$E_\alpha(\Omega_N) := \frac{1}{N^2} \sum_{i=1}^{N} \sum_{j=1}^{N} T_\alpha(x_i - x_j) + A_{d,\alpha}$$
$$= c_{d,\alpha} \sum_{\ell=1}^{\infty} \frac{\Gamma(\ell - \alpha/2)}{\Gamma(\ell + d + \alpha/2)} \sum_{m=1}^{q_\ell} \left[\frac{1}{N} \sum_{j=1}^{N} Y_{\ell,m}(x_j) \right]^2.$$

The function $U_\alpha(x, \Omega_N)$ can be considered as the difference between the Riesz α-potentials of the rotationally invariant measure σ and Q_N, the normalized counting measure supported on Ω_N. Also, the sum $\frac{1}{N}\sum_{j=1}^{N}|x-x_j|^\alpha$ is the classical α-mean of the distances from x to the points of Ω_N. The double sum $\sum_{i=1}^{N}\sum_{j=1}^{N}T_\alpha(x_i-x_j)$ is the N-point discrete Riesz α-energy functional of Ω_N. Likewise, $E_\alpha(\Omega_N)$ is the difference between the normalized energy functionals of the two measures σ and Q_N.

The following inequality is immediate:

$$E_\alpha(\Omega_N) \le \|U_\alpha(x,\Omega_N)\|_\infty. \tag{12}$$

Using the reproducing kernel Hilbert space structure, the authors of [28] have shown that the above inequality can be turned around "half-way".

Proposition 2. *Let $0<\alpha<2$. Then the following inequality holds true:*

$$\|U_\alpha(x,\Omega_N)\|_\infty \le \sqrt{A_{d,\alpha}}\,(E_\alpha(\Omega_N))^{1/2}.$$

Let $x_1,\ldots,x_N$ be N distinct points in $\mathbb{S}^d$. In the Hilbert space $\mathcal{N}_\alpha$, it is easy to see that the functional ψ defined by

$$\mathcal{N}_\alpha \ni f \mapsto \psi(f) := \int_{\mathbb{S}^d} f(x)d\sigma(x) - N^{-1}\sum_{j=1}^{N} f(x_j)$$

is linear and continuous. By the Riesz representation theorem, there exists a unique $\xi \in \mathcal{N}_\alpha$ such that

$$\psi(f) = \langle f,\xi\rangle, \quad f\in\mathcal{N}_\alpha.$$

The reproducing kernel structure of $\mathcal{N}_\alpha$ allows us to easily identify $\xi(x)$ as:

$$\xi(x) = A_{d,\alpha} + N^{-1}\sum_{j=1}^{N} T_\alpha(x-x_j), \quad x\in\mathbb{S}^d.$$

The following result follows as a direct consequence.

Proposition 3. *For each $f\in\mathcal{N}_\alpha$, we have*

$$\left|\int_{\mathbb{S}^d} f(x)d\sigma(x) - N^{-1}\sum_{j=1}^{N} f(x_j)\right| \le \|f\|_{\mathcal{N}_\alpha}(E_\alpha(\Omega_N))^{1/2}.$$

Proof. Using the Cauchy-Schwarz inequality, we have

$$\left|\int_{\mathbb{S}^d} f(x)d\sigma(x) - N^{-1}\sum_{j=1}^{N} f(x_j)\right| = |\langle f,\xi\rangle| \le \|f\|_{\mathcal{N}_\alpha}\|\xi\|_{\mathcal{N}_\alpha}.$$

We complete the proof by noting that $\|\xi\|_{\mathcal{N}_\alpha} = (E_\alpha(\Omega_N))^{1/2}$.

Let $0 < \alpha < 2$. Much attention has been devoted in the literature to the estimation of the quantities

$$\mathcal{E}(N,\alpha) := \min_{\Omega_N \subset \mathbb{S}^d} E_\alpha(\Omega_N),$$

in which the minimum is taken over all possible subsets of N distinct points in $\mathbb{S}^d$. We refer the readers to [43, 44, 45] and the references therein. If $\Omega_N^{(\alpha)} := \{x_1^{(\alpha)}, x_2^{(\alpha)}, \ldots, x_N^{(\alpha)}\}$ is such that

$$E_\alpha(\Omega_N^{(\alpha)}) = \mathcal{E}(N,\alpha),$$

then $\Omega_N^{(\alpha)}$ is called an (N,α)-minimal energy configuration. Here we use the super index α to emphasize the dependence of such configuration on α. In the remainder of this article, we summarize the results that are obtained in [28] by using Theorems 4 and 5 to estimate the discrepancies of the normalized counting measures supported on (N,α)-minimal energy configurations.

Wagner derived a variety of estimates for $U_\alpha(\cdot, \Omega_N)$ as well as the energy functionals $E_\alpha(\Omega_N)$ for a wide range of α. Here we quote two of his estimates for α in the range $0 < \alpha < 2$. Here we are primarily concerned with the order of estimates, and we will extensively engage in the use of Vinogradov's symbol $\ll (\gg)$.

Proposition 4. *Let $0 < \alpha < 2$. There exists a set Ω_N^* of N points in $\mathbb{S}^d$ such that*

$$\|U_\alpha(x, \Omega_N^*)\|_\infty \ll N^{-\frac{d+\alpha}{d}}.$$

Proposition 5. *Let $0 < \alpha < 2$. Then the following inequality holds true:*

$$\mathcal{E}(N,\alpha) \gg N^{-\frac{d+\alpha}{d}}.$$

The orders of the estimates given in Propositions 4 and 5 are sharp. For the special case $\alpha = 1$, Wagner [45] accredited the result of Proposition 4 to Stolarsky [36]. The result of Proposition 5 for the special case $\alpha = 1$ was first proved by Beck [2].

Wagner obtained several upper bounds estimates for $\mathcal{E}(N,\alpha)$ by using those derived for $\|U_\alpha(\cdot, \Omega_N)\|_\infty$ and Inequality (12). Proposition 2 shows that one can reverse the process by using the energy functionals $E_\alpha(\Omega_N)$ to control $\|U_\alpha(\cdot, \Omega_N)\|_\infty$. In a broader sense, the result of Theorems 4 can be considered as a successful example in this application. Furthermore, numerical experiments show that Proposition 2 yields very favorable estimates for $U_\alpha(x, \Omega_N)$ when the point set Ω_N is uniformly distributed. The close connection to interpolation and approximation in native spaces is also evident. Further investigation of these problems needs to be carried out in the realm of this general methodology. Here we present two estimates (obtained in [28]) for discrepancy $D(N)$) using Propositions 4 and 5 and Theorems 4 and 5.

Proposition 6. *For every set Ω_N of N distinct points in $\mathbb{S}^d$, the following inequality holds true:*

$$D(N) \gg N^{-\frac{d+1}{2d}}.$$

We remark that the order of the lower bound estimate for $D(N)$ is sharp up to a logarithmic factor; see [3]. Proposition 6 shows that Theorem 5 is capable of obtaining lower bound of near-optimal orders for discrepancy $D(N)$.

Proposition 7. *Let Ω_N^* be a set of N distinct points in $\mathbb{S}^d$ such that*

$$E_1(\Omega_N^*) \ll N^{-\frac{(d+1)}{d}}.$$

Then the discrepancy $D(N)$ of Ω_N^* satisfies the following inequality:

$$D(N) \ll N^{-\frac{(d+1)}{d(d+2)}}.$$

In particular, the above discrepancy estimate holds true for each $(N,1)$-minimal energy configuration.

Using the above result, the authors of [28] find another way of getting an discrepancy estimate that Brauchart [5] has obtained recently for the case $\alpha = 0$. For further information, readers may also find it helpful to consult [4]. The following is a detailed account of their derivation. For α in the range $0 < \alpha < 1$, one drops the multiplier $\ell^{-(1-\alpha)}(0 < \alpha < 1)$ from the right hand side of Inequality 6. Doing so makes the right hand side of the inequality bigger. One thus gets the following inequality:

$$D(N) \ll \left[\sum_{\ell=1}^{\infty} \ell^{-(d+\alpha)} \sum_{m=1}^{q_\ell} \left(\frac{1}{N} \sum_{j=1}^{N} Y_{\ell,m}(x_j) \right)^2 \right]^{\frac{1}{d+2}}. \tag{13}$$

Up to a constant depending only on d, what is inside of the bracket is exactly $E_\alpha(\Omega_N)$. Wagner (see Proposition 4.5 in the current paper) proved that there exists an Ω_N (independent of α) such that

$$E_\alpha(\Omega_N) \ll N^{-(d+\alpha)/d}.$$

For such an Ω_N, one applies Inequality (13) to get the following discrepancy estimate

$$D(N) \ll N^{-(d+\alpha)/d(d+2)}.$$

Letting $\alpha \downarrow 0$, one gets (for $\alpha = 0$) that

$$D(N) \ll N^{-1/(d+2)},$$

which is what Brauchart [5] has obtained for the minimal logarithmic energy points ($\alpha = 0$).

Using Propositions 3 and 4, we derive the following result.

Proposition 8. *If $\{x_1^{(\alpha)}, x_2^{(\alpha)}, \ldots, x_N^{(\alpha)}\}$ is an (N,α)-minimal energy configuration, then for each $f \in \mathcal{N}_\alpha$, there exists a constant $C > 0$ independent of f and N, such that*

$$\left| \int_{\mathbb{S}^d} f(x) d\sigma(x) - N^{-1} \sum_{j=1}^{N} f(x_j) \right| \leq C \|f\|_{\mathcal{N}_\alpha} N^{-\frac{d+\alpha}{2d}}.$$

As an interesting comparison to Proposition 8, we present the following result proved in [40].

Proposition 9. *Let ϕ be an SBF. If $\{x_1^{(\phi)}, x_2^{(\phi)}, \ldots, x_N^{(\phi)}\}$ is a set of N distinct points in $\mathbb{S}^d$, such that*

$$\sum_{j=1}^{N} \sum_{k=1}^{N} \phi\left(x_j^{(\phi)} \cdot x_k^{(\phi)}\right) = \inf_{\Omega_N} \sum_{j=1}^{N} \sum_{k=1}^{N} \phi\left(x_j \cdot x_k\right),$$

where the infimum is taken over all $\Omega_N := \{x_1, \ldots, x_N\}$, N distinct points in $\mathbb{S}^d$. Then for each $f \in \mathcal{N}_\phi$, the native space of ϕ, there exists a constant $C > 0$ independent of f and N, such that

$$\left| \int_{\mathbb{S}^d} f(x) d\sigma(x) - N^{-1} \sum_{j=1}^{N} f(x_j^{(\phi)}) \right| \leq C \|f\|_{\mathcal{N}_\alpha} N^{-1/2}.$$

6 Generalisations to Other Manifolds

In this section we indicate how one might generalize such results to other compact homogeneous manifolds. For the compact two-point homogeneous spaces, of which the sphere is the most straightforward example, but include the projective spaces, we expect to be able to reproduce the results presented in the previous sections, but this is a matter for future research.

Let $M \subset \mathbb{R}^{d+k}$ be a d-dimensional embedded compact homogeneous C^∞ manifold; i.e. there is a compact group G of isometries of $\mathbb{R}^{d+k}$ such that for some $\eta \in M$ (often referred to as the pole) $M = \{g\eta : g \in G\}$. A kernel $\kappa : M \times M \to \mathbb{R}$ is termed *zonal* (or G-invariant) if $\kappa(x,y) = \kappa(gx, gy)$ for all $g \in G$ and $x, y \in M$. This kernel plays the part of the spherical kernels on the sphere. Since the maps in G are isometries of Euclidean space, they preserve both Euclidean distance and the (arc-length) metric $d(\cdot,\cdot)$ induced on the components of M by the Euclidean metric. Thus the distance kernel $d(x,y)$ is zonal, as are all the radial functions, $\phi(d(x,y))$, which are kernels that depend only on the distance between x and y.

The manifold carries a unique normalized G-invariant measure which we call σ. Then, we can define the inner product of real functions

$$\langle f, g\rangle = \int_M fg d\sigma.$$

We assume that

$$\int_M x d\sigma(x) = 0. \tag{14}$$

First let $H_0 = P_0$ be the constants, and H_1 be the set of linear functionals on M. These are essentially the linear polynomials. Let $P_1 = H_0 \bigcup H_1$. Then, inductively we define

$$P_\ell = P_{\ell-1} \bigcup \{p_{\ell-1}p_1 : \quad p_{\ell-1} \in P_{\ell-1}, p_1 \in P_1\}, \quad \ell \geq 2.$$

We can then break the polynomials into orthogonal pieces (called harmonic polynomials):

$$H_\ell = P_\ell \bigcap P_{\ell-1}^\perp, \quad \ell \geq 2,$$

where orthogonality is with respect to the inner product defined above (it is clear from (14) that H_1 is orthogonal to H_0). This is a nice definition of the polynomials and harmonic polynomials since it is intrinsic to the manifold and does not require the restriction of polynomials from any ambient space.

Let $\{Y_{\ell,1}, \cdots, Y_{\ell,q_\ell}\}$ be an orthonormal basis for H_ℓ. Then

$$r_\ell(x, y) = \sum_{j=1}^{q_\ell} Y_{\ell,j}(x) Y_{\ell,j}(y), \quad x, y \in M,$$

is the reproducing kernel for H_ℓ. It is relatively straight forward to show that r_ℓ is zonal, i.e.

$$r_\ell(gx, gy) = r_\ell(x, y) \quad g \in G, \ x, y, \in M.$$

Thus $r_\ell(x, x)$ is constant. If we integrate this constant on M we get

$$\int_M r_\ell(x, x)\, d\sigma(x) = \sum_{j=1}^{q_\ell} \int_M Y_{\ell,j}(x) Y_{\ell,j}(x)\, d\sigma(x) = q_\ell.$$

Hence,

$$\sum_{j=1}^{q_\ell} Y_{\ell,j}(x) Y_{\ell,j}(x) = q_\ell. \tag{15}$$

We are interested in kernels with the following expansions

$$\kappa(x, y) = \sum_{\ell=0}^{\infty} \hat{\kappa}(\ell) r_\ell(x, y).$$

If for each $x, y \in M$ there is a $g \in G$ such that $gx = y$ then in [22], it is shown that all zonal kernels have such an expansion.

We can decompose an arbitrary function $f \in L^2(M)$:

$$f = \sum_{\ell=0}^{\infty} f_\ell,$$

where

$$f_\ell(x) = \langle f, r_\ell(\cdot, x) \rangle.$$

We can define a spherical cap in M in exactly the same way as in Section 1, i.e. $C(x,r) = \{y \in M : |y - x| \leq r\}$, where the distance is the distance in the ambient space (everything can be reproduced in terms of geodesic distance on the manifold, but this is more convenient). Then, we have the following analogue to Definition 1,

Definition 2. *For each $N \geq 1$, let $\{x_{N,1}, \ldots, x_{N,N}\}$ be a set of N points in M. This sequence is uniformly distributed in M if for each spherical cap $C(x,r)$, we have*

$$\lim_{N \to \infty} \frac{\#\{x_{N,j} : x_{N,j} \in C(x,r)\}}{N} = \sigma(C(x,r)).$$

In Damelin et al. [8] the equivalence of statements 2 and 3 of following counterpart of Theorem 2 is proved:

Theorem 6. *Let $\{x_{N,1}, \ldots, x_{N,N}\}_{N=1}^{\infty}$ be a triangular array of points in M. Then the following three statements are equivalent:*

1. *The points $x_{N,1}, \ldots, x_{N,N}$ are uniformly distributed in M.*
2. *For each fixed integer $\ell \geq 1$, and each fixed m, $1 \leq m \leq q_\ell$, we have*

$$\lim_{N \to \infty} \frac{1}{N} \sum_{j=1}^{N} Y_{\ell,m}(x_{N,j}) = 0. \tag{16}$$

3. *For every continuous function f on M, we have*

$$\lim_{N \to \infty} \frac{1}{N} \sum_{j=1}^{N} f(x_{N,j}) = \int_M f(x) d\sigma(x).$$

The equivalence of 1 and 3 is a simple consequence of duality.

If we assume that the $\hat{\kappa}(\ell) > 0$ for all $\ell \geq 0$, then we can define a native space for κ

$$\mathcal{N}_\kappa := \left\{ f \in L^2(M) : \|f\|_\kappa < \infty \right\},$$

where $\|\cdot\|_\kappa$ is the norm associated with the inner product

$$\langle f, g \rangle_\kappa = \sum_{\ell=0}^{\infty} \hat{\kappa}^{-1}(\ell) \langle f_\ell, g_\ell \rangle.$$

Of course, the crucial property here is that for every $f \in \mathcal{N}_\kappa$,

$$f(x) = \langle f, \kappa(x, \cdot)\rangle, \quad x \in M.$$

Let us define

$$A_\kappa = \int_M \kappa(x, y) d\sigma(y).$$

This is a constant since, for any $g \in G$, using the fact that κ is a spherical kernel,

$$\begin{aligned}
\int_M \kappa(gx, y) d\sigma(y) &= \int_M \kappa(x, g^{-1}y) d\sigma(y) \\
&= \int_M \kappa(x, y) d\sigma(gy) \\
&= \int_M \kappa(x, y) d\sigma(y).
\end{aligned}$$

In the second step we used the volume preserving change of variable $y \to gy$, and the final equality follows from the G-invariance of the measure σ.

Define the sequence of functions

$$T_{\kappa,N} = \frac{1}{N} \sum_{j=1}^{N} \kappa(x, x_{N,j}), \quad x \in M.$$

The only explicit property of the spherical harmonics required for the proof of Theorem 3 is the specialization of (15) to the sphere. Thus we can follow exactly the same proof as in [40]) to obtain the following result:

Theorem 7. *Let*

$$\kappa = \sum_{\ell=0}^{\infty} \hat{\kappa}(\ell) r_\ell(x, y),$$

where $\hat{\kappa}(\ell) > 0$, $\ell = 0, 1, \cdots$, *and* $\sum_{\ell=1}^{\infty} \hat{\kappa}(\ell) q_\ell < \infty$. *Then we have the following equivalent statements:*

1. *The triangular array* $\{x_{N,1}, \ldots, x_{N,N}\}_{N=1}^{\infty}$ *are uniformly distributed in* M.
2. *The following limit holds true:* $\lim_{N\to\infty} \|T_{\kappa,N} - A_\phi\|_{\mathcal{N}_\kappa} = 0$.
3. *For every fixed p,* $0 < p \le \infty$, *we have* $\lim_{N\to\infty} \|T_{\kappa,N} - A_\phi\|_p = 0$.
4. *The following limit holds true:*

$$\lim_{N\to\infty} \sum_{\ell=1}^{\infty} |\hat{\kappa}(\ell)|^2 \sum_{m=1}^{q_\ell} \left(\frac{1}{N} \sum_{j=1}^{N} Y_{\ell,m}(x_{N,j}) \right)^2 = 0.$$

Exactly as in the discussion following Proposition 2, for $\Omega_N=\{x_1,\ldots,x_N\}$, a set of N distinct points in M the continuous linear functional ψ defined by

$$\mathcal{N}_\kappa \ni f \mapsto \psi(f) := \int_{\mathbb{S}^d} f(x)d\sigma(x) - N^{-1}\sum_{j=1}^{N} f(x_j)$$

has representer

$$\xi = A_\kappa + T_{\kappa,N}.$$

If we now define

$$E_\kappa(\Omega_N) = \|\xi\|_{\mathcal{N}_\kappa} = A_\kappa + \frac{1}{N^2}\sum_{j=1}^{N}\sum_{k=1}^{N}\kappa(x_j,x_k),$$

we have the following result, with proof identical to that of Proposition 3:

Proposition 10. *For each $f\in\mathcal{N}_\kappa$, we have*

$$\left|\int_M f(x)d\sigma(x) - N^{-1}\sum_{j=1}^{N} f(x_j)\right| \le \|f\|_{\mathcal{N}_\kappa}(E_\kappa(\Omega_N))^{1/2}.$$

We close this section with a generalization of Proposition 9, in which we reprove a result from [16].

Theorem 8. *Let κ be a zonal positive definite kernel. If $\{x_1^{(\kappa)},x_2^{(\kappa)},\ldots,x_N^{(\kappa)}\}$ is a set of N distinct points in $\mathbb{S}^d$, such that*

$$\sum_{j=1}^{N}\sum_{k=1}^{N}\kappa\left(x_j^{(\kappa)},x_k^{(\kappa)}\right) = \inf_{\Omega_N}\sum_{j=1}^{N}\sum_{k=1}^{N}\kappa\left(x_j,x_k\right),$$

where the infimum is taken over all $\Omega_N := \{x_1,\ldots,x_N\}$, N distinct points in M, then for each $f\in\mathcal{N}_\kappa$, the native space of κ, there exists a constant $C>0$ independent of f and N, such that

$$\left|\int_M f(x)d\sigma(x) - N^{-1}\sum_{j=1}^{N} f(x_j^{(\kappa)})\right| \le C\|f\|_{\mathcal{N}_\kappa}N^{-1/2}.$$

Proof. First we have from [8], that the uniform measure $d\sigma$ minimizes the energy integral

$$E_\kappa(\mu) = \int_M\int_M \kappa(x,y)d\mu(x)d\mu(y),$$

is minimized over all probability measures. However,

$$\int_M\int_M \kappa(x,y)d\sigma(x)d\sigma(y) = A_\kappa.$$

Hence, for any set of N points $x_1, \cdots, x_N$, writing

$$\mu^{(\kappa)} = \frac{1}{N}\delta_{x_i^{(\kappa)}},$$

and

$$\mu = \frac{1}{N}\delta_{x_i},$$

we have

$$\begin{aligned} A_\kappa &\leq E_\kappa(\mu^{(\kappa)}) \\ &= \frac{1}{N^2}\sum_{j=1}^{N}\sum_{k=1}^{N}\kappa\left(x_j^{(\kappa)}, x_k^{(\kappa)}\right) \\ &\leq \frac{1}{N^2}\sum_{j=1}^{N}\sum_{k=1}^{N}\kappa\left(x_j, x_k\right). \end{aligned}$$

If we integrate the right hand side inequality with respect to $d\sigma(x_j)$ and $d\sigma(x_k)$ we get $N(N-1)$ integrals all in the same form

$$\int_M \int_M \kappa(x_j, x_k) d\mu(x_j) d\mu(x_k) = A_\kappa.$$

On the diagonal we have a constant $\kappa(x, x)$ (remember it is independent of x due to the zonal nature of the kernel) and hence from integrating these contributions to the sum we obtain $N\kappa(x, x)$, for some fixed $x \in M$. Thus we have

$$\begin{aligned} A_\kappa &\leq \frac{1}{N^2}\sum_{j=1}^{N}\sum_{k=1}^{N}\kappa\left(x_j^{(\kappa)}\right) \\ &\leq \frac{N(N-1)}{N^2}A_\kappa + \frac{\kappa(x,x)}{N} \\ &< A_\kappa + \frac{\kappa(x,x)}{N}, \end{aligned} \tag{17}$$

for any fixed $x \in M$.

Now, using the reproducing property of κ, for $f \in \mathcal{N}_\kappa$ and $x \in M$, we have

$$\begin{aligned}\left|\int_M f(x)d\sigma(x) - \frac{1}{N}\sum_{j=1}^N f\left(x_j^{(\kappa)}\right)\right| &= \left|\left\langle f, A_\kappa - \frac{1}{N}\sum_{j=1}^N \kappa\left(x_j^{(\kappa)},\cdot\right)\right\rangle\right| \\ &\le \|f\|_{\mathcal{N}_\kappa}\left\|A_\kappa - \frac{1}{N}\sum_{j=1}^N \kappa\left(x_j^{(\kappa)},\cdot\right)\right\|_{\mathcal{N}_\kappa} \\ &= \left(A_\kappa - \frac{1}{N^2}\sum_{j=1}^N\sum_{k=1}^N \kappa\left(x_j^{(\kappa)}\right)\right)^{1/2} \|f\|_{\mathcal{N}_\kappa} \\ &\le \left(\frac{\kappa(x,x)}{N}\right)^{1/2}\|f\|_{\mathcal{N}_\kappa},\end{aligned}$$

by using (17). This completes the proof.

References

1. B. Baxter and S. Hubbert: Radial basis functions for the sphere. In: *Progress in Multivariate Approximation*, International Series of Numerical Mathematics, vol. 137, Birkhäuser, Basel, 2001, 33–47.
2. J. Beck: On the sum of distances between N points on a sphere. Mathematika **31**, 1984, 33–41.
3. J. Beck and W.W.L. Chen: *Irregularities of Distribution*. Cambridge Tracts in Math., vol. 89, Cambridge University Press, 1987.
4. J.S. Brauchart: Note on a generalized invariance principle and its relevance for cap discrepancy and energy. In: *Modern Developments in Multivariate Approximation*, International Series of Numerical Mathematics, vol. 145, Birkhäuser, Basel, 2003, 41–55.
5. J.S. Brauchart: Optimal logarithmic energy points on the unit sphere. Math. Comp. **77**, 2008, 1599–1613.
6. W. zu Castell and F. Filbir: Radial basis functions and corresponding zonal series expansions on the sphere. J. Approx. Theory **134**, 2005, 65–79.
7. D. Chen, V.A. Menegatto, and X. Sun: A necessary and sufficient condition for strictly positive definite functions on spheres. Proc. Amer. Math. Soc. **131**, 2003, 2733–2740.
8. S.B. Damelin, J. Levesley, and X. Sun: Energy estimates and the Weyl criterion on homogeneous manifolds. In: *Algorithms for Approximation*, A. Iske and J. Levesley (eds.), Springer, Berlin, 2007, 359–367.
9. M. Drmota and R.F. Tichy: *Sequences, Discrepancies and Applications*. Lecture Notes in Mathematics, 1651. Springer-Verlag, Berlin, 1997.
10. N. Dyn, F.J. Narcowich, and J.D. Ward: Variational principles and Sobolev-type estimates for generalized interpolation on a Riemannian manifold. Constr. Approx. **15**, 1999, 175–208.
11. P. Erdős and P. Turán: On a problem in the theory of uniform distribution I and II. Indag. Math. **10**, 1948, 370–378 and 406–413.
12. I.M. Gel'fand and N.Ya. Vilenkin: *Generalized Functions*, vol. 4. Academic Press, New York and London, 1964.

13. P.J. Grabner: Erdős-Turán type discrepancy bounds. Mh. Math. **111**, 1991, 127–135.
14. D.P. Hardin and E.B. Saff: Minimal Riesz energy point configurations for rectifiable d-dimensional manifolds. Adv. Math. **193**, 2005, 174–204.
15. D.P. Hardin and E.B. Saff: Discretizing manifolds via minimum energy points. Notices of Amer. Math. Soc. **51**(10), 2004, 1186–1194.
16. D.P. Hardin, E.B. Saff, and H. Stahl: The support of the logarithmic equilibrium measure on sets of revolution. J. Math. Phys. **48**, 2007, 022901 (14pp).
17. J.F. Koksma: Een algemeene stelling uit de theorie der gelijkmatige verdeeling modulo 1. Mathematica B (Zutphen) **11** (1941/1943), 7–11.
18. A.B.J. Kuijlaars and E.B. Saff: Asymptotics for minimal discrete energy on the sphere. Trans. Amer. Math. Soc. **350**, 1998, 523–538.
19. L. Kuipers and H. Niederreiter: *Uniform Distribution of Sequences*. John Wiley & Sons, 1974.
20. N.S. Landkov: *Foundations of Modern Potential Theory*. Springer-Verlag, Berlin, Heidelberg, New York, 1972.
21. W.J. LeVeque: An inequality connected with the Weyl's criterion for uniform distribution. Proc. Symp. Pure Math. **129**, 1965, 22–30.
22. J. Levesley and D.L. Ragozin: The fundamentality of translates of spherical functions on compact homogeneous spaces. Journal of Approximation Theory **103**, 2000, 252–268.
23. X.-J. Li and J. Vaaler: Some trigonometric extremal functions and the Erdős-Turán type inequalities. Indiana University Mathematics Journal **48**(1), 1999, 183–236.
24. H.L. Montgomery: *Ten Lectures on the Interface between Analytic Number Theory and Harmonic Analysis*. CBMS Regional Conference Series in Mathematics, no. 84, American Mathematical Society, Providence, RI, 1990.
25. C. Müller: *Spherical Harmonics*. Lecture Notes in Math. 17, Springer-Verlag, Berlin, 1966.
26. F.J. Narcowich, X. Sun, and J.D. Ward: Approximation power of RBFs and their associated SBFs: a connection. Adv. Comput. Math. **27**(1), 2007, 107–124.
27. F.J. Narcowich, X. Sun, J. Ward, and H. Wendland: Direct and inverse Sobolev error estimates for scattered data interpolation via spherical basis functions. Found. Comput. Math. **7**(3), 2007, 369–390.
28. F.J. Narcowich, X. Sun, J.D. Ward, and Z. Wu: LeVeque type inequalities and discrepancy estimates for minimal energy configurations on spheres. Preprint.
29. F.J. Narcowich and J.D. Ward: Scattered data interpolation on spheres: error estimates and locally supported basis functions. SIAM J. Math. Anal. **33**, 2002, 1393–1410.
30. H. Niederreiter: *Random Number Generation and Quasi Monte Carlo Methods*. CBMS-NSF Regional Conference Series in Applied Mathematics **63**, SIAM, Philadelphia, 1992.
31. G. Polya and G. Szegő: On the transfinite diameters (capacity constant) of subsets in the plane and in space. J. für Reine und Angew. Math. **165**, 1931, 4–49 (in German).
32. A. Ron and X. Sun: Strictly positive definite functions on spheres in Euclidean spaces. Math. Comp. **65**(216), 1996, 1513–1530.
33. C. Runge: Über empirische Funktionen und die Interpolation zwischen äquidistanten Ordinaten. Zeitschrift für Mathematik und Physik **46**, 1901, 224–243.

34. I.J. Schoenberg: Metric spaces and completely monotone functions. Ann. of Math. **39**, 1938, 811–841.
35. I.J. Schoenberg: Positive definite functions on spheres. Duke Math. J. **9**, 1942, 96–108.
36. K.B. Stolarsky: Sums of distance between points on a sphere II. Proc. Amer. Math. Soc. **41**, 1973, 575–582.
37. E. Stein and G. Weiss: *Introduction to Fourier Analysis on Euclidean Space*. Princeton University Press, Princeton, 1971.
38. F.E. Su: A LeVeque-type lower bound for discrepancy. In: *Monte Carlo and Quasi-Monte Carlo Methods 1998*, H. Niederreiter and J. Spanier (eds.), Springer-Verlag, 2000, 448–458.
39. F.E. Su: *Methods for Quantifying Rates of Convergence on Groups*. Ph.D. thesis, Harvard University, 1995.
40. X. Sun and Z. Chen: Spherical basis functions and uniform distribution of points on spheres. J. Approx. Theory **151**(2), 2008, 186–207.
41. J. Vaaler: Some extremal functions in Fourier analysis. Bulletin (New Series) of The American Mathematical Society **12**, 1985, 183–216.
42. J. Vaaler: A refinement of the Erdős-Turán inequality. In: *Number Theory with an Emphasis on the Markoff Spectrum*, A.D. Pollington, W. Moran, (eds.), Marcel Dekker, 1993, 163–270.
43. G. Wagner: On the means of distances on the surface of a sphere (lower bounds). Pacific J. Math. **144**, 1990, 389–398.
44. G. Wagner: On the means of distances on the surface of a sphere (upper bounds). Pacific J. Math. **153**, 1992, 381-396.
45. G. Wagner: On a new method for constructing good point sets on spheres. Discrete & Comput. Geom. **9**, 1993, 111–129.
46. G.N. Watson: *A Treatise on the Theory of Bessel Functions*. 2nd edition, Cambridge University Press, London, 1966.
47. H. Weyl: Über die Gleichverteilung von Zahlen modulo 1. Math. Ann. **77**, 1916, 313–352.
48. Y. Xu and E.W. Cheney: Strictly positive definite functions on spheres. Proc. Amer. Math. Soc. **116**, 1992, 977–981.

Part II

Contributed Research Papers

Modelling Clinical Decay Data Using Exponential Functions

Maurice G. Cox

National Physical Laboratory, Teddington TW11 0LW, UK

Summary. Monitoring of a cancer patient, following initial administration of a drug, provides a time sequence of measured response values constituting the level in serum of the relevant enzyme activity. The ability to model such clinical data in a rigorous way leads to (a) an improved understanding of the biological processes involved in drug uptake, (b) a measure of total absorbed dose, and (c) a prediction of the optimal time for a further stage of drug administration. A class of mathematical decay functions is studied for modelling such activity data, taking into account measurement uncertainties associated with the response values. Expressions for the uncertainties associated with the biological processes in (a) and with (b) and (c) are obtained. Applications of the model to clinical data from two hospitals are given.

1 Introduction

Following initial administration of a drug, a cancer patient is monitored by taking a time sequence of measured response values constituting the level in serum of the relevant enzyme activity. The drug administered may be a radiopharmaceutical or a fusion protein consisting of a tumour-targeting antibody linked to an enzyme product. Three requirements in clinical drug administration and related research are

1. determination of the half-lives of the biological decay processes specific to the patient concerned,
2. a measure of total absorbed dose, and
3. prediction of the optimal time for a further stage of drug administration.

This paper studies a class of mathematical decay functions for modelling such data, which is typically very sparse. The functions are composed of a sum of terms involving exponentials and are required to possess feasible properties that relate to the biological processes specific to the patient concerned. The model is used to help address the above requirements.

Key to this work is the consideration of the uncertainties associated with the measured activity values. Evaluations of these uncertainties are available

E.H. Georgoulis, A. Iske, J. Levesley (eds.), *Approximation Algorithms for Complex Systems*, Springer Proceedings in Mathematics 3, DOI: 10.1007/978-3-642-16876-5_8,
© Springer-Verlag Berlin Heidelberg 2011

based on a knowledge of the measurement process involved. Generally, random errors dominate the measurement, there being negligible contributing systematic errors. Thus, the quantities involved can be regarded as mutually independent, that is, there are negligible covariance effects associated with the data. Standard uncertainties, representing standard deviations of quantities regarded as random variables (of which the measured activity values are realizations) characterized by probability distributions, are stated relative to those values in percentage terms.

Estimates of the exponential model parameters, namely, the half-lives and the amplitudes (initial activities) of the exponential terms, are primary results from modelling the data. As a consequence of the uncertainties associated with the measured activity values, there will be uncertainty associated with these parameter estimates.

Regarding the other quantities indicated, a measure of total absorbed dose is given by the product of the area under the curve of the model (from time zero to infinity) and a known constant, and a prediction of the optimal time for a further stage of drug administration is given by the time point that corresponds to a prescribed activity value. These quantities can be expressed in terms of the model parameters. In turn, estimates of these derived quantities have associated uncertainties.

Because of the magnitude of the relative standard uncertainties associated with the measured activity values, typically of the order of 10 % or 20 %, and, as a result of the sparsity of the data, the parameter estimates and estimates of the derived quantities can have appreciable associated uncertainties. It is important that those involved in drug administration and planning appreciate the magnitude of these uncertainties, which should be taken into account when making inferences from results from the model.

Following an introduction to the nature of the available activity data, the problems of estimating the model parameters and of obtaining estimates of the derived quantities are formulated (Section 2).

An approach to the solution is detailed (Section 3), focusing on determining a good approximation to the global solution of the problem of estimating the model parameters. Best estimates of the parameters are then given by solving a non-linear least-squares (NLS) problem in a reduced parameter space in the region of this approximate solution. A model-data consistency check is given, and only if it is satisfied is it reasonable to evaluate the uncertainties associated with the parameter estimates and estimates of derived quantities.

Results are provided for three examples (Section 4). In the first two examples, comprising clinical data from an ADEPT therapy [8], a predicted time is required corresponding to a particular activity value. The third example comprises clinical data from a ^{90}Y-DOTATOC therapy [10]. In all examples, the area under the curve is determined.

Scope for further work (Section 5) considers the use of contextual information to support the sparse data, experimental design issues, and the improved and additional information available from probability density functions for the various measurands, and conclusions are given (Section 6).

2 Problem Formulation

2.1 Raw Data and Associated Uncertainties

The clinical data for any specific patient consists of points (t_i, y_i), $i = 1, \ldots, m$, and standard measurement uncertainties $u(y_i)$ associated with the y_i. The t_i are an increasing set of time values, typically recorded in hour units (h), regarded as known with negligible uncertainty. The y_i are the corresponding measured activity values in Uml^{-1}, where U is the enzyme unit, namely the amount of enzyme that catalyzes the reaction of 1 mmol of substrate per minute. (The unit U is used in radiation dosimetry in hospitals, but is not an SI unit. In terms of SI, $\mathrm{U} = \mu\mathrm{mol\,s}^{-1} = 16.67\,\mathrm{nkat}$.)

Each $u(y_i)$ corresponds to the standard deviation of a quantity Y_i (the activity at time t_i) regarded as a random variable characterized by a probability distribution, of which y_i is a realization. The $u(y_i)$ are supplied by the hospital concerned, using knowledge of the measuring system that provided the y_i.

Typical data are shown in Figure 1 (top). A cross denotes a measured activity value, and the accompanying bar represents ± 1 standard uncertainty associated with the value. Since it is difficult to see all such bars, the data is also given in Figure 1 (bottom) on a log scale for the activity variable. The uncertainty bars have identical lengths on this scale, if, as here, the standard uncertainties on the original scale are proportional to the measured values.

2.2 Model Function and Feasibility

The model function of which the measured values constitute outcomes is a linear combination of exponential terms involving n biological process [11]:

$$f(\mathbf{A}, \mathbf{T}, t) = e^{-\lambda_p t} \sum_{j=1}^{n} A_j e^{-t/T_j}. \tag{1}$$

In expression (1), $\mathbf{A} = (A_1, \ldots, A_n)^\top$ and $\mathbf{T} = (T_1, \ldots, T_n)^\top$ respectively denote initial activities and time constants of these processes, and λ_p is the physical decay constant for the radionuclide used in the measuring system. The A_j and the T_j are to be positive for the function to have a meaningful interpretation. The number of terms n is in general unknown.

For any particular data set, estimates of the parameters $\mathbf{A}$ and $\mathbf{T}$ are to be determined such that the function is consistent with the data. Moreover, the function is to be *minimally consistent*, namely n, the number of terms, is to be as small as possible (Section 2.4).

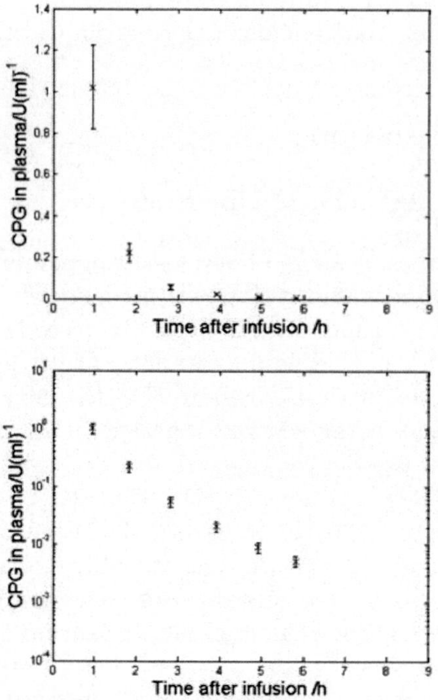

Fig. 1. Data set and bars representing ±1 standard uncertainty, and (bottom) the same, but with activity values on a log scale.

Correction for the effect of the radionuclide gives the adjusted model:

$$f(\mathbf{A}, \mathbf{T}, t) = \sum_{j=1}^{n} A_j e^{-t/T_j}. \tag{2}$$

It will henceforth be assumed that such adjustment has been made, which causes correlation to be associated with the corrected data. This correlation is negligible for practical purposes. See Appendix A.

2.3 Derived Quantities

The required quantities can be derived from the parameters in the model:

1. The biological half-lives, given by

$$(t_{1/2})_j = T_j \ln 2, \quad j = 1, \ldots, n,$$

used to explain behavioural response in the body, and constituting parameters used in comparing the effect of two radiopharmaceuticals;

2. The product of the cumulated activity Q, namely, the area under the curve from the time of initial administration (time zero),

$$Q = \sum_{j=1}^{n} A_j T_j, \tag{3}$$

and the appropriate S-value, as a value for the total absorbed dose in grays (Gy) [5, 7]. The S-value is a conversion constant obtained from tables [14] for the radionuclide used;

3. The time t_0 corresponding to an activity y_0, a prescribed threshold value, given by the equation

$$\sum_{j=1}^{n} A_j e^{-t_0/T_j} = y_0.$$

Here, t_0 is the optimal time at which to administer either the prodrug within the ADEPT scheme (Appendix B) or stem-cell support to counter the toxic effects of treatment to bone marrow. An iterative procedure for determining t_0 is given in Appendix C, where it is shown that this equation has a unique solution to which the procedure will converge for a meaningful value of y_0.

2.4 Objective

The task is to estimate the parameters n, $\mathbf{A}$ and $\mathbf{T}$ of the model function (2) subject to the restrictions that

$$A_j \geq 0, \quad T_j \geq 0, \qquad j = 1, \ldots, n, \tag{4}$$

and that n is to be as small as possible subject to the model being consistent with the data.

For any particular choice of n such that $2n \leq m$, the measure of consistency is based on the sum of squares of the weighted deviations of the activity values y_i from the respective modelled activity values $f(\mathbf{A}, \mathbf{T}, t_i)$, namely, those corresponding to estimates

$$\widehat{\mathbf{A}} = (\widehat{A}_1, \ldots, \widehat{A}_n)^\top, \qquad \widehat{\mathbf{T}} = (\widehat{T}_1, \ldots, \widehat{T}_n)^\top$$

of $\mathbf{A}$ and $\mathbf{T}$, where the weights are taken as the reciprocals of the standard uncertainties $u(y_i)$ associated with the y_i:

$$F(\widehat{\mathbf{A}}, \widehat{\mathbf{T}}) = \sum_{i=1}^{m} \left(\frac{y_i - f(\widehat{\mathbf{A}}, \widehat{\mathbf{T}}, t_i)}{u(y_i)} \right)^2 . \tag{5}$$

If this sum is no greater than a critical value, the model is considered consistent with the data. The critical value is chosen to be the 95th percentile of the chi-squared distribution with $m - 2n$ degrees of freedom (the number of data points minus the number of model parameters), based on regarding the Y_i as characterized by Gaussian probability distributions [3].

If $m = 2n$, the number of model parameters is identical to the number of data points, and the model is regarded as consistent with the data only if $F(\widehat{\mathbf{A}}, \widehat{\mathbf{T}}) = 0$, that is the model function $f(\widehat{\mathbf{A}}, \widehat{\mathbf{T}}, t)$ passes exactly through (interpolates) the data points, namely

$$f(\widehat{\mathbf{A}}, \widehat{\mathbf{T}}, t_i) = y_i, \quad i = 1, \ldots, m.$$

The best feasible least-squares solution, given n, is provided by the values $\widehat{A}_j$ and $\widehat{T}_j$ of the A_j and the T_j that solve the problem

$$\min_{\mathbf{A},\mathbf{T}} F(\mathbf{A}, \mathbf{T}) \quad \text{subject to} \quad A_j \geq 0, \quad T_j \geq 0, \quad j = 1, \ldots, n. \tag{6}$$

For some data sets there might not exist a feasible solution to this problem for any n. Even if a feasible solution existed, it might not be consistent with the data. Such cases need to be identified, since other possibilities would need to be considered, perhaps a decision by the clinician involved.

For a given n such that $2n \leq m$, the problem (6) constitutes the minimization with respect to $\mathbf{A}$ and $\mathbf{T}$ of a sum of squares of non-linear functions subject to non-negativity constraints on the variables (parameters) [5].

3 Solution Approach

3.1 Analysis

Problem non-linearity implies that there might be local solutions in addition to the global solution to the problem. The global solution is that for which $F(\mathbf{A}, \mathbf{T})$ (expression (5)) is least over all feasible values of $\mathbf{A}$ and $\mathbf{T}$. A local solution is to be avoided, since it is generally inferior to the global solution, possibly considerably so, in terms of the value of $F(\mathbf{A}, \mathbf{T})$, that is in terms of closeness of the model to the data. Typical algorithms for NLS problems in general determine a local solution. Such algorithms are iterative in nature, producing a sequence of generally improving approximations to a solution, starting from an initial approximation. The quality of the initial approximation influences the number of iterations taken and particularly the solution obtained. The global solution is more likely to be obtained if it is close to the initial approximation. One of the focusses of this paper is the provision of a good initial approximation.

Consider, for some prescribed value of n such that $2n \leq m$, the solution to the constrained problem (6). If the solution values of A_j and T_j are non-zero for all j, this solution is identical to that of the unconstrained problem

$$\min_{\mathbf{A},\mathbf{T}} F(\mathbf{A}, \mathbf{T}).$$

If, however, one (or more) of the constraints (4) is active at the solution, that is A_j or T_j (or both) is zero for some j, then the corresponding term $A_j \exp(-t/T_j)$ in the model function (2) is also zero. In this case, a formally identical solution is possible for a smaller value of n.

The problem of determining a minimum of $F(\mathbf{A}, \mathbf{T})$ can be re-formulated using variable projection [9]. Given a particular (vector) value of $\mathbf{T}$, the best choice of $\mathbf{A}$ is that which minimizes F with respect to the n parameters constituting $\mathbf{A}$, for that $\mathbf{T}$. This problem is one of linear least squares. So, formally expressing $\mathbf{A}$ as a function of $\mathbf{T}$, namely

$$\mathbf{A} = \mathbf{A}(\mathbf{T}) = (A_1(\mathbf{T}), \ldots, A_n(\mathbf{T}))^\top,$$

the problem (6) becomes

$$\min_{\mathbf{T}} F(\mathbf{A}(\mathbf{T}), \mathbf{T}) \quad \text{subject to} \quad T_j \geq 0, \quad A_j(\mathbf{T}) \geq 0, \quad j = 1, \ldots, n.$$

3.2 Initial Parameter Approximation

The provision of a good initial approximation, as stated in Section 3.1, influences whether the global solution is obtained. Consider the use of the model, and the data and information available concerning the data, to provide such an initial approximation. This information is as follows. The number m of data points is small. For instance, for 25 sets of patient data provided by the Royal Free and University College Medical School, $3 \leq m \leq 9$, and for six sets provided by the Royal Marsden Hospital and the Institute of Cancer Research, $3 \leq m \leq 8$. The values of the time constants T_j are dictated by the biological processes involved in clearance of the infused drug. Although these values vary appreciably across patients, it is possible to prescribe, using knowledge of previous studies, *a priori* values $T_{\min}$ and $T_{\max}$ such that

$$0 < T_{\min} \leq T_1 < \cdots < T_n \leq T_{\max}, \tag{7}$$

corresponding to a particular permutation of terms in the model. Hence, judging by the above patient data, the number n of exponential terms in the model that can be identified will generally be smaller than five, say, and certainly no larger than the integer part of $m/2$. Moreover, it is feasible to carry out a search over a discretization of the interval containing the possible values of the T_j.

Such a discretization constitutes the vertices $(T_{i_1}, \ldots, T_{i_n})$ of an n-dimensional rectangular mesh (50 vertices in each dimension have generally been found adequate) with the exception that vertices that do not satisfy

$$T_{\min} \leq T_{i_1} < \cdots < T_{i_n} \leq T_{\max} \tag{8}$$

are excluded. The resulting discretization avoids coincident time constants, and ensures that time constants corresponding to any one vertex are distinct

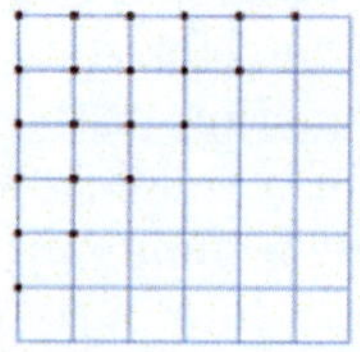

Fig. 2. Mesh for $n = 2$, with points marked by small circles satisfying the inequalities (8).

from those corresponding to all other vertices. Appendix D gives an algorithm for generating the discretization, and Figure 2 shows such a discretization.

For each n-dimensional discretization point, relating to a specific set of n feasible time constants $\widetilde{\mathbf{T}}$, the corresponding amplitudes $\widetilde{\mathbf{A}}$ are obtained by solving the linear least-squares (LLS) problem

$$\min_{\mathbf{A}} g(\mathbf{A}) = \sum_{i=1}^{m} \left(\frac{y_i - f(\mathbf{A}, \widetilde{\mathbf{T}}, t_i)}{u(y_i)} \right)^2 . \qquad (9)$$

Over all points for which $\widetilde{\mathbf{A}}$ is feasible ($\widetilde{\mathbf{A}} > \mathbf{0}$), select as an initial approximation to the time constants $\mathbf{T}$ that point for which g is least.

If the solution is characterized by reasonably separated time constants, relative to the chosen discretization, the discretization points that neighbour the selected discretization point would have larger values of g, and therefore by continuity g would have a minimum in that neighbourhood. Therefore, it would be expected that a NLS algorithm would converge from this initial approximation to the global solution if that solution lay within the boundary of the mesh and the mesh were sufficiently fine.

3.3 Model Parameter Estimation

A popular NLS algorithm (Gauss-Newton) to minimize $F(\mathbf{A}(\mathbf{T}), \mathbf{T})$ generates a succession of iterates, each iterate obtained as the previous iterate updated by the solution to a LLS problem. This problem is given by approximating F by a form that involves the Jacobian $\mathbf{J}(\mathbf{T}) = \{\partial F_i / \partial T_j\}$ of dimension $m \times n$ evaluated at the current iterate.

The process starts with the initial approximation determined as in Section 3.2. Iteration is carried out with respect to $\mathbf{T}$, the corresponding $\mathbf{A} = \mathbf{A}(\mathbf{T})$ at each stage being determined by solving the LLS problem (9). The solution values are denoted by $\widehat{\mathbf{T}}$ and $\widehat{\mathbf{A}}$.

In all cases of clinical data analyzed (31 data sets), the solution obtained as described starting from the initial approximation in Section 3.2 is feasible, and the benefits of variable projection, stated by Golub and Pererya [9], are realized with respect to convergence and robustness of the procedure.

3.4 Consistency of Model and Data

A solution is acceptable if it is feasible and consistent with the data. Feasibility is assured by the initial approximation procedure (Section 3.2). Consistency is assessed by computing the value of $F(\widehat{\mathbf{A}}, \widehat{\mathbf{T}})$ (expression (5)). If this value, χ^2_{obs}, known as the observed chi-squared value, satisfies [3]

$$\Pr\left\{\chi^2(\nu) > \chi^2_{\text{obs}}\right\} \geq 0.05, \tag{10}$$

where $\chi^2(\nu)$ is the chi-squared distribution for ν degrees of freedom, with $\nu = m - 2n$ (the number of data points less the number of model parameters), the solution is considered acceptable. Satisfaction of inequality (10) means that χ^2_{obs} lies in the left-most 95 % of the distribution of $\chi^2(\nu)$. Note that $m > 2n$ for this test to be carried out. However, if $m = 2n$ and the solution *interpolates* the data, viz. $f(\widehat{\mathbf{A}}, \widehat{\mathbf{T}}, t_i) = y_i$, $i = 1, \ldots, m$, the solution is also considered acceptable (Section 2.4), and uncertainties can still be propagated.

The *reduced chi-squared value* $\widetilde{\chi}^2_{\text{obs}}$ is the ratio of χ^2_{obs} and the 95th percentile of the chi-squared distribution. Consistency is indicated by $\widetilde{\chi}^2_{\text{obs}} \leq 1$.

Also see Appendix E.

3.5 Uncertainties Associated with Parameter Estimates and Estimates of Derived Quantities

When the model is consistent with the data, uncertainties associated with the parameter estimates and with estimates of derived quantities can be evaluated.

The NLS algorithm provides a covariance matrix $\boldsymbol{U}_{\widehat{\boldsymbol{T}}}$ associated with the estimates $\widehat{\mathbf{T}}$ of the parameters $\mathbf{T}$. This matrix is of dimension $n \times n$, containing on its diagonal the squares of the standard uncertainties associated with the components of $\widehat{\mathbf{T}}$, and in its off-diagonal positions the covariances associated with pairs of the components of $\widehat{\mathbf{T}}$. Let $\mathbf{D}$ be the diagonal of $\boldsymbol{U}_{\widehat{\boldsymbol{T}}}$. Then the correlation matrix associated with $\widehat{\mathbf{T}}$ is

$$\mathbf{D}^{-1/2}\boldsymbol{U}_{\widehat{\boldsymbol{T}}}\mathbf{D}^{-1/2}.$$

For a derived scalar quantity $Z = \phi(\mathbf{A}, \mathbf{T}) = \phi(\mathbf{A}(\mathbf{T}), \mathbf{T})$, let $u(z)$ be the standard uncertainty associated with the estimate $z = \phi(\mathbf{A}(\widehat{\mathbf{T}}), \widehat{\mathbf{T}})$ of Z, and define the row vector

$$\mathbf{c}^{\top} = \left[\frac{\partial Z}{\partial T_1}, \ldots, \frac{\partial Z}{\partial T_n}\right]\Bigg|_{\boldsymbol{T} = \widehat{\boldsymbol{T}}}.$$

The quantity $u(z)$ is evaluated using the law of propagation of uncertainty [1]:

$$u^2(z) = \mathbf{c}^\top U_{\widehat{\boldsymbol{T}}}\mathbf{c}.$$

The covariance matrix associated with parameter estimates $\widehat{\mathbf{A}}$ and $\widehat{\mathbf{T}}$ is

$$\mathbf{U}_{\widehat{\mathbf{p}}} = \left[\mathbf{J}^\top(\widehat{\mathbf{p}})\mathbf{U}_{\mathbf{y}}^{-1}\mathbf{J}(\widehat{\mathbf{p}})\right]^{-1},$$

where $\widehat{\mathbf{p}}$ is the solution value of $\mathbf{p} = \left(\mathbf{A}^\top, \mathbf{T}^\top\right)^\top$ and $\mathbf{J}(\widehat{\mathbf{p}})$ is the Jacobian matrix $\partial f(\mathbf{A}, \mathbf{T}, \mathbf{t})/\partial \mathbf{p}$ of dimension $m \times 2n$, where $\mathbf{t} = (t_1, \ldots, t_m)^\top$, evaluated at $\widehat{\mathbf{p}} = (\widehat{\mathbf{A}}^\top, \widehat{\mathbf{T}}^\top)^\top$. $\mathbf{U}_{\widehat{\mathbf{p}}}$ is calculated as $\mathbf{R}^{-1}\mathbf{R}^{-\top}$, with $\mathbf{R}$ the Cholesky factor of $\mathbf{K}$, where the ith row of $\mathbf{K}$, $i = 1, \ldots, m$, is that of $\mathbf{J}(\widehat{\mathbf{p}})$ scaled by $1/u(y_i)$.

4 Results

Three examples are given, for each of which clinical data is modelled by a function of the form (2) subject to restrictions (4), the intention being to obtain the number of terms n as the smallest satisfying the chi-squared test (10). This number is determined by using successively $1, 2, \ldots$ terms in the model until either the chi-squared test is satisfied or the number of terms is too great for the m items of data to enable a unique solution to be obtained. In the latter case, no consistent solution is possible and the matter would be referred to the clinician concerned.

In the first two examples, comprising clinical data from an ADEPT therapy [8], a time is predicted corresponding to a particular activity value. The third example comprises clinical data from a ^{90}Y-DOTATOC therapy [10]. In all three examples, the area under the curve is determined to obtain total absorbed dose. In a case of inconsistency, a data point judged to be discrepant is excluded from the data set, and the data re-analyzed. Such a decision would in practice be made by the clinician involved.

In the figures for the examples in this section are shown, where appropriate,

1. the provided activity data (crosses),
2. bars representing ± 1 standard uncertainty (20 % for ADEPT, 10 % for ^{90}Y-DOTATOC) associated with the measured activity values,
3. the best-fitting model (central, blue line) for the data, obtained as the consistent feasible model with fewest exponential terms,
4. a ± 1 standard uncertainty band (outer, red lines) for the model obtained by propagation of the activity data uncertainties, and
5. a predicted time (circle) corresponding to a threshold of 0.002 Uml^{-1} (horizontal line).

4.1 Example 1

The first example (Figure 3) has five data points. For $n = 1$, $\widetilde{\chi}^2_{\text{obs}} = 0.34$ and thus the model is consistent with the data. The biological half-life

$(t_{1/2})_1$ is estimated as $(\widehat{t_{1/2}})_1 = 0.79\,\text{h}$, with associated standard uncertainty $u((\widehat{t_{1/2}})_1) = 0.06\,\text{h}$, expressed as $0.79(0.06)\,\text{h}$. The estimate of the area Q in formula (3) is $\widehat{Q} = 3.3\,\text{Uhml}^{-1}$, with associated relative standard uncertainty $\widetilde{u}(\widehat{Q}) = 93\,\%$, that is, $3.3(93\,\%)\,\text{Uhml}^{-1}$. Corresponding to activity value $y_0 = 0.002\,\text{Uml}^{-1}$ is the time $t_0 = 8.3(0.5)\,\text{h}$ for prodrug administration.

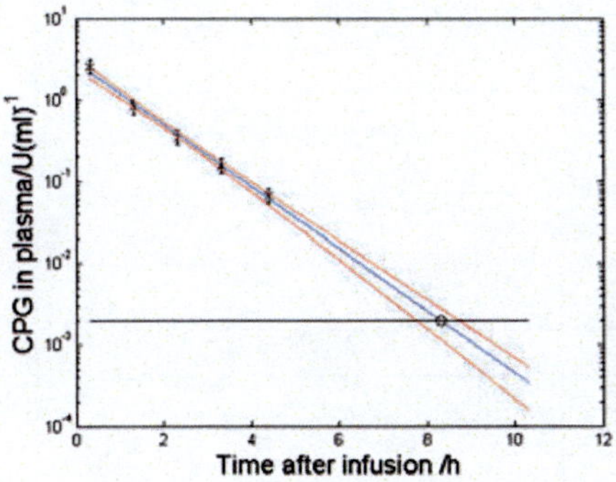

Fig. 3. Data and model for example 1, with standard uncertainty band.

4.2 Example 2

The second example has six data points. For $n = 1$ and 2, $\widetilde{\chi}^2_{\text{obs}} = 4.62$ and 2.33, respectively. There is no feasible solution for $n = 3$, which is not surprising since the fifth data point has a larger activity value than the fourth, and the number of model parameters equals the number of data points (six). The best feasible solution is shown in Figure 4 (top), corresponding to $n = 2$.

Excluding the fifth data point gives for $n = 1$ and 2, $\widetilde{\chi}^2_{\text{obs}} = 4.25$ and 0.03, respectively. The model for $n = 2$ is consistent with the reduced data set. The solution for $n = 2$ is shown in Figure 4 (bottom). Estimates of the two half-lives are $0.34(0.11)\,\text{h}$ and $1.35(0.95)\,\text{h}$, and the correlation coefficient associated with these estimates is $\rho = 0.82$. The estimated area $\widehat{Q} = 0.89(26\,\%)\,\text{Uhml}^{-1}$. For activity value $0.002\,\text{Uml}^{-1}$, $t_0 = 6.4(1.2)\,\text{h}$. Observe the rapid widening of the uncertainty band with t.

4.3 Example 3

The third example has seven data points. For $n = 1$ and 2, $\widetilde{\chi}^2_{\text{obs}} = 1.37$ and 0.13, respectively. The model for $n = 2$ is consistent with the data and is shown in Figure 5. Estimates of the half-lives are $2.0(2.1)\,\text{h}$ and $36(26)\,\text{h}$, and the associated correlation coefficient $\rho = 0.90$. The estimated area $\widehat{Q} = 1752(0.3\,\%)\,\text{Uhml}^{-1}$.

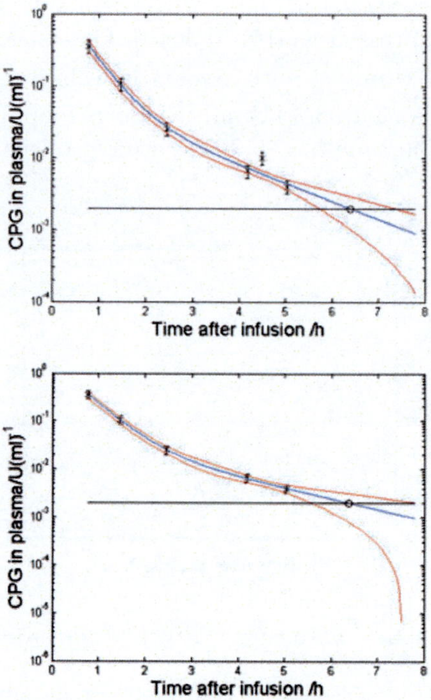

Fig. 4. Data and model for example 2, and (below) excluding the fifth point, with standard uncertainty band.

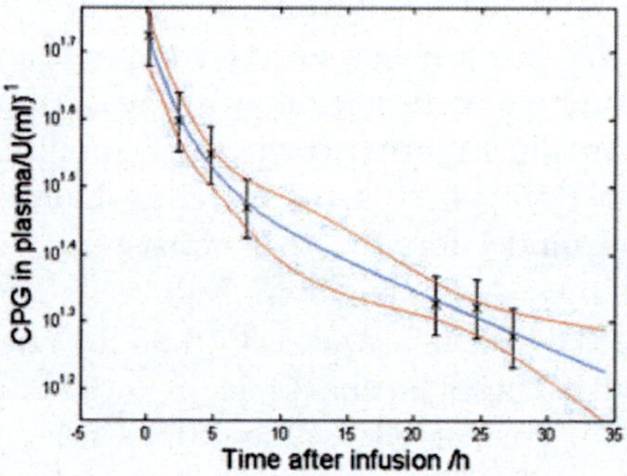

Fig. 5. Data and model, with standard uncertainty band for example 3.

5 Scope for Further Work

5.1 Contextual Information

Contextual information such as historical data could be used to improve results such as those in Section 4, which are based only on the clinical data for the individual patient concerned. Historical data relates to groups of previous patients. Such information has limited value in the case of an individual, since biological processes differ appreciably across patients. However, since individual patient data is sparse, the aggregation of such data and more general information should lead to better results than the consideration of individual patient data alone.

Consider a group of patients that have been given identical therapy to the patient whose data is being clinically analyzed. Suppose historical data comprising values $\widehat{T}_j$, p in number, of each of the T_j, $j = 1, \ldots, p$, for a particular value of n, are available for those patients. Let the average of these $\widehat{T}_j$ be denoted by $\widehat{T}^{(0)}$ and the covariance matrix associated with $\widehat{T}^{(0)}$ by $\mathbf{S}$. Then, taking account of this information in estimating $\mathbf{T}$ for the current patient can be accomplished by

$$\min_{\mathbf{T}} \left[F(\mathbf{A}(\mathbf{T}), \mathbf{T}) + \left(\mathbf{T} - \widehat{\mathbf{T}}^{(0)}\right)^{\top} \mathbf{S}^{-1} \left(\mathbf{T} - \widehat{\mathbf{T}}^{(0)}\right) \right]. \tag{11}$$

There are now $m + p$ 'data points' and, still, $2n$ parameters, so the degrees of freedom $\nu = m + p - 2n$. As a consequence of the increased degrees of freedom, fewer data are needed to estimate the exponential parameters. For example, two half-lives can be estimated from three and even two meaningful data points. In addition, the approach should help to regularize the solution. A regularization parameter γ can be inserted before the term $\left(\mathbf{T} - \widehat{\mathbf{T}}^{(0)}\right)^{\top} \mathbf{S}^{-1} \left(\mathbf{T} - \widehat{\mathbf{T}}^{(0)}\right)$ in expression (11). By this means smaller weight $(0 < \gamma < 1)$ can be given to historical data to reflect the fact that that data are not specific to the current patient.

Early trials with the approach appear promising. In particular, when such additional data is not needed, the solution is essentially unchanged.

5.2 Experimental Design Issues

The uncertainty associated with an estimate of the area Q under the curve is influenced by the number m of data points, and the locations in time of those points and the activity values at those points. An important question is "Which choice of m locations minimizes this uncertainty?", which is difficult to answer, since this choice depends on the unknown model parameter values.

One way to proceed would be to use the contextual information in Section 5.1 to provide a model curve defined by average time constants and average initial activities, and the covariance matrix associated with the computed model parameter estimates. As a consequence, optimal time locations could be determined for this 'average' curve. The time locations could be used for a particular patient data set.

Arguably a better way to proceed would be adaptively. Measure the activity corresponding to the first time point so defined, and use the historical data together with this data point to refine the estimate of the curve. Then, in terms of the first time point and this estimate, the second time point could be defined, the activity measured there, and so on. The procedure proceeds until an adequate number of time points were available.

Night-time measurement might be a problem, in which case there would be periods of time in which hospital staff might not be available to take measurement. (Figure 5 illustrates data obtained only in daylight hours: there is a distinct gap between two groups of activity values taken in daytime.) The procedure would be subject to the necessary time constraints.

The procedure could also be applied to time prediction. The time points so determined would be expected to be appreciably different from those for area determination. There would be a compromise between optimal time points for area determination and for time prediction if estimates of both derived quantities were required.

Other experimental design issues include the balance between conflicting aspects such as patient trauma (from taking too many measured values), equipment availability, and the benefits of improved modelling.

5.3 Probability Density Functions

The uncertainty evaluation carried out here is based on an application of the GUM uncertainty framework [1], which comprises the following stages. A mathematical model of a measurement of a single (scalar) quantity is expressed generically as a functional relationship $Y = f(\mathbf{X})$, where Y is a scalar output quantity and $\mathbf{X}$ represents the N input quantities $(X_1, \ldots, X_N)^\top$. The X_i and Y are regarded as random variables.

Each input quantity X_i is regarded as a random variable, characterized by a probability density function (PDF) and summarized by its expectation and standard deviation. These PDFs are constructed from available knowledge of the quantities. The expectation is taken as the best estimate x_i of X_i and the standard deviation as the standard uncertainty $u(x_i)$ associated with x_i. This information is propagated, using the law of propagation of uncertainty, through a first-order Taylor series approximation to the model to provide the standard uncertainty $u(y)$ associated with y, the estimate y of Y given by evaluating the model at the x_i. A coverage interval for Y is provided based on taking the PDF for Y as Gaussian, namely, invoking the central limit theorem.

In the context of the current application, the generic input quantities X_i correspond to the activity quantities Y_i, and the generic output quantity Y to (a) one of the half-lives T_j, (b) the area Q under the curve of the model, or (c) the time T_0 for further drug administration.

Although the above explicitly specifies the measurand Y as a measurement function, that is, as a formula in terms of the X_i, (a) to (c) above constitute measurement models, that is, implicit models of the form $h(Y, \mathbf{X}) = 0$. This is because the model (a) constitutes the solution that cannot be written down explicitly to a NLS problem, and the models for the derived quantities (b) and (c) use the output quantities from (a), the model parameters, as input quantities. In its own right, model (b) is an explicit model, whereas model (c) is implicit. The best-practice guide [4] gives advice on handling explicit and implicit models. If the model is non-linear, as here, or the PDF for Y is not Gaussian, the GUM uncertainty framework is approximate. The larger are the uncertainties associated with the estimates x_i, the poorer the quality of the approximation can be expected to be.

Cases where assumptions break down appear in the examples in Section 4. In some examples the standard uncertainties associated with estimates of derived quantities are comparable in magnitude to the estimates themselves, and the quantities realized by these estimates are positive. These quantities cannot be distributed as Gaussian variables as assumed by the GUM uncertainty framework: a Gaussian variable covers both positive and negative values.

A PDF provides richer information than an estimate and the associated standard uncertainty alone regarding a quantity of concern. This statement holds particularly when the PDF departs appreciably from Gaussian, for instance has marked asymmetry or broad tails. Initial calculations, using a Monte Carlo method for the propagation of distributions [2], suggest that appreciable asymmetry can exist in the PDF for the measurands corresponding to the predicted time and the area under the curve. By appealing to the Principle of Maximum Entropy [12] these calculations are based on characterizing the quantities realized by the activity data by Gaussian PDFs. Positivity of the model parameters can be taken into account when applying the Monte Carlo method using the treatment of Elster [6].

The model is to be used for predictive purposes, namely, to provide the time t_0 corresponding to a prescribed activity value y_0 of the model function. A probability distribution with PDF $g_{T_0}(\tau_0)$ that characterizes T_0, regarded as a random variable, is to be inferred. This distribution describes the possible values that can reasonably be attributed to the measurand T_0 given the information available. Once $g_{T_0}(\tau_0)$ is available, questions can be asked such as (a) what is the best estimate t_0 of T_0 and what is the standard uncertainty $u(t_0)$ associated with t_0, and (b) at what time should the prodrug be administered?

The answer to the first question is, in accordance with the GUM, that t_0 should be taken as the expectation $\mathrm{E}(T_0)$ of T_0 and $u(t_0)$ as the square root of the variance $\mathrm{V}(T_0)$ of T_0.

The answer to the second question needs more careful consideration. It relates to the degree of assurance that the threshold has been reached before administering the prodrug. If the prodrug is administered too early, the level of toxicity arising from the therapy might be too great. If the prodrug is administered too late, the level of toxicity would be acceptable, but the therapy would be less effective. Therefore, there is a balance of risk associated with the time of administration. Armed with the distribution $g_{T_0}(\tau_0)$, the specialist should be better placed to make a decision than would be the case in the absence of $g_{T_0}(\tau_0)$. For instance, to control the risk of the level of toxicity being too great, only 10 % of the possible values of T_0 given by the PDF might be allowed to exceed the threshold, in which case the 90th percentile would be chosen. On the other hand, to control the risk of the treatment being relatively ineffective, 90 % of the possible values of T_0 might be required to exceed the threshold, in which case the 10th percentile would be chosen.

6 Conclusions

The problems of estimating model parameters and obtaining estimates of derived quantities are considered for the modelling of sparse clinical data by a series of exponential terms. The approach to a solution is detailed by first determining a good approximation to the global solution of the problem of estimating the model parameters. Best estimates of the parameters are then given by solving a NLS problem in the region of this approximation. Variable projection is used to reduce the dimensionality of the parameter space. A model-data consistency check is given, and only if it is satisfied is it reasonable to evaluate the uncertainties associated with the parameter estimates and estimates of derived quantities.

Results are provided for three examples, two comprising clinical data from an ADEPT therapy and one from a ^{90}Y-DOTATOC therapy. In two of these examples, a derived quantity corresponds to the predicted time for prodrug administration in the ADEPT scheme, and, in all three examples, it corresponds to the area under the curve, used in determining total absorbed dose. Uncertainty propagation is carried out to evaluate the uncertainties associated with estimates of the model parameters and the derived quantities.

Possible extensions to the work described here are considered.

Acknowledgement

This work constituted part of NPL's Strategic Research programme of the National Measurement System of the UK's Department for Business, Innovation and Skills. Provision and analysis of the clinical data greatly benefited from collaboration with the Royal Free and University College Medical School and the Royal Marsden Hospital Medical School. Dr Peter Harris reviewed drafts of this paper, which benefited greatly from a anonymous referee's report.

Appendix A: Correction for Radioactive Decay

The data for analysis is corrected for physical effects (Section 2.2), exemplified in the case of a radiopharmaceutical by the natural decay of the radionuclide used. The correction, to all measured activity values y_i, involves the estimated half-life of the radionuclide. Since this estimated half-life has an associated uncertainty, the corrected values have associated covariance. If such correction is made, the quantity Y becomes the quantity $\widetilde{Y}$ according to $\widetilde{Y} = Ye^{\lambda_p t}$. Thus,

$$\frac{\partial \widetilde{Y}}{\partial Y} = e^{\lambda_p t}, \qquad \frac{\partial \widetilde{Y}}{\partial \lambda_p} = tYe^{\lambda_p t}.$$

It therefore follows, applying the law of propagation of uncertainty [1], that the standard uncertainty $u(\widetilde{y}_i)$ associated with $\widetilde{y}_i$ and the covariance $\mathrm{cov}(\widetilde{y}_i, \widetilde{y}_j)$ associated with $\widetilde{y}_i$ and $\widetilde{y}_j$ are given by

$$u^2(\widetilde{y}_i) = v_i^2 u^2(y_i) + t_i^2 y_i^2 v_i^2 u^2(\widehat{\lambda}_p), \quad \mathrm{cov}(\widetilde{y}_i, \widetilde{y}_j) = t_i t_j y_i y_j v_i v_j u^2(\widehat{\lambda}_p), \quad (12)$$

where $v_i = e^{\widehat{\lambda}_p t_i}$ and $u(\widehat{\lambda}_p)$ is the standard uncertainty associated with $\widehat{\lambda}_p$, the estimated half life. For instance, for the ^{90}Y-DOTATOC data sets, the radionuclide is ^{90}Y, with half life 64.057 h, with standard uncertainty 0.016 h. Analyzing the ^{90}Y-DOTATOC examples with and without the covariance correction in expressions (12), $i = 1, \ldots, m$, $j = 1, \ldots, m$, to the covariance matrix made no practical difference. The reason is that the standard uncertainty associated with the ^{90}Y half life is four orders of magnitude smaller than the standard uncertainties associated with the activity data. Comparable results are expected for other radionuclides.

Appendix B: Antibody-Directed Enzyme Prodrug Therapy (ADEPT)

Antibody-directed enzyme prodrug therapy (ADEPT) [8] aims to overcome shortcomings of existing treatment of common tumours such as colorectal and gastric cancer by selective generation of a high concentration of drug in tumour while sparing healthy tissue. It is a two-stage system, in which a fusion protein consisting of a tumour-targeting antibody linked to an enzyme is given intravenously. Following adequate clearance from healthy tissue, prodrug is given and converted to active drug in tumour by the targeted enzyme. ADEPT has potential to overcome drug resistance with minimal toxicity.

To optimize performance, ADEPT requires delivery of effective concentrations of enzyme to the tumour followed by administration of a potentially effective prodrug dose when enzyme levels in the blood are low enough to avoid systemic prodrug activation.

The optimal time point for giving prodrug is a balance defined by safe levels of the tumour-targeting antibody-enzyme complex in serum and sufficient amounts in tumour for effective prodrug conversion. The determination of this time point constitutes a potential challenge since molecules that clear rapidly tend to accumulate less in tumour [13].

The form of chemotherapy relating to the considerations of this paper is a multistage targeted therapy. This approach to administering chemotherapy has been found to be effective in reducing the toxic effects of the treatment on normal tissues whilst delivering a localised toxic effect to tumour cells.

The reaction between the fusion protein and the prodrug produces a toxic drug, which causes damage to tumour cells. In order to control the level of toxicity arising from the therapy, the prodrug should be administered when the plasma concentration of the fusion protein falls below a prescribed threshold value. This threshold concentration is some four orders of magnitude smaller than the initial fusion product concentration. A process that enabled clinicians to estimate reliably the time for prodrug administration from a sparse series of plasma concentration measurement data would have great potential value for analyzing the toxic impact of targeted chemotherapy on patients.

Appendix C: Time Prediction for Prodrug Administration

The time point t_0 for prodrug administration is the value of t satisfying $F(\widehat{\mathbf{A}}, \widehat{\mathbf{T}}, t) = y_0$, where y_0 is the activity threshold (a value smaller than the sum of the initial activities). Thus t_0 is the solution of the non-linear algebraic equation

$$G(t) \equiv F(\widehat{\mathbf{A}}, \widehat{\mathbf{T}}, t) - y_0 = \sum_{j=1}^{n} \widehat{A} e^{-t/\widehat{T}_j} - y_0 = 0.$$

For $\widehat{A}_j > 0$ and $\widehat{T}_j > 0$, and $t \geq 0$, $G'(t) < 0$ and $G''(t) > 0$. Hence $G(t)$ is convex and monotonically decreasing, and has at most one zero. Since

$$0 < y_0 < \sum_{j=1}^{n} \widehat{A}_j, \tag{13}$$

$G(0)$ and $G(\infty)$ take opposite signs, and so $G(t) = 0$ has a solution.

Taking inequality (13) as holding, the required t_0 can be obtained by Newton-Raphson iteration starting from an initial approximation $t_0^{(0)}$:

$$t_0^{(r)} = t_0^{(r-1)} - G(t_0^{(r-1)})/G'(t_0^{(r-1)}), \qquad r = 1, 2, \ldots. \tag{14}$$

Consider an initial approximation $t_0^{(0)}$ satisfying $G(t_0^{(0)}) > 0$. That the iteration will then converge to the required solution can be seen as follows. Taylor's theorem with remainder term gives

$$G(t+\Delta t) = G(t) + \Delta t G'(t) + \frac{1}{2}(\Delta t)^2 G''(t+\theta\Delta t), \qquad 0 \le \theta \le 1.$$

A general step of the Newton-Raphson process (14) can be expressed as $\Delta t = -G(t)/G'(t)$, and so

$$G(t+\Delta t) = G(t) - \frac{G(t)}{G'(t)}G'(t) + \frac{1}{2}(\Delta t)^2 G''(t+\theta\Delta t) = \frac{1}{2}(\Delta t)^2 G''(t+\theta\Delta t).$$

But the term $\frac{1}{2}(\Delta t)^2 G''(t+\theta\Delta t) > 0$, since $G''(t) > 0$. Thus $G(t_0^{(1)}) > 0$, and by induction the $t_0^{(r)}$ form an increasing sequence.

Practical convergence of this process implemented in floating-point arithmetic is given by terminating the iteration when this increasing property fails to hold, namely when the computed value of $t_0^{(r)}$ fails to exceed that of $t_0^{(r-1)}$.

An initial approximation that is always satisfactory is $t_0^{(0)} = 0$.

The standard uncertainty $u(t_0)$ associated with t_0 is given by considering the model

$$G(\mathbf{A}, \mathbf{T}, T_0) \equiv \sum_{j=1}^{n} A_j e^{-T_0/T_j} - y_0 = 0. \tag{15}$$

Expression (15) constitutes an implicit model, and so setting $\mathbf{p} = (\mathbf{A}^\top, \mathbf{T}^\top)^\top$ and applying the results in reference [4] gives an expression for $u^2(t_0)$:

$$\left. \frac{\left(\dfrac{\partial G}{\partial \mathbf{p}}\right)^\top \mathbf{U}_\mathbf{p} \dfrac{\partial G}{\partial \mathbf{p}}}{\left(\dfrac{\partial G}{\partial T_0}\right)^2} \right|_{T_0 = t_0, \mathbf{p} = \widehat{\mathbf{p}}} .$$

Appendix D: Generating all Meshpoints

To search over a discrete mesh satisfying inequalities (7), a count over the set of indices defined by the mesh is carried out. Such counting is given by the following code fragment, which provides the index set of the next in the sequence of index sets representing a choice of n distinct items from L items, the number of vertices. Starting with $c = (1, 2, \ldots, n-1, n)$, where n is the number of time constants, the code fragment generates all required vertices:

```
n = length(c);
k = n;
while c(k) == L - n + k
  k = k - 1;
end
c(k) = c(k) + 1;
c(k+1:n) = c(k)+1:c(k)+n-k;
```

The code is based on simulating a digital counter, with the following properties: (1) the counter has n digits lying between 1 and L; (2) no digit is repeated; (3) when the counter is regarded as displaying an n-digit set of numbers in arithmetic to base L, the sequence produced is strictly increasing.

Appendix E: Model-Data Consistency

Consistency of model and data is required to use the model to infer information about the processes that underpin the data. The test of consistency used in Section 3.4 is the chi-squared test. It makes the assumption that the model deviations, and hence the model values, are realizations of Gaussian variables, and so the sum of their squares is a realization of a chi-squared distribution with $m - 2n$ degrees of freedom.

A distinction should be drawn between making an individual statistical test and applying a statistical test on a routine basis to a succession of models determined from clinical data. The test relates to the possible values that could be attributed to the test statistic and the actual ('observed') value in a particular case. If the observed value is considered extreme (in the tail of the distribution of possible values), it is reasonable to conclude that the model determined in that case is inconsistent with the model. However, if, as here, the tail probability corresponding to values that are considered extreme is 5 %, a value often used in statistical testing, it can be expected that over very many sets of patient data, of the order of 1 in 20 of the corresponding models would be judged inconsistent with the data. This statement is made on the basis of statistical variation alone.

It is hence important that such cases should not *automatically* be judged as displaying model-data inconsistency, but be referred to an expert for consideration. Graphical aids are helpful in this regard.

References

1. BIPM, IEC, IFCC, ILAC, ISO, IUPAC, IUPAP, and OIML. Evaluation of measurement data — guide to the expression of uncertainty in measurement. Joint Committee for Guides in Metrology, JCGM 100:2008. `www.bipm.org/utils/common/documents/jcgm/JCGM_100_2008_E.pdf`.
2. BIPM, IEC, IFCC, ILAC, ISO, IUPAC, IUPAP, and OIML. Evaluation of measurement data — supplement 1 to the "Guide to the expression of uncertainty in measurement" — propagation of distributions using a Monte Carlo method. Joint Committee for Guides in Metrology, JCGM 101:2008. `www.bipm.org/utils/common/documents/jcgm/JCGM_101_2008_E.pdf`.
3. M.G. Cox: The evaluation of key comparison data. Metrologia **39**, 2002, 589–595.
4. M.G. Cox and P.M. Harris: *SSfM Best Practice Guide No. 6, Uncertainty Evaluation*. Technical Report DEM-ES-011, National Physical Laboratory, Teddington, UK, 2006. `http://publications.npl.co.uk/npl_web/pdf/dem_es11.pdf`.

5. A. Divoli, A. Spinelli, S. Chittenden, D. Dearnaley, and G. Flux: Whole-body dosimetry for targeted radionuclide therapy using spectral analysis. Cancer Biol. Radiopharm. **20**, 2005, 66–71.
6. C. Elster: Calculation of uncertainty in the presence of prior knowledge. Metrologia **44**, 2007, 111–116.
7. G.D. Flux, M.J. Guy, R. Beddows, M. Pryor, and M.A. Flower: Estimation and implications of random errors in whole-body dosimetry for targeted radionuclide therapy. Phys. Med. Biol. **47**, 2002, 3211–3223.
8. R. Francis and R.H.J. Begent: Monoclonal antibody targeting therapy: an overview. In *Targeted Therapy for Cancer* K. Syrigos and K. Harrington (eds.), Oxford University Press, Oxford, 2003, 29–46.
9. G. Golub and V. Pereyra: Separable nonlinear least squares: the variable projection method and its applications. Inverse Problems **19**, 2003, R1–R26.
10. C. Hindorf, S. Chittenden, L. Causer, V.J. Lewington, and H. Mäcke: Dosimetry for ^{90}Y-DOTATOC therapies in patients with neuroendocrine tumors. Cancer Biol. Radiopharm. **22**, 2007, 130–135.
11. IUPAC compendium of chemical terminology. `http://goldbook.iupac.org/B00658.html`.
12. E.T. Jaynes: Where do we stand on maximum entropy? In *Papers on Probability, Statistics, and Statistical Physics*, R.D. Rosenkrantz (ed.), Kluwer Academic, Dordrecht, The Netherlands, 1989, 210–314.
13. A. Mayer, R.J. Francis, S.K. Sharma, B. Tolner, C.J.S. Springer, J. Martin, G.M. Boxer, J. Bell, A.J. Green, J.A. Hartley, C. Cruickshank, J. Wren, K.A. Chester, and R.H.J. Begent: A phase I study of single administration of antibody-directed enzyme prodrug therapy with the recombinant anti-carcinoembryonic antigen antibody-enzyme fusion protein MFECP1 and a bis-iodo phenol mustard prodrug. Clin. Cancer Res. **12**, 2006, 6509–6516.
14. M.G. Stabin and J.A. Siegel: Physical models and dose factors for use in internal dose assessment. Health Phys. **85**, 2003, 294.

Towards Calculating the Basin of Attraction of Non-Smooth Dynamical Systems Using Radial Basis Functions

Peter Giesl

Department of Mathematics, University of Sussex, Falmer, BN1 9QH, UK

Summary. We consider a special type of non-smooth dynamical systems, namely $\dot{x} = f(t,x)$, where $x \in \mathbb{R}$, f is t-periodic with period T and non-smooth at $x = 0$. In [6] a sufficient Borg-like condition to determine a subset of its basin of attraction was given. The condition involves a function W and its partial derivatives; the function W is t-periodic and non-smooth at $x = 0$.

In this article, we describe a method to approximate this function W using radial basis functions. The challenges that W is non-smooth at $x = 0$ and a time-periodic function are overcome by introducing an artificial gap in x-direction and using a time-periodic kernel. The method is applied to an example which models a motor with dry friction.

1 Introduction

Non-smooth dynamical systems arise in a number of applications, for example in mechanical systems with dry friction, cf. Section 6. Compared to smooth systems, non-smooth systems show different behaviour; for example, solutions are in general not unique with respect to backward time.

We consider a special type of non-smooth dynamical systems, namely $\dot{x} = f(t,x)$, where $x \in \mathbb{R}$, f is t-periodic with period T and non-smooth at $x = 0$, i.e. f can be split into the two smooth functions $f^{\pm} \colon \mathbb{R} \times \mathbb{R}_0^{\pm}$ and the natural phase space is the cylinder $S_T^1 \times \mathbb{R}$.

In [6] a sufficient condition for existence, uniqueness and exponential stability of a periodic orbit was given, which at the same time determines a subset of its basin of attraction. The condition involves a function W, which is t-periodic and non-smooth at $x = 0$. W serves as a weight function to measure the distance between adjacent trajectories. W has to satisfy the following three conditions with constants $\nu, \epsilon > 0$ in a positively invariant subset K of the phase space $S_T^1 \times \mathbb{R}$, see also Theorem 1;

1. $f_x(t,x) + W'(t,x) \leq -\nu$ for all $(t,x) \in K$ with $x \neq 0$,
2. $\frac{f^-(t,0)}{f^+(t,0)} e^{W^-(t,0) - W^+(t,0)} \leq e^{-\epsilon}$ for all $(t,0) \in K$ with $f^-(t,0) < 0$,

E.H. Georgoulis, A. Iske, J. Levesley (eds.), *Approximation Algorithms for Complex Systems*, Springer Proceedings in Mathematics 3, DOI: 10.1007/978-3-642-16876-5_9,
© Springer-Verlag Berlin Heidelberg 2011

3. $\frac{f^+(t,0)}{f^-(t,0)} e^{W^+(t,0)-W^-(t,0)} \leq e^{-\epsilon}$ for all $(t,0) \in K$ with $f^+(t,0) > 0$,

where $W'(t,x) = W_x(t,x)f(t,x) + W_t(t,x)$ denotes the orbital derivative, i.e. the derivative along solutions.

Let us explain the meaning of these conditions, for details cf. [6]: Condition 1 means that adjacent trajectories approach each other with respect to the weighted distance. Conditions 2 and 3 relate to the jumps in the functions f and W at points $(t,0)$, where the solution changes sign from + to − (Condition 2) or from − to + (Condition 3). If we compare two adjacent trajectories near these points $(t,0)$, then the weighted distance has two contributions, on the one hand the weight function changes from W^+ to W^- or the other way round, on the other hand the two solutions have different signs for a small time interval and thus are determined by the right-hand sides f^+ and f^-, respectively. Conditions 2 and 3 ensure that those two contributions together result in a decreasing weighted distance between the two trajectories.

In this article, we will approximate the function W using radial basis functions and thus we can determine the basin of attraction of a periodic orbit in non-smooth systems. The error estimates for radial basis function approximation require a smooth target function, whereas the function W is non-smooth. Thus, we introduce an artificial gap $S^1_T \times (-v,0)$ of size $v > 0$ and consider the smooth function $V \colon S^1_T \times \mathbb{R} \to \mathbb{R}$, which satisfies

$$V(t,x) = W^+(t,x) \text{ for } x \geq 0$$
$$\text{and } V(t,x-v) = W^-(t,x) \text{ for } x \leq 0.$$

The above conditions for W will be transformed into similar conditions for V. Note that the conditions are linear (differential) operators, and thus V can be approximated using meshless collocation with radial basis functions. Since we seek to approximate functions which are periodic with respect to one variable, we use the approach in [11] which involves a time-periodic positive kernel of the form

$$\Phi(t,x) = \sum_{k\in\mathbb{Z}} \Psi(t+kT,x)$$

where Ψ is a positive definite kernel in $\mathbb{R}^2$. Note that we will use kernels Ψ associated with Wendland's compactly supported radial basis functions, such that the sum becomes finite.

Let us give an overview over the contents: In Section 2 we give an introduction to non-smooth dynamical systems, discussing Filippov solutions, periodic orbits and their basins of attractions as well as a sufficient condition for their calculation. In Section 3 we summarise the results on collocation of a time-periodic function. In Section 4 we introduce the method to approximate the non-smooth function W by using an artificial gap and a related function V. In Section 5 we discuss the collocation matrix for the approximation of V and in Section 6 we apply the method to an example. In an appendix we consider Wendland's compactly supported radial basis functions and give explicit formulae for this choice.

2 Non-Smooth Dynamical Systems

2.1 Filippov Solutions

Non-smooth dynamical systems arise in mechanical applications, e.g. through dry friction. A model for a motor with dry friction is the equation $\dot{x} = \sin \omega t - m\,\mathrm{sign} x$, where x denotes the angular velocity of a rod, $\sin \omega t$ is the periodic momentum and $-m\,\mathrm{sign} x$ models the dry friction, where $m > 0$, cf. Section 6. This is an example for a more general class of non-smooth dynamical systems that we consider in this article, namely

$$\dot{x} = f(t, x), \tag{1}$$

where $x \in \mathbb{R}$, f is t-periodic with minimal period $T > 0$ and non-smooth at $x = 0$. The cylinder $S^1_T \times \mathbb{R}$, where S^1_T denotes the circle of circumference T, is both the phase space of the dynamical system and the domain of f. We denote the solution $x(t)$ of the initial value problem (1) together with the initial condition $x(t_0) = x_0$ by $\varphi(t, t_0, x_0) := x(t)$.

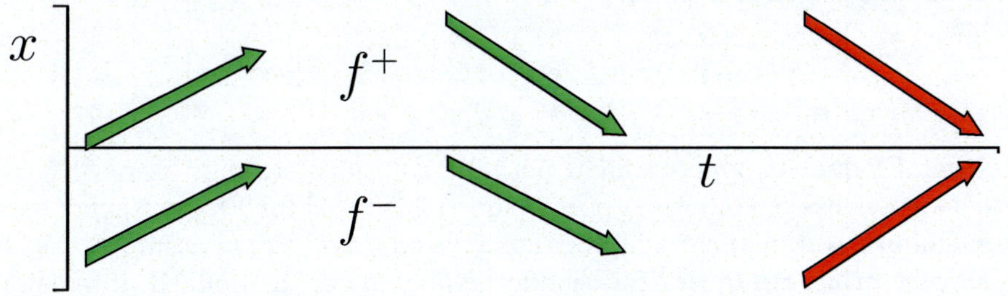

Fig. 1. The figure shows the three possible cases of signs for f^+ and f^-. Left: $f^+, f^- > 0$ and the solution moves from the negative half-plane ($x < 0$) to the positive ($x > 0$). Middle: $f^+, f^- < 0$ and the solution moves from the positive half-plane ($x > 0$) to the negative ($x < 0$). Right: $f^+ < 0$, $f^- > 0$. After the solution intersects with the t-axis, it stays on it, i.e. $x(t) = 0$; this is called a sliding motion. Note that the remaining case $f^+ > 0$, $f^- < 0$ is excluded by definition.

While solutions of smooth ordinary differential equations are unique both in forward and backward time, the situation changes if f is not smooth. Solutions of (1) are defined in the sense of Filippov [4], and we assume that at least one of the following inequalities holds: $f^+(t, 0) < 0$ or $f^-(t, 0) > 0$, where $f^+(t, 0) = \lim_{x \searrow 0} f(t, x)$, $f^-(t, 0) = \lim_{x \nearrow 0} f(t, x)$. This implies that we have unique sclutions in forward time, but not necessarily in backward time. We illustrate the three different situations of the signs of f^+ and f^- in Figure 1: in the first case, the solution moves from the negative half-plane ($x < 0$) to the positive ($x > 0$), in the second case the other way round. The solution is still continuous, but not differentiable when it crosses the t-axis.

In the third case, the solution intersects with the line $x = 0$ from below or above, and then the solution follows the t-axis, i.e. $x(t) = 0$. Note that this is neither the solution of $\dot{x} = f^+(t, 0)$ nor of $\dot{x} = f^-(t, 0)$, but of $\dot{x} = 0$; this is called a *sliding motion*. In this case we have non-uniqueness in backward time since we have no information about when the sliding motion has started.

2.2 Periodic Solution

We are interested in periodic solutions and their stability. The definitions are similar to the smooth case.

Definition 1 (Periodic orbit). *A* periodic solution with period T *of (1) is a solution* $\varphi(t, t_0, x_0)$*, such that* $\varphi(t + T, t_0, x_0) = \varphi(t, t_0, x_0)$ *holds for all* $t \in \mathbb{R}$. *The set* $\Omega = \{(t, \varphi(t, t_0, x_0)) \in S^1_T \times \mathbb{R}\}$ *is called a* periodic orbit.

A periodic orbit $\Omega = \{(t, \varphi(t, t_0, x_0)) \in S^1_T \times \mathbb{R}\}$ *is called* exponentially stable *with exponent* $-\nu$, *if it is*

- orbitally stable, *i.e. for all* $\epsilon > 0$ *there is a* $\delta > 0$ *such that*
$$\operatorname{dist}((t \bmod T, \varphi(t, t_0, y_0)), \Omega) < \epsilon$$
for all $|y_0 - x_0| \leq \delta$ *and all* $t \geq 0$,
- exponentially attractive, *i.e. for all* $\iota > 0$ *there are* $\delta' > 0$ *and* $C > 0$, *such that*
$$|\varphi(t, t_0, y_0) - \varphi(t, t_0, x_0)| \leq Ce^{(-\nu+\iota)t}|y_0 - x_0|$$
holds for all y_0 *with* $|y_0 - x_0| \leq \delta'$ *and all* $t \geq 0$.

Note that Floquet exponents for a periodic solution (see e.g. [2] or [13] for Floquet exponents of smooth systems, and [14] and [7] for Floquet exponents of non-smooth systems) can only be defined under additional assumptions on the periodic orbit. However, if a Floquet exponent can be defined, it is equal to the maximal exponent $-\nu$ in Definition 1.

Definition 2 (Basin of attraction). *The* basin of attraction $A(\Omega)$ *of an exponentially stable periodic orbit* $\Omega \subseteq S^1_T \times \mathbb{R}$ *is the set*

$$A(\Omega) := \left\{(t_0, x_0) \in S^1_T \times \mathbb{R} \mid \operatorname{dist}((t \bmod T, \varphi(t, t_0, x_0)), \Omega) \stackrel{t \to \infty}{\longrightarrow} 0\right\}.$$

In [6] the following sufficient condition for existence, uniqueness and asymptotic stability of a periodic orbit was given, which, at the same time, shows that K is a subset of its basin of attraction. This is a generalisation of Borg's criterion [1]. Note that Condition 1 of the following Theorem 1 means that adjacent solutions $\varphi(t, t_0, x_0)$ and $\varphi(t, t_0, y_0)$, where $|x_0 - y_0|$ is small, approach each other. More precisely, the weighted distance

$$d(t) = e^{W(t, \varphi(t, t_0, x_0))}|\varphi(t, t_0, x_0) - \varphi(t, t_0, y_0)|$$

is decreasing exponentially. The Conditions 2 and 3 take account of intervals where the solutions $\varphi(t, t_0, x_0)$ and $\varphi(t, t_0, y_0)$ have different signs; for more details cf. [6].

Theorem 1 ([6], Theorem 3.3). *Consider $\dot{x} = f(t,x)$, where $f \in C^1(\mathbb{R} \times (\mathbb{R} \setminus \{0\}), \mathbb{R})$ and $f(t,x) = f(t+T,x)$ for all $(t,x) \in \mathbb{R} \times \mathbb{R}$. Moreover assume that each of the functions $f^{\pm}(t,x) := f(t,x)$ with $x > (<) 0$ can be extended to a continuous function $f^{\pm}(t,x)$ up to $x = 0$. Let the same hold for the derivatives $f_x^{\pm}(t,x)$. Finally, let $f^+(t,0) - f^-(t,0)$ be a C^1-function with respect to t, and assume that for all $t \in [0,T]$ at least one of the following inequalities holds: $f^+(t,0) < 0$ or $f^-(t,0) > 0$.*

Assume furthermore that $W^{\pm} \colon \mathbb{R} \times \mathbb{R}_0^{\pm}$ are continuous functions with $W^{\pm}(t+T,x) = W^{\pm}(t,x)$ for all $(t,x) \in \mathbb{R} \times \mathbb{R}_0^{\pm}$. Let the orbital derivatives $W' = (W^{\pm})'$ exist, be continuous functions in $\mathbb{R} \times \mathbb{R}^{\pm}$, and be extendable to $\mathbb{R} \times \mathbb{R}_0^{\pm}$ in a continuous way. Note that the orbital derivative is defined by $W'(t,x) = W_x(t,x) f(t,x) + W_t(t,x)$.

Let $K \subseteq S_T^1 \times \mathbb{R}$ be a nonempty, connected, positively invariant and compact set, such that the following three conditions hold with constants $\nu, \epsilon > 0$;

1. *$f_x(t,x) + W'(t,x) \leq -\nu < 0$ for all $(t,x) \in K$ with $x \neq 0$,*
2. *$\frac{f^+(t,0)}{f^-(t,0)} e^{W^+(t,0) - W^-(t,0)} \leq e^{-\epsilon} < 1$ for all $(t,0) \in K$ with $f^+(t,0) > 0$,*
3. *$\frac{f^-(t,0)}{f^+(t,0)} e^{W^-(t,0) - W^+(t,0)} \leq e^{-\epsilon} < 1$ for all $(t,0) \in K$ with $f^-(t,0) < 0$.*

Then there is one and only one periodic orbit Ω with period T in K. Ω is exponentially stable with exponent $-\nu$ and for its basin of attraction we have the inclusion $K \subseteq A(\Omega)$.

The goal of this paper is to construct such a function W using radial basis functions. The main problem is that the error estimates for radial basis functions require a smooth target function whereas the function W is non-smooth. This problem will be addressed in Section 4.

3 Collocation by Radial Basis Functions

Radial basis functions are a powerful tool to approximate solutions of linear PDEs. In this article, we will use the symmetric approach for this generalised interpolation which was developed in [3, 5, 15, 19], and also see [18]. A generalisation with application to dynamical systems, in particular the construction of Lyapunov functions for equilibria in autonomous systems, can be found in [8, 10]. The generalised interpolation of a function $V(t,x)$ which is periodic with respect to the time variable t with application to the construction of Lyapunov functions for a periodic orbit in time-periodic systems using meshless collocation was developed in [11, 12]. We will describe the collocation of a time-periodic function in the following; for details we refer the reader to [11].

Consider the linear operator L, which can, for example, be a differential operator. We assume that it is of order ≤ 1, i.e. it involves only derivatives up to order one. We also restrict ourselves to the case $x \in \mathbb{R}$, although the theory is available for general $x \in \mathbb{R}^n$. We wish to approximately solve the following equation for V

$$LV(t,x) = g(t,x), \tag{2}$$

where g is a given function.

We define a reproducing kernel Hilbert space with a positive definite kernel. This will ensure that the interpolation problem leads to a system of linear equations with a positive definite matrix and thus has a unique solution. We take into account that the functions are periodic with respect to t. For simplicity we restrict ourselves to the period $T = 2\pi$ in this section and denote $S^1_{2\pi}$ by S^1.

We give the following definition of a positive definite, periodic function.

Definition 3 ([11], Definition 3.6). *A function $\Phi : S^1 \times \mathbb{R} \to \mathbb{R}$, periodic in t, is called positive definite if for all choices of pairwise distinct points $(t_j, x_j) \in S^1 \times \mathbb{R}$, $1 \le j \le N$, and all $\alpha \in \mathbb{R}^N \setminus \{0\}$, we have*

$$\sum_{j,k=1}^{N} \alpha_j \alpha_k \Phi(t_j - t_k, x_j - x_k) > 0.$$

Positive definite functions are often characterised using Fourier transform. Since our functions are periodic in their t argument, the appropriate form of the Fourier transform of a function $\Phi : S^1 \times \mathbb{R} \to \mathbb{R}$ is defined by

$$\widehat{\Phi}_\ell(\omega) := (2\pi)^{-2} \int_0^{2\pi} \int_{\mathbb{R}} \Phi(t,x) e^{-i\ell t} e^{-ix\omega}\, dx\, dt,$$

where $\ell \in \mathbb{Z}$ and $\omega \in \mathbb{R}$. This is a discrete Fourier transform with respect to t and a continuous one with respect to x. The inverse Fourier transform is then given by

$$\Phi(t,x) = \sum_{\ell \in \mathbb{Z}} \int_{\mathbb{R}} \widehat{\Phi}_\ell(\omega) e^{i(\ell t + x\omega)}\, d\omega.$$

If the Fourier transform $\widehat{\Phi}_\ell(\omega)$ is positive for all $\ell \in \mathbb{Z}$ and all $\omega \in \mathbb{R}$, then the function Φ is positive definite, cf. [11, Lemma 3.7].

In [11] it was also shown that a t-periodic positive definite kernel can be constructed from a positive definite function $\Psi \colon \mathbb{R}^2 \to \mathbb{R}$ by making it periodic in the first argument:

$$\Phi(t,x) = \sum_{k \in \mathbb{Z}} \Psi(t + 2\pi k, x). \tag{3}$$

Note that this sum is finite if Ψ has compact support.

The associated reproducing kernel Hilbert space for a kernel $\Phi(t,x)$ with positive Fourier transform $\widehat{\Phi}_\ell(\omega)$, where $\ell \in \mathbb{Z}$ and $\omega \in \mathbb{R}$, can be defined by

$$\mathcal{N}_\Phi(S^1 \times \mathbb{R}) := \left\{ g : \sum_{\ell \in \mathbb{Z}} \int_{\mathbb{R}} \frac{|\widehat{g}_\ell(\omega)|^2}{\widehat{\Phi}_\ell(\omega)} d\omega < \infty \right\}.$$

The space is a Hilbert space with the inner product

$$(g,h)_{\mathcal{N}_\Phi} := \sum_{\ell\in\mathbb{Z}} \int_{\mathbb{R}} \frac{\widehat{g}_\ell(\omega)\overline{\widehat{h}_\ell(\omega)}}{\widehat{\Phi}_\ell(\omega)} d\omega.$$

Now, suppose we are given a kernel possessing a Fourier transform $\widehat{\Phi}_\ell(\omega)$ behaving like

$$c_1(1+\ell^2+\omega^2)^{-\tau} \leq \widehat{\Phi}_\ell(\omega) \leq c_2(1+\ell^2+\omega^2)^{-\tau} \tag{4}$$

with $0 < c_1 \leq c_2$. Then, according to [11, Lemma 3.5], see also [11, Section 3.3], the associated function space $\mathcal{N}_\Phi(S^1 \times \mathbb{R})$ is norm equivalent to the Sobolev space $\widetilde{W}_2^\tau(S^1 \times \mathbb{R})$ of functions which are periodic in t; for the definition of this space see [11, Section 3.1].

Typical kernels satisfying (4) are Wendland's compactly supported radial basis functions $\Psi(\widetilde{x}) = \psi_{l,k}(\|\widetilde{x}\|)$, cf. Definition 4. Here, $\widetilde{x} = (t,x) \in \mathbb{R} \times \mathbb{R}$, $k \in \mathbb{N}$ is the smoothness index of the compactly supported Wendland function and $l = k+2$. Then (4) holds for the kernel (3) with $\tau = k + 3/2$.

Definition 4 (Wendland functions, [16, 17]). *Let $l \in \mathbb{N}$, $k \in \mathbb{N}_0$. We define by recursion*

$$\psi_{l,0}(r) = (1-r)_+^l$$

$$\text{and } \psi_{l,k+1}(r) = \int_r^1 s\psi_{l,k}(s)\,ds$$

for $r \in \mathbb{R}_0^+$. Here we set $x_+ = x$ for $x \geq 0$ and $x_+ = 0$ for $x < 0$.

In the following we will describe the approach for the approximation of the solutions of (2). First, points $\widetilde{x}_j := (t_j, x_j) \in S^1 \times \mathbb{R}$, $j = 1, \ldots, N$ are chosen and the linear functionals $\lambda_j := \delta_{(t_j,x_j)} \circ L$ are defined. If these functionals are linearly independent over the reproducing kernel Hilbert space $\mathcal{N}_\Phi$, then the following theorem holds true.

Theorem 2 ([18], Theorem 16.1). *Suppose $\mathcal{N}_\Phi$ is a reproducing kernel Hilbert space with reproducing kernel Φ. Suppose further, that there are linearly independent linear functionals $\lambda_1, \ldots, \lambda_N \in \mathcal{N}_\Phi^*$. Then, to every $V \in \mathcal{N}_\Phi$, there exists one and only one norm-minimal generalized interpolant s_V, i.e. s_V is the unique solution to*

$$\min\{\|s\|_{\mathcal{N}_\Phi} : s \in \mathcal{N}_\Phi \text{ with } \lambda_j(s) = \lambda_j(V)\}.$$

Moreover, s_V has the representation

$$s_V(\tilde{x}) = \sum_{j=1}^N \beta_j \lambda_j^{\tilde{y}} \Phi(\tilde{x} - \tilde{y}), \tag{5}$$

where the coefficients are determined by solving the linear system $\lambda_i(s_V) = \lambda_i(V)$, $1 \leq i \leq N$ and $\tilde{x}, \tilde{y} \in S^1 \times \mathbb{R}$.

Remark 1. The linear system is given by $A\beta = \gamma$, where $A = (a_{jl})_{j,l=1,\dots,N}$ with $a_{jl} = \lambda_j^{\tilde{x}} \lambda_l^{\tilde{y}} \Phi(\tilde{x} - \tilde{y})$, $\tilde{x}_j = (t_j, x_j) \in S^1 \times \mathbb{R}$, and $\gamma = (\gamma_i)_{i=1,\dots,N}$ with $\gamma_i = \lambda_i(V) = g(t_i, x_i)$ according to (2). Since A is positive definite if Φ is a positive definite kernel, the linear system $A\beta = \gamma$ has a unique solution β which determines s_V by (5).

Hence, we have to show that the linear functionals $\lambda_j := \delta_{(t_j,x_j)} \circ L$ are linearly independent over a sufficiently smooth Sobolev space. Then the error analysis from [11] for $LV - Ls_V$ which vanishes at the collocation points yields the following result in Theorem 3.

To measure the quality of our approximants we will use *mesh norms.* Let $\widetilde{K} \subseteq S^1 \times \mathbb{R}$ and assume that the points $\widetilde{X} = \{\tilde{x}_j := (t_j, x_j) \in S^1 \times \mathbb{R}, j = 1\dots, N\}$ lie in $\widetilde{K}$, i.e. $\widetilde{X} \subseteq \widetilde{K}$. The quantity $h_{\widetilde{X},\widetilde{K}} = \sup_{\tilde{x}\in\widetilde{K}} \min_{\tilde{x}_j\in X} \|\tilde{x} - \tilde{x}_j\|$ measures how well $\widetilde{X}$ is distributed over $\widetilde{K}$. However, since $\widetilde{K}$ is periodic in the t variable, it is more natural to use the measure

$$\widetilde{h}_{\widetilde{X},\widetilde{K}} := \sup_{\tilde{x}\in\widetilde{K}} \min_{\tilde{x}_j\in\widetilde{X}} \|\tilde{x} - \tilde{x}_j\|^c$$

where the "cylinder"-norm is defined by $\|\tilde{x}\|^c = ((t \bmod 2\pi)^2 + x^2)^{1/2}$ and $t \bmod 2\pi \in [-\pi, \pi)$.

Theorem 3. *Denote by k the smoothness index of the compactly supported Wendland function. Let $k > 1$. Set $\tau = k + 3/2$ and $\sigma = \lceil \tau \rceil$. Consider the kernel $\Phi(t,x) = \sum_{k\in\mathbb{Z}} \Psi(t + 2\pi k, x)$, where $\Psi(\tilde{x}) = \psi_{l,k}(c\|\tilde{x}\|)$, $c > 0$ and $\psi_{l,k}$ is defined in Definition 4. Assume that the functionals $\lambda_j := \delta_{(t_j,x_j)} \circ L \in \mathcal{N}_\Phi^*$, $j = 1, \dots, N$ are linearly independent and of order at most 1. Furthermore, assume that the solution V of*

$$LV(t,x) = g(t,x)$$

satisfies $V \in C^\sigma(S^1 \times \mathbb{R}, \mathbb{R})$.

Then the reconstruction s_V of V with respect to the set $\widetilde{X} \subseteq \widetilde{K} := \{(t,x) \in S^1 \times \mathbb{R}\}$, where $\widetilde{K}$ has a C^σ-boundary, satisfies

$$\|LV - Ls_V\|_{L_\infty(\widetilde{K})} \leq C \widetilde{h}_{\widetilde{X},\widetilde{K}}^{k-\frac{1}{2}} \|V\|_{\widetilde{W}_2^{k+3/2}(\widetilde{K})}.$$

The proof of this theorem is similar to the proof of [11, Corollary 3.21].

4 Approximation of the Weight Function

4.1 Artificial Gap

The main problem of applying meshless collocation to approximate the function W of Theorem 1 in Section 2, is that W is non-smooth at $x = 0$, but the error estimates for approximation with radial basis functions require that the approximated function W is smooth.

We solve this problem by introducing an artificial gap of width $v > 0$ between W^+ and W^- allowing for a smooth function, cf. Figure 2. More precisely, we define a smooth function $V\colon S^1_T \times \mathbb{R} \to \mathbb{R}$ in the following way.

Definition 5 (of V and B). *Fix $v > 0$. Assume that the set $K \subseteq S^1_T \times \mathbb{R}$ and the function $W\colon K \to \mathbb{R}$ are given, where W is non-smooth at $x = 0$. Define $B = B^+ \cup B^-$, where $B^+ = \{(t,x) \in K \mid x \geq 0\}$ and $B^- = \{(t, x-v) \mid (t,x) \in K, x \leq 0\}$.*

Further define $V\colon B \subseteq S^1_T \times \mathbb{R} \to \mathbb{R}$ by

$$V(t,x) := W^+(t,x) \text{ for } x \geq 0, x \in K, \tag{6}$$

$$V(t,x-v) := W^-(t,x) \text{ for } x \leq 0, x \in K. \tag{7}$$

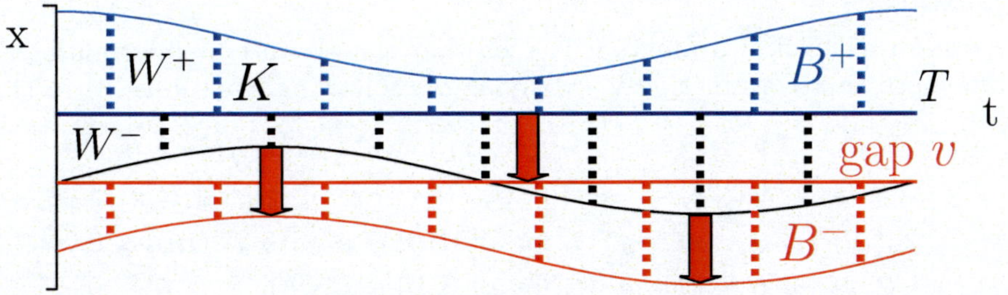

Fig. 2. The functions W^+ and W^- are defined in $x > 0$, $x < 0$, respectively. The set $K \subseteq S^1_T \times \mathbb{R}$ is transformed by introducing an artificial gap in x-direction of size v. The previous area $K^- = \{(t,x) \in K \mid x \leq 0\}$ is shifted to $B^- := \{(t, x-v) \mid (t,x) \in K, x \leq 0\}$. The function V on B^- is defined by the corresponding values of W in K^- before the shift.

In the newly introduced area $S^1_T \times (-v, 0)$, the function V can bridge the jump between $W^-(t,0)$ and $W^+(t,0)$. As a consequence, there is a smooth function V satisfying (6)-(7).

We translate the conditions on W in Theorem 1 into the corresponding ones for V. The three conditions become the following four conditions 1.

Conditions 1 *Let $\nu, \epsilon > 0$ and assume that*

1. $f^+_x(t,x) + V_t(t,x) + V_x(t,x) f^+(t,x) \leq -\nu$ *for all* $(t,x) \in B$ *with* $x > 0$,
2. $f^-_x(t,x+v) + V_t(t,x) + V_x(t,x) f^-(t,x+v) \leq -\nu$ *for all* $(t,x) \in B$ *with* $x < -v$,
3. $V(t,0) - V(t,-v) + \ln\left(\frac{f^+(t,0)}{f^-(t,0)}\right) \leq -\epsilon$ *for all* $(t,0) \in B$ *with* $f^+(t,0) > 0$,
4. $V(t,-v) - V(t,0) + \ln\left(\frac{f^-(t,0)}{f^+(t,0)}\right) \leq -\epsilon$ *for all* $(t,0) \in B$ *with* $f^-(t,0) < 0$.

Note that due to the assumptions of Theorem 1, $f^+(t,0) > 0$ implies $f^-(t,0) > 0$ and thus $\frac{f^+(t,0)}{f^-(t,0)} > 0$. Similarly, $f^-(t,0) < 0$ implies $f^+(t,0) < 0$ and thus $\frac{f^-(t,0)}{f^+(t,0)} > 0$.

The existence and smoothness of such functions have been discussed in [7] under some additional assumptions on the periodic orbit which ensure that a Floquet exponent $-\nu_0$ can be defined. Note that in non-smooth systems, the Floquet exponent may also be $-\nu_0 = -\infty$, but here we assume that the Floquet exponent $-\nu_0$ exists, is finite and negative.

The construction of W in [7] and thus also of V starts on the periodic orbit. Here, ν and ϵ of Conditions 1 have to satisfy a certain relation. More precisely, if we define $0 < \nu < \nu_0$ then ϵ is defined by

$$\epsilon = \frac{(\nu_0 - \nu)T}{L}, \tag{8}$$

where T denotes the period and L the number of changes of sign of the periodic orbit in one period. We will discuss how to obtain these quantities in practice in Section 4.2.

In order to find a function V such that the above four inequalities hold, we approximate a function V satisfying the following equations (9) to (12),

$$LV(t,x) = -\nu - f_x^+(t,x) \text{ for all } (t,x) \in B, x > 0, \tag{9}$$
$$L^v V(t,x) = -\nu - f_x^-(t,x+v) \text{ for all } (t,x) \in B, x < -v, \tag{10}$$
$$V(t,0) - V(t,-v) = g^+(t) \text{ for all } (t,0) \in B \text{ with } f^+(t,0) > 0, \tag{11}$$
$$V(t,0) - V(t,-v) = g^-(t) \text{ for all } (t,0) \in B \text{ with } f^-(t,0) < 0, \tag{12}$$

where the functions $g^+(t)$ and $g^-(t)$ are specified later in Section 4.3; at this point we expect $g^+(t) = -\epsilon - \ln\left(\frac{f^+(t,0)}{f^-(t,0)}\right)$ and $g^-(t) = \epsilon + \ln\left(\frac{f^-(t,0)}{f^+(t,0)}\right)$.

Note that the left-hand sides are all linear operators of order at most 1 applied to V. The first-order differential operators L and L^v in (9) and (10) are defined by

$$LV(t,x) = V_t(t,x) + V_x(t,x)f^+(t,x), \tag{13}$$
$$L^v V(t,x) = V_t(t,x) + V_x(t,x)f^-(t,x+v). \tag{14}$$

4.2 Determination of ν and ϵ

Before we start the approximation, we use a simple Euler method to approximate the periodic orbit $\tilde{x}(t)$. The solution $y(t)$ of the first variation equation $\frac{d}{dt}y(t) = f_x(t,\tilde{x}(t))y(t)$ along the periodic orbit together with the jumps when crossing $x = 0$ gives us an approximation of the Floquet exponent $-\nu_0$ through, cf. [7, Equation (3)],

$$-\nu_0 T = \int_0^T f_x(\tau,\tilde{x}(\tau))\,d\tau + \sum_{i=1}^{L} \ln L_i,$$

where T denotes the period and L the number of changes of sign of the periodic orbit in one period. L_i is defined by $L_i = \frac{f^+(t_i,0)}{f^-(t_i,0)}$ if the periodic orbit changes sign at t_i from $-$ to $+$ and by $L_i = \frac{f^-(t_i,0)}{f^+(t_i,0)}$ if the periodic orbit changes sign at t_i from $+$ to $-$. Note that a sliding motion, cf. Figure 1, cannot occur on the periodic orbit, because this would lead to the Floquet exponent $-\nu_0 = -\infty$ which we have excluded. If $L = 0$, then the periodic orbit is completely in one area of sign and thus in a smooth system; we do not consider this case further.

Then we choose $0 < \nu < \nu_0$ and

$$\epsilon := \frac{(\nu_0 - \nu)T}{L} > 0$$

in accordance with (8). The nearer we choose ν to the Floquet exponent ν_0, the smaller we can choose ϵ, i.e. 3 and 4 in Conditions 1 become less restrictive.

4.3 Jump Conditions and Breakpoint

Concerning the two equations (11) and (12), we expect $g^+(t) = -\epsilon - \ln\left(\frac{f^+(t,0)}{f^-(t,0)}\right)$ and $g^-(t) = \epsilon + \ln\left(\frac{f^-(t,0)}{f^+(t,0)}\right)$. But towards the boundary of the interval where $f^+(t,0) > 0$ we have $f^+(t,0) \to 0$, i.e. $-\ln\left(\frac{f^+(t,0)}{f^-(t,0)}\right) \to \infty$.

The choice of ν and ϵ implies that we have to use the above functions at points where the periodic orbit crosses the axis $x = 0$; for other points the function g^+ and g^- can be chosen differently, if the inequalities in 3 and 4 of Conditions 1 are satisfied.

We choose a number $b \in \mathbb{R}^+$, the *breakpoint*, and define

$$g^+(t) = \begin{cases} -\epsilon - \ln\left(\frac{f^+(t,0)}{f^-(t,0)}\right), & \text{if } \frac{f^+(t,0)}{f^-(t,0)} > b, \\ -\epsilon - \ln b, & \text{otherwise,} \end{cases}$$

$$g^-(t) = \begin{cases} \epsilon + \ln\left(\frac{f^-(t,0)}{f^+(t,0)}\right), & \text{if } \frac{f^-(t,0)}{f^+(t,0)} > b, \\ \epsilon + \ln b & \text{otherwise.} \end{cases}$$

Note that then the inequality in 3 of Conditions 1 is satisfied since for $\frac{f^+(t,0)}{f^-(t,0)} \leq b$ we have $g^+(t) + \ln\left(\frac{f^+(t,0)}{f^-(t,0)}\right) \leq -\epsilon - \ln b + \ln b = -\epsilon$; a similar argument holds for 4 of Conditions 1.

The breakpoint b should be small enough so that g^+ and g^- are defined according to the first line in the definition above at points where the periodic orbit crosses the t-axis. On the other hand, it should be defined according to the second line in the definition above at the beginning and end of intervals where $f^+(t,0) > 0$, $f^-(t,0) < 0$, respectively, cf. the example in Section 6.

5 Approximation of V

5.1 Collocation Matrix

We fix a positive definite radial basis function $\Psi\colon \mathbb{R}^2 \to \mathbb{R}$, given by $\Psi(t,x) = \psi(\|(t,x)\|)$, where $\psi\colon \mathbb{R} \to \mathbb{R}$. In this article, we use $\psi = \psi_{k+2,k}$ where the function $\psi_{k+2,k}$ is a Wendland function, cf. Definition 4, which has compact support. From this function we construct a positive definite function which is T-periodic in its first argument by setting

$$\Phi(t,x) = \sum_{k\in\mathbb{Z}} \Psi(t+kT,x). \tag{15}$$

Note that the sum is finite, since Ψ has compact support.

We choose pairwise different points $(t_1,x_1),\ldots,(t_{n^+},x_{n^+}) \in B$ with $x_j \geq 0$ for $1 \leq j \leq n^+$ and $(t_{n^++1},x_{n^++1}),\ldots,(t_{n^++n^-},x_{n^++n^-}) \in B$ with $x_j \leq -v$ for $n^++1 \leq j \leq n^++n^-$. Furthermore, we choose points $(\tau_1,0),\ldots,(\tau_{N^+},0)$ with $f^+(\tau_j,0) > 0$ and $(\tau_{N^++1},0),\ldots,(\tau_{N^++N^-},0)$ with $f^-(\tau_j,0) < 0$. We always assume that $t_j,\tau_j \in [0,T]$ and denote $N := N^+ + N^-$.

The ansatz for $s_V\colon S^1_T \times \mathbb{R} \to \mathbb{R}$ as an approximation for $V(t,x)$ satisfying (9) to (12) is given by, cf. (5),

$$\begin{aligned} s_V(\tilde{x}) &= \sum_{j=1}^{n^+} c_j(\delta_{(t_j,x_j)} \circ L)^{\tilde{y}} \Phi(\tilde{x}-\tilde{y}) + \sum_{j=n^++1}^{n^++n^-} c_j(\delta_{(t_j,x_j)} \circ L^v)^{\tilde{y}} \Phi(\tilde{x}-\tilde{y}) \\ &\quad + \sum_{j=1}^{N} d_j[\Phi(\tilde{x}-(\tau_j,0)) - \Phi(\tilde{x}-(\tau_j,-v))], \end{aligned} \tag{16}$$

where the linear operators L and L^v were defined in (13) and (14), and $\tilde{x} = (t,x)$ and $\tilde{y} = (s,y)$.

The coefficients c and d are determined by solving the linear system given by the conditions, cf. (9) to (12),

$$M \begin{pmatrix} c \\ d \end{pmatrix} = \begin{pmatrix} \alpha \\ \beta \\ \gamma \\ \delta \end{pmatrix} \tag{17}$$

where $M = \begin{pmatrix} M_{11} & M_{12} & M_{13} \\ M_{21} & M_{22} & M_{23} \\ M_{31} & M_{32} & M_{33} \end{pmatrix}$ is a symmetric matrix with

$$
\begin{aligned}
(M_{11})_{ij} &= (\delta_{(t_i,x_i)} \circ L)^{\tilde{x}} (\delta_{(t_j,x_j)} \circ L)^{\tilde{y}} \Phi(\tilde{x} - \tilde{y}), \\
&\qquad \text{where } i,j = 1,\ldots,n^+, \\
(M_{12})_{ij} &= (\delta_{(t_i,x_i)} \circ L)^{\tilde{x}} (\delta_{(t_j,x_j)} \circ L^v)^{\tilde{y}} \Phi(\tilde{x} - \tilde{y}), \\
&\qquad \text{where } i = 1,\ldots,n^+, j = n^+ + 1,\ldots,n^+ + n^-, \\
(M_{13})_{ij} &= (\delta_{(t_i,x_i)} \circ L)^{\tilde{x}} [\Phi(\tilde{x} - (\tau_j, 0)) - \Phi(\tilde{x} - (\tau_j, -v))], \\
&\qquad \text{where } i = 1,\ldots,n^+, j = 1,\ldots,N, \\
(M_{22})_{ij} &= (\delta_{(t_i,x_i)} \circ L^v)^{\tilde{x}} (\delta_{(t_j,x_j)} \circ L^v)^{\tilde{y}} \Phi(\tilde{x} - \tilde{y}), \\
&\qquad \text{where } i,j = n^+ + 1,\ldots,n^+ + n^-, \\
(M_{23})_{ij} &= (\delta_{(t_i,x_i)} \circ L^v)^{\tilde{x}} [\Phi(\tilde{x} - (\tau_j, 0)) - \Phi(\tilde{x} - (\tau_j, -v))], \\
&\qquad \text{where } i = n^+ + 1,\ldots,n^+ + n^-, j = 1,\ldots,N, \\
(M_{33})_{ij} &= 2\Phi(\tau_i - \tau_j, 0) - \Phi(\tau_i - \tau_j, v) - \Phi(\tau_i - \tau_j, -v), \\
&\qquad \text{where } i,j = 1,\ldots,N.
\end{aligned}
$$

Detailed formulae of the matrix elements can be found in the Appendix A.

The right-hand sides are given by, cf. (9) to (12),

$$
\begin{aligned}
\alpha_i &= -\nu - f_x^+(t_i, x_i), i = 1\ldots,n^+, \\
\beta_i &= -\nu - f_x^-(t_{n^++i}, x_{n^++i} + v), i = 1\ldots,n^-, \\
\gamma_i &= g^+(\tau_i), i = 1\ldots,N^+, \\
\delta_i &= g^-(\tau_i), i = N^+ + 1\ldots,N^+ + N^-.
\end{aligned}
$$

The approximant s_V is given by (16). For more detailed formulae of s_V and the Conditions 1 that we need to check, see Appendix B.

5.2 Error Analysis

The linear functionals are linearly independent provided that the points are pairwise different. This relies on the fact the the operators L and L^v have no singular points, cf. (13) and (14), due to the coefficient 1 of V_t. One can show the linear independence for the operators in (11) and (12) similarly to the proof in [9, Proposition 3.9]; see also [8, Proposition 3.29] for a combination of different linear operators. Theorem 3 thus guarantees that the approximant s_V satisfies the inequalities 1. to 4. of Conditions 1 provided that the meshes are dense enough and V is smooth enough.

6 Example: Dry Friction

We consider the example [6, Section 4.1], namely

$$\dot{x} = \sin t - m \operatorname{sign}(x).$$

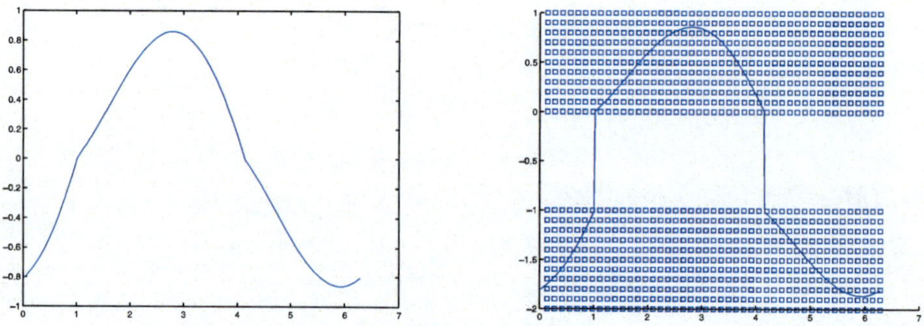

Fig. 3. Left: The periodic orbit in the interval $[0, 2\pi]$. There are two points where the periodic orbit changes sign. Right: The periodic orbit after the introduction of the gap of size $v = 1$ together with the $n^+ + n^- = 990$ points (t_j, x_j) used for the generalised interpolation with respect to the operators L and L^v.

The term $-m\,\mathrm{sign}(x)$ models dry friction and is non-smooth at $x = 0$. We choose the value $m = 0.35$ which gives an exponentially stable periodic orbit with Floquet exponent $-\nu_0 = -0.2840$ and $L = 2$, i.e. two changes of sign in one period $T = 2\pi$, cf. Figure 3. We choose $\nu = 0.8 \cdot \nu_0 = 0.2272$ which results in $\epsilon = 0.1785$ using (8). We use equally distributed points $(\tau_j, 0)$, dividing $[0, T]$ into 40 pieces and deciding for each point whether it satisfies $f^+(\tau_i, 0) > 0$, $f^-(\tau_i, 0) < 0$ or neither. We arrive at $N^+ = N^- = 15$. We used the breakpoint $b = 0.3$, which results in the vector

$$\gamma = (1.0255, 1.0255, 0.9068, 0.7478, 0.6518, 0.5938, 0.5624, 0.5524,\\ 0.5624, 0.5938, 0.6518, 0.7478, 0.9068, 1.0255, 1.0255).$$

The equal numbers at the beginning and end of γ indicate that we cut the values just at the end, and that we used the correct definition on the periodic orbit.

The gap introduced is of width $v = 1$. The points for the orbital derivative for $x \geq 0$ are chosen on a grid (t_j, x_j) where $x_j \in \{0, \delta, 2\delta, \ldots, 1\}$ and t_j are 45 equally distributed in the interval $(0, T]$ and $\delta = \frac{1}{10}$, which results in $n^+ = 495$.

The points for the orbital derivative for $x < 0$ are chosen on a grid $(t_j, x_j - v)$ where $x_j \in \{-1, \ldots, -2\delta, -\delta, 0\}$ and t_j are 45 equally distributed in the interval $(0, T]$ and $\delta = \frac{1}{10}$, which results in $n^- = 495$.

We use Wendland's compactly supported radial basis function $\psi_{4,2}(0.7 \cdot r)$, where $\psi_{4,2}(r) = (1-r)^6(35r^2 + 18r + 3)$. The choice of the scaling factor 0.7 in the radial basis function corresponds to the distance of the grid points: if the factor is too large, then the support of the radial basis function centered at a grid point only includes this one grid point and the approximation is bad; if it is too small then the collocation matrix has a bad condition number.

Figure 3 shows the periodic orbit which crosses the t-axis twice in one period (left). The right-hand side figure shows the same periodic orbit after the introduction of the artificial gap together with the points (t_j, x_j) used for the collocation of the differential operators L and L^v. Figure 4 (left) shows 3 and 4 of Conditions 1 for the approximated function s_V: all values are negative, and they are bounded by approximately $-\epsilon$. Towards the beginning and end of the intervals where $f^+(t,) > 0$, $f^-(t,0) < 0$, respectively, the values drop below $-\epsilon$ as a consequence of the introduced breakpoint.

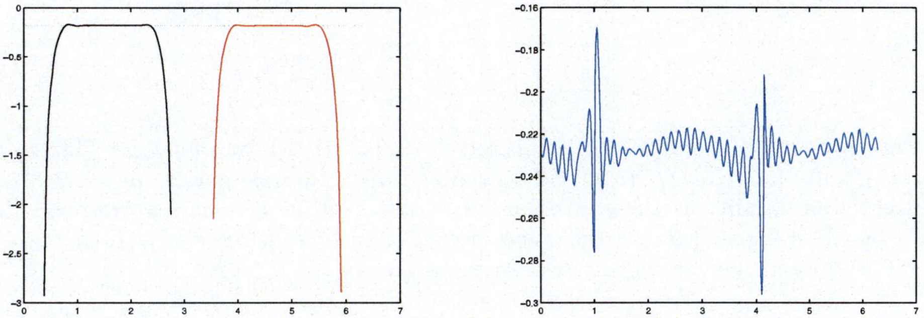

Fig. 4. Left: 3 and 4 of Conditions 1 for s_V are checked. The points in black on the t-axis are the points t where $f^+(t,0) > 0$, the values are $s_V(t,0) - s_V(t,-v) + \ln\left(\frac{f^+(t,0)}{f^-(t,0)}\right)$ and they are negative, bounded approximately by $-\epsilon = -0.1785$. The points in red on the t-axis are the points t where $f^-(t,0) < 0$, the values are $s_V(t,-v) - s_V(t,0) + \ln\left(\frac{f^-(t,0)}{f^+(t,0)}\right)$ and they are negative, bounded approximately by $-\epsilon = -0.1785$. Right: 1 and 2 of Conditions 1 are evaluated along the periodic orbit. The values are negative and approximately $-\nu = -0.2272$; note that the largest variation occurs at the points where the periodic orbit changes sign, cf. Figure 3.

Figures 4 (right) and 5 illustrate 1 and 2 of Conditions 1. In Figure 5 the values of $f_x^+(t,x) + (s_V)_t(t,x) + (s_V)_x(t,x)f^+(t,x)$ are shown in the positive plane $S_T^1 \times \mathbb{R}^+$ (left) and the values of $f_x^-(t,x) + (s_V)_t(t,x-v) + (s_V)_x(t,x-v)f^-(t,x)$ are shown in the negative plane $S_T^1 \times \mathbb{R}^-$ (right). The values are all negative, mostly they are approximately $-\nu = -0.2272$. Larger variations occur near the non-smooth axis $x = 0$. Figure 4 (right) shows these value along the periodic orbit; here, we observe the same tendency: the values are negative, approximately $-\nu$ and the largest variations occur at the points where the periodic orbit crosses the t-axis.

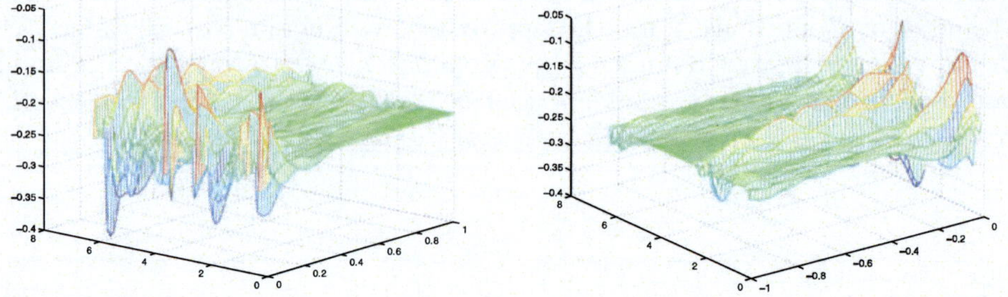

Fig. 5. Left: 1 of Conditions 1 is checked on $[0,T]\times(0,1]$: the values are $f_x^+(t,x)+(s_V)_t(t,x)+(s_V)_x(t,x)f^+(t,x)$ and they are negative, approximately $-\nu=-0.2272$. Right: 2 of Conditions 1 is checked on $[0,T]\times[-1,0)$ for W, which corresponds to $[0,T]\times[-1-v,-v)$ for V; the values are $f_x^-(t,x)+(s_V)_t(t,x-v)+(s_V)_x(t,x-v)f^-(t,x)$ and they are negative, approximately $-\nu=-0.2272$.

Appendix A: Formulae for the Collocation Matrix

To determine the collocation matrix in more detail, we define $F^\pm(t,x) := (1,f^\pm(t,x))$. We set $\tilde{x}=(t,x)$, $\tilde{y}=(\tau,y)$ and $\tilde{x}_j := (t_j,x_j)$; we assume that $T=2\pi$ and $t_j\in[0,2\pi]$.

We use $\Phi(t,x)=\sum_{k\in\mathbb{Z}}\psi(\|(t+k2\pi,x)\|)$, cf. (15) and set $\psi_1(r) := \frac{1}{r}\frac{d\psi(r)}{dr}$, and $\psi_2(r) := \frac{1}{r}\frac{d\psi_1(r)}{dr}$ if $r>0$ and $\psi_2(0):=0$. In the following table we present the Wendland function $\psi_{4,2}$ used for the example in Section 6 together with ψ_1 and ψ_2, cf. also [8, Appendix B.1].

	$\psi_{4,2}(cr)$
$\psi(r)$	$(1-cr)_+^6[35(cr)^2+18cr+3]$
$\psi_1(r)$	$-56c^2(1-cr)_+^5[1+5cr]$
$\psi_2(r)$	$1680c^4(1-cr)_+^4$

We use the formulae of Section 5.1 to calculate

$$
\begin{aligned}
&(M_{11})_{ij}\\
&= (\delta_{\tilde{x}_i} \circ L)^{\tilde{x}} (\delta_{\tilde{x}_j} \circ L)^{\tilde{y}} \Phi(\tilde{x} - \tilde{y})\\
&= (\delta_{\tilde{x}_i} \circ L)^{\tilde{x}} \sum_{k \in \mathbb{Z}} \langle \nabla_{(\tau,y)} \psi(\|(t - \tau + 2\pi k, x - y)\|), F^+(\tau, y)\rangle \big|_{(\tau,y)=(t_j,x_j)}\\
&= (\delta_{\tilde{x}_i} \circ L)^{\tilde{x}} \sum_{k \in \mathbb{Z}} \psi_1(\|(t - t_j + 2\pi k, x - x_j)\|)\\
&\quad \times \langle (t_j - t - 2\pi k, x_j - x), F^+(t_j, x_j)\rangle\\
&= \sum_{k \in \mathbb{Z}} \Big[- \psi_2(\|(t_i - t_j + 2\pi k, x_i - x_j)\|)\\
&\quad \times \langle (t_i - t_j + 2\pi k, x_i - x_j), F^+(t_i, x_i)\rangle \langle (t_i - t_j + 2\pi k, x_i - x_j), F^+(t_j, x_j)\rangle\\
&\quad - \psi_1(\|(t_i - t_j + 2\pi k, x_i - x_j)\|) \langle F^+(t_i, x_i), F^+(t_j, x_j)\rangle \Big].
\end{aligned}
$$

Assume that $\operatorname{supp} \psi \subseteq \{(t, x) \mid \|(t, x)\| \leq R\}$. Then the sum is empty for $|k| > \frac{R}{2\pi} + 1$, since then

$$
\begin{aligned}
\|(t_i - t_j + 2\pi k, x_i - x_j)\| &\geq |t_i - t_j + 2\pi k|\\
&\geq 2\pi |k| - |t_i - t_j|\\
&> R + 2\pi - 2\pi,
\end{aligned}
$$

because $t_j, t_k \in [0, 2\pi]$. Thus, the sum is finite, and we have with $\kappa := \lfloor \frac{R}{2\pi} \rfloor + 1$

$$
\begin{aligned}
&(M_{11})_{ij}\\
&= \sum_{k=-\kappa}^{\kappa} \Big[- \psi_2(\|(t_i - t_j + 2\pi k, x_i - x_j)\|) \langle (t_i - t_j + 2\pi k, x_i - x_j), F^+(t_i, x_i)\rangle\\
&\quad \times \langle (t_i - t_j + 2\pi k, x_i - x_j), F^+(t_j, x_j)\rangle\\
&\quad - \psi_1(\|(t_i - t_j + 2\pi k, x_i - x_j)\|) \langle F^+(t_i, x_i), F^+(t_j, x_j)\rangle \Big].
\end{aligned}
$$

In a similar way we obtain

$$(M_{12})_{ij} = \sum_{k=-\kappa}^{\kappa} \Big[-\psi_2(\|(t_i - t_j + 2\pi k, x_i - x_j)\|)$$
$$\times\langle(t_i - t_j + 2\pi k, x_i - x_j), F^+(t_i, x_i)\rangle$$
$$\times\langle(t_i - t_j + 2\pi k, x_i - x_j), F^-(t_j, x_j + v)\rangle$$
$$-\psi_1(\|(t_i - t_j + 2\pi k, x_i - x_j)\|)\langle F^+(t_i, x_i), F^-(t_j, x_j + v)\rangle \Big],$$

$$(M_{13})_{ij} = \sum_{k=-\kappa}^{\kappa} \Big[\psi_1(\|(t_i - \tau_j + 2\pi k, x_i)\|)\langle F^+(t_i, x_i), (t_i - \tau_j + 2\pi k, x_i)\rangle$$
$$-\psi_1(\|(t_i - \tau_j + 2\pi k, x_i + v)\|)\langle F^+(t_i, x_i), (t_i - \tau_j + 2\pi k, x_i + v)\rangle \Big],$$

$$(M_{22})_{ij} = \sum_{k=-\kappa}^{\kappa} \Big[-\psi_2(\|(t_i - t_j + 2\pi k, x_i - x_j)\|)$$
$$\times\langle(t_i - t_j + 2\pi k, x_i - x_j), F^-(t_i, x_i + v)\rangle$$
$$\times\langle(t_i - t_j + 2\pi k, x_i - x_j), F^-(t_j, x_j + v)\rangle$$
$$-\psi_1(\|(t_i - t_j + 2\pi k, x_i - x_j)\|)\langle F^-(t_i, x_i + v), F^-(t_j, x_j + v)\rangle \Big],$$

$$(M_{23})_{ij} = \sum_{k=-\kappa}^{\kappa} \Big[\psi_1(\|(t_i - \tau_j + 2\pi k, x_i)\|)\langle F^-(t_i, x_i + v), (t_i - \tau_j + 2\pi k, x_i)\rangle$$
$$-\psi_1(\|(t_i - \tau_j + 2\pi k, x_i + v)\|)$$
$$\times\langle F^-(t_i, x_i + v), (t_i - \tau_j + 2\pi k, x_i + v)\rangle \Big].$$

Appendix B: The Formula and Conditions for the Approximant

Recall that $(c, d)^T$ denotes the solution of (17). In a similar way as above, we can calculate the approximant $s_V(t, x)$ from (16),

$$
\begin{aligned}
s_V(t,x) = \sum_{k=\kappa_1(t)}^{\kappa_2(t)} \Bigg[& \sum_{j=1}^{n^+} c_j \psi_1(\|(t-t_j+2\pi k, x-x_j)\|) \\
& \times \langle (t_j - t - 2\pi k, x_j - x), F^+(t_j, x_j)\rangle \\
& + \sum_{j=n^+ +1}^{n^+ + n^-} c_j \psi_1(\|(t-t_j+2\pi k, x-x_j)\|) \\
& \times \langle (t_j - t - 2\pi k, x_j - x), F^-(t_j, x_j + v)\rangle \\
& + \sum_{j=1}^{N} d_j [\psi(\|(t-\tau_j+2\pi k, x)\|) - \psi(\|(t-\tau_j+2\pi k, x+v)\|)] \Bigg],
\end{aligned}
$$

where $\kappa_1(t) := \lceil \frac{-R-t}{2\pi} \rceil$ and $\kappa_2(t) := \lfloor \frac{R-t}{2\pi} \rfloor + 1$. $s_V(t,x)$ is a periodic function by construction and thus it suffices to calculate the values for $t \in [0, 2\pi]$; here we obtain again $\min_{t\in[0,2\pi]} \kappa_1(t) = -\kappa$ and $\max_{t\in[0,2\pi]} \kappa_2(t) = \kappa$.

The Conditions 1 that we have to check for the approximant s_V involve the orbital derivative. For $x > 0$ we have, cf. (9):

$$
\begin{aligned}
&(L \circ s_V)(t,x) \\
&= \sum_{k=\kappa_1(t)}^{\kappa_2(t)} \Bigg[\sum_{j=1}^{n^+} c_j \{ -\psi_2(\|(t-t_j+2\pi k, x-x_j)\|) \\
&\times \langle (t-t_j+2\pi k, x-x_j), F^+(t,x)\rangle \langle (t-t_j+2\pi k, x-x_j), F^+(t_j,x_j)\rangle \\
&-\psi_1(\|(t-t_j+2\pi k, x-x_j)\|)\langle F^+(t,x), F^+(t_j,x_j)\rangle \} \\
&+ \sum_{j=n^+ +1}^{n^+ + n^-} c_j \{ -\psi_2(\|(t-t_j+2\pi k, x-x_j)\|) \\
&\times \langle (t-t_j+2\pi k, x-x_j), F^+(t,x)\rangle \langle (t-t_j+2\pi k, x-x_j), F^-(t_j,x_j+v)\rangle \\
&-\psi_1(\|(t-t_j+2\pi k, x-x_j)\|)\langle F^+(t,x), F^-(t_j,x_j+v)\rangle \} \\
&+ \sum_{j=1}^{N} d_j [\psi_1(\|(t-\tau_j+2\pi k, x)\|)\langle F^+(t,x), (t-\tau_j+2\pi k, x)\rangle \\
&-\psi_1(\|(t-\tau_j+2\pi k, x+v)\|)\langle F^+(t,x), (t-\tau_j+2\pi k, x+v)\rangle] \Bigg].
\end{aligned}
$$

For $x < -v$ we have:

$$
\begin{aligned}
&(L^v \circ s_V)(t,x)\\
&= \sum_{k=\kappa_1(t)}^{\kappa_2(t)} \Bigg[\sum_{j=1}^{n^+} c_j\{-\psi_2(\|(t-t_j+2\pi k, x-x_j)\|)\\
&\quad \times\langle(t-t_j+2\pi k, x-x_j), F^-(t,x+v)\rangle\langle(t-t_j+2\pi k, x-x_j), F^+(t_j,x_j)\rangle\\
&\quad -\psi_1(\|(t-t_j+2\pi k, x-x_j)\|)\langle F^-(t,x+v), F^+(t_j,x_j)\rangle\}\\
&\quad + \sum_{j=n^++1}^{n^++n^-} c_j\{-\psi_2(\|(t-t_j+2\pi k, x-x_j)\|)\\
&\quad \times\langle(t-t_j+2\pi k, x-x_j), F^-(t,x+v)\rangle\\
&\quad \times\langle(t-t_j+2\pi k, x-x_j), F^-(t_j,x_j+v)\rangle\\
&\quad -\psi_1(\|(t-t_j+2\pi k, x-x_j)\|)\langle F^-(t,x+v), F^-(t_j,x_j+v)\rangle\}\\
&\quad +\sum_{j=1}^{N} d_j\left[\psi_1(\|(t-\tau_j+2\pi k, x)\|)\langle F^-(t,x+v), (t-\tau_j+2\pi k, x)\rangle\right.\\
&\quad \left.-\psi_1(\|(t-\tau_j+2\pi k, x+v)\|)\times\langle F^-(t,x+v), (t-\tau_j+2\pi k, x+v)\rangle\right]\Bigg].
\end{aligned}
$$

References

1. G. Borg: A condition for the existence of orbitally stable solutions of dynamical systems. Kungl. Tekn. Högsk. Handl. **153**, 1960.
2. C. Chicone: *Ordinary Differential Equations with Applications.* Springer, 1999.
3. G.E. Fasshauer: Solving partial differential equations by collocation with radial basis functions. In *Surface Fitting and Multiresolution Methods*, A.L. Méhauté, C. Rabut, and L. L. Schumaker (eds.), Vanderbilt University Press, 1997, 131–138.
4. A. Filippov: *Differential Equations with Discontinuous Righthand Sides.* Kluwer, 1988.
5. C. Franke and R. Schaback: Convergence order estimates of meshless collocation methods using radial basis functions. Adv. Comput. Math. **8**, 1998, 381–399.
6. P. Giesl: The basin of attraction of periodic orbits in nonsmooth differential equations. ZAMM Z. Angew. Math. Mech. **85**, 2005, 89–104.
7. P. Giesl: Necessary condition for the basin of attraction of a periodic orbit in non-smooth periodic systems. Discrete Contin. Dyn. Syst. **18**, 2007, 355–373.
8. P. Giesl: *Construction of Global Lyapunov Functions using Radial Basis Functions.* Lecture Notes in Mathematics **1904**, Springer-Verlag, Heidelberg, 2007.
9. P. Giesl: On the determination of the basin of attraction of discrete dynamical systems. J. Difference Equ. Appl. **13**, 2007, 523–546.
10. P. Giesl and H. Wendland: Meshless collocation: error estimates with application to dynamical systems. SIAM J. Numer. Anal. **45**, 2007, 1723–1741.
11. P. Giesl and H. Wendland: Approximating the basin of attraction of time-periodic ODEs by meshless collocation. Discrete Contin. Dyn. Syst. **25**, 2009, 1249–1274.

12. P. Giesl and H. Wendland: Approximating the basin of attraction of time-periodic ODEs by meshless collocation of a Cauchy problem. Discrete Contin. Dyn. Syst. Supplement, 2009, 259–268.
13. Ph. Hartman: *Ordinary Differential Equations*. Wiley, New York, 1964.
14. B. Michaeli: *Lyapunov-Exponenten bei nichtglatten dynamischen Systemen*. PhD Thesis, University of Köln, 1999 (in German).
15. F.J. Narcowich and J.D. Ward: Generalized Hermite interpolation via matrix-valued conditionally positive definite functions. Math. Comput. **63**, 1994, 661–687.
16. H. Wendland: Piecewise polynomial, positive definite and compactly supported radial functions of minimal degree. Adv. Comput. Math. **4**, 1995, 389–396.
17. H. Wendland: Error estimates for interpolation by compactly supported radial basis functions of minimal degree. J. Approx. Theory **93**, 1998, 258–272.
18. H. Wendland: *Scattered Data Approximation*. Cambridge Monographs on Applied and Computational Mathematics, Cambridge University Press, Cambridge, UK, 2005.
19. Z. Wu: Hermite-Birkhoff interpolation of scattered data by radial basis functions, Approximation Theory Appl. **8**, 1992, 1–10.

Stabilizing Lattice Boltzmann Simulation of Fluid Flow past a Circular Cylinder with Ehrenfests' Limiter

Tahir S. Khan and Jeremy Levesley

Department of Mathematics, University of Leicester, LE1 7RH, UK

Summary. In this study, two-dimensional fluid flow around a circular cylinder for different laminar and turbulent regimes has been analyzed. We will show that introduction of Ehrenfests' coarse-graining idea in lattice Boltzmann method can stabilize the simulation for high Reynolds numbers without any grid refinement in the vicinity of circular cylinder where sharp gradients occur. A Strouhal-Reynolds number relationship from low to high Reynolds number has been captured and found satisfactory in agreement with other numerical and experimental simulations.

1 Introduction

The laminar and turbulent unsteady viscous flow around the circular cylinder has been a fundamental fluid mechanics problem due to its wide variety of applications in engineering such as submarines, bridge piers, towers, pipelines and off shore structures etc. Numerous experimental and numerical investigations [15, 17, 18, 19, 20, 21, 22, 26] have been carried out to understand the complex dynamics of the cylinder wake flow over the last century. The governing dimensionless parameter for the idealized disturbance-free flow around a nominally two-dimensional cylinder is the Reynolds number $Re = UD/\nu$ where U is the free-stream velocity, D the cylinder diameter and ν the kinematic viscosity. It has been observed both experimentally and numerically that as the Reynolds number increases, flow begins to separate behind the cylinder causing vortex shedding which leads to a periodic flow known as Von Karman vortex street. Recently, Zdravkovich [29] has compiled almost all the experimental and numerical simulation data on the flow past circular cylinders and classified this phenomenon into five different regimes based on the Reynolds numbers. For 30 to $48 < Re <$ 180 to 200, there is a laminar vortex shedding in the wake of the cylinder. Transition from laminar to turbulence occurs in the region 180 to $200 < Re <$ 350 to 400. In the region $Re >$ 350 to 400 the wake behind circular cylinder becomes completely turbulent.

During last two decades the lattice Boltzmann method (LBM) [23, 27] has emerged as an alternative to the conventional computational fluid dynamics

E.H. Georgoulis, A. Iske, J. Levesley (eds.), *Approximation Algorithms for Complex Systems*, Springer Proceedings in Mathematics 3, DOI: 10.1007/978-3-642-16876-5_10,
© Springer-Verlag Berlin Heidelberg 2011

(CFD) methods (finite difference, finite volume, finite elements and spectral methods). Unlike the traditional CFD tools, which are based on the discretization of continuous partial differential equations (Navier-Stokes for fluid dynamics), the LBM is based on evolution equations for the mesoscopic Boltzmann densities from which macroscopic quantities can be calculated. The main advantages of the LBM are simple programming, parallel computation and ease of implementation of boundary conditions for complex geometries. Among the different variants of the LBM in use, are multiple relaxation lattice Boltzmann method, finite volume lattice Boltzmann method, interpolation-supplemented lattice Boltzmann method, entropic lattice Boltzmann method [2, 12, 13, 14, 25] and recently introduced lattice Boltzmann method with Ehrenfests' step [3, 4, 5, 6]. Despite successful LBM simulations of various fluid flows of engineering interests, it has been observed that the LBM exhibits numerical instabilities in low viscosity regimes. The reasons for these instabilities are lack of positivity and deviations of the populations from the quasi-equilibrium states. On the curved boundary of cylinder the interpolation-based shemes [7, 11, 16, 28] play an important role to improve the numerical stability. The stability of the LBM has been improved in entropic lattice Boltzmann method (ELBM) through compliance with an H-theorem [24] which ensures the positivity of the distribution function. As an alternative and versatile approach, the LBM with Ehrenfests' steps has been able to control the deviations of the populations from quasi-equilibrium states by fixing a tolerance value for the difference in microscopic and macroscopic entropy. When this tolerance value is exceeded the populations are returned to their quasi-equilibrium states.

Both models ELBM and LBM with Ehrenfests' steps have efficiently simulated turbulent flow past a square cylinder [2, 5]. In the present work we have tested the efficiency of the second method i.e., LBM with Ehrenfests' steps for the flow past a circular cylinder. A main feature of the unsteady cylinder wake is the global parameter Strouhal number which is defined as $St = Lf_\omega/U_\infty$ where f_ω is the vortex shedding frequency, L is the characteristic length scale (diameter of the cylinder here) and U_∞ is the free-stream fluid velocity (velocity at the inlet here).The main goal of this work is to visualize the laminar and turbulent unsteady flow fields for different Reynolds number and to find a Strouhal-Reynolds number relationship. The focus would be the two-dimensional vortex shedding behind a circular cylinder and we will show that introduction of Ehrenfests' steps can stabilize the LBM simulation of flow past circular cylinder for quite a high Reynolds number up to $Re = 20,000$. Numerical results presented here are compared with other experimental and numerical results [9, 10, 17, 18, 19, 20, 21, 22, 26, 29] and the agreements are found satisfactory.

The work is organized as follows: In Sec. 2, a brief description of the LBM is presented. In Sec. 3, Ehrenfests' coarse-graining idea is introduced. In Sec. 4, the computational set up for the flow is defined. In Sec. 5, the boundary conditions are explained and finally in Sec. 6, we present the results of our numerical experiment and their comparisons with other results.

2 Lattice Boltzmann Method

The Boltzmann equation

$$\frac{\partial f}{\partial t} + \mathbf{v} \cdot \nabla f = Q(f),$$

is a kinetic transport equation where $f = f(\mathbf{x}, \mathbf{v}, t)$ is the distribution function to find the probability of a particle moving with velocity $\mathbf{v}$ at site $\mathbf{x}$ and time t and $Q(f)$ is the collisional integral which represents the interactions of the populations f. The developments made for the solutions of the Boltzmann equation are focused on finding simpler expressions for the complicated collisional integral $Q(f)$. One of the approximations of $Q(f)$ is the well known Bhatnagar-Gross-Krook (BGK) collisional integral,

$$Q(f) = -\omega(f - f^{eq}).$$

This represents the relaxation towards local equilibrium f^{eq} defined below, on a time scale $\tau = 1/\omega$ which results in a viscous behavior of the model. It has been shown through Chapman-Enskog expansion [23] that the resulting macrodynamics are the Navier-Stokes equations to second-order in τ. On the other hand, in [3] the authors demonstrate that using Ehrenfests' coarse-graining idea,the lattice Boltzmann iteration delivers macroscopic dynamics which are the Navier-Stokes equations with viscosity $\Delta t/2$, to order Δt^2 where Δt is the time step. The difference between the two approaches is that in the first approach, the kinetic equation acts as an intermediary between macroscopic transport equations and LBM simulation whereas in the second approach, Navier-Stokes equations are the result of free-flight dynamics followed by equilibration.

A discretization of the velocity space into a finite set of velocities $\mathbf{v} = (\mathbf{e}_0, \mathbf{e}_1, ..., \mathbf{e}_{n-1})$ and associated distribution function $f = (f_0, f_1, ..., f_{n-1})$ results in the discrete Boltzmann equation, known as lattice Boltzmann equation (LBE),

$$\frac{\partial f_i}{\partial t} + \mathbf{e}_i \cdot \nabla f_i = -\omega(f_i - f_i^{eq}), \qquad i = 0, 1, ..., n-1.$$

Further discretization of LBE in time and space leads to the BGK lattice Boltzmann equation,

$$f_i(\mathbf{x} + \mathbf{e}_i \Delta t, t + \Delta t) - f_i(\mathbf{x}, t) = -\frac{1}{\tau}(f_i - f_i^{eq}).$$

In order to recover macroscopic fluid dynamics (Navier-Stokes) equations, the set of discrete velocities is selected in such a way that it must satisfy the mass, momentum and energy conservation. This requires the following constraints on the local equilibrium distribution:

$$\rho = \sum_i f_i^{eq},$$

$$\mathbf{u} = \frac{1}{\rho}\sum_i f_i^{eq}\mathbf{e}_i,$$

where the explicit expression of the local equilibrium [1] has the following form:

$$f_i^{eq} = w_i\rho\prod_{j=1}^{2}(2-\sqrt{1+3u_j^2})\left(\frac{2u_j+\sqrt{1+3u_j^2}}{1-u_j}\right)^{e_{i,j}/c},$$

where j is the index of the spatial directions, so $e_{i,j}$ represents the jth component of $\mathbf{e}_i$, and w_i are the weighting factors defined below. The second order expansion gives the following polynomial quasiequilibria:

$$f_i^{eq} = w_i\rho\left[1+\frac{3}{c^2}(\mathbf{e}_i\cdot\mathbf{u})+\frac{9}{2c^2}(\mathbf{e}_i\cdot\mathbf{u})^2-\frac{3}{2c^2}(\mathbf{u}\cdot\mathbf{u})\right].$$

For the two-dimensional case, the lattice which exhibits rotational symmetry to ensure the conservation constraints is D2Q9 as shown in Figure 1. The discrete velocities for this lattice are defined as:

$$\mathbf{e}_i = \begin{cases} (0,0), & i=0,\\ (c\cos[(i-1)\pi/2], c\sin[(i-1)\pi/2]), & i=1,2,3,4,\\ (\sqrt{2}c\cos[(i-5)\pi/2+\pi/4], \sqrt{2}c\sin[(i-5)\pi/2+\pi/4]), & i=5,6,7,8, \end{cases}$$

where $c = \Delta x/\Delta t$, Δx and Δt are lattice constant and the time step size, respectively. The weights w_i are given by

$$w_i = \begin{cases} \frac{4}{9}, & i=0,\\ \frac{1}{9}, & i=1,2,3,4,\\ \frac{1}{36}, & i=5,6,7,8. \end{cases}$$

With this model the macroscopic variables are given by:

$$\rho = \sum_i f_i,$$

$$\mathbf{u} = \frac{1}{\rho}\sum_i f_i\mathbf{e}_i.$$

The speed of sound of this model is

$$c_s = \frac{c}{\sqrt{3}}.$$

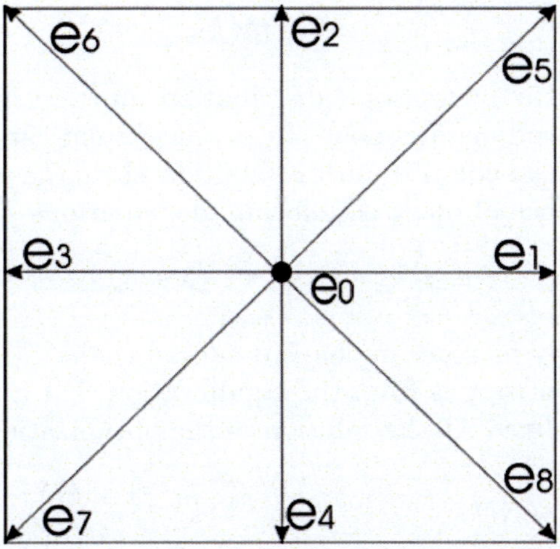

Fig. 1. Two-dimensional D2Q9 lattice

The viscosity of the model is given by

$$\nu = c_s^2(\tau - \frac{1}{2}).$$

The two computational steps for the LBM are:

$$\begin{array}{ll} \textit{Collision}: & \tilde{f}_i(\mathbf{x},t) = f_i(\mathbf{x},t) - \frac{1}{\tau}[f_i(\mathbf{x},t) - f_i^{eq}(\mathbf{x},t)], \\ \textit{Streaming}: & f_i(\mathbf{x} + \mathbf{e}_i\Delta t, t + \Delta t) = \tilde{f}_i(\mathbf{x},t), \end{array}$$

where f_i and $\tilde{f}_i$ denote the pre-collision and post-collision distribution functions, respectively.

3 Ehrenfests' Coarse-Graining

The introduction of Ehrenfests' coarse-graining idea to the LBM [3, 4, 5, 6, 8] can be understood as follows: First consider the continuous Boltzmann Equation (1) as a combination of two alternating operations, the free-flight and the collision as shown in the Figure 2. The free-flight operator is simply a linear map $\Theta_\tau : f(\mathbf{x}, \mathbf{v}, t) \rightarrow f(\mathbf{x} - \mathbf{v}\tau, \mathbf{v}, \tau)$ describing the distribution of particles in phase space as a shift transformation of the conservative dynamics and can be expressed by the equation

$$f(\mathbf{x}, \mathbf{v}, t + \tau) = f(\mathbf{x} - \mathbf{v}\tau, \mathbf{v}, t),$$

where τ is the fixed coarse-graining time. For the collision operator, the velocity space $\mathbf{v}$ is discretized as $\mathbf{v}_i$. This collision operator does not affect the macroscopic variables of the particle distribution. Defining a linear map

$$M = m(f),$$

which transforms the microscopic distribution function into the macroscopic variables, the hydrodynamic moments of the system can be retrieved [3]. If f_0 is an initial quasi-equilibrium distribution then the Ehrenfests' chain is defined as a sequence of quasi-equilibrium distributions $f_0, f_1, ...$, where

$$f_\lambda = f^{eq}_{m[\Theta_\tau(f_{\lambda-1})]}, \qquad \lambda = 1, 2, \ldots.$$

There is no entropy increase in the Ehrenfests' chain due to mechanical motion, the gain in entropy is from the equilibration. For a given entropy functional $S(f)$ and a fixed M the solution of the optimization problem

$$f^{eq}_M = \arg\max \{S(f) : m(f) = M\},$$

is unique. For the Boltzmann entropy

$$S(f) = -\int\int f \log f d\mathbf{v} d\mathbf{x},$$

the quasi-equilibrium is the Maxwellian distribution

$$f^{eq}_M = \frac{\rho^2}{2\pi P} \exp(-\frac{\rho}{P}(\mathbf{v} - \mathbf{u})^2).$$

In [3], the authors have shown that by introducing Ehrenfests' steps after free-flight, we can recover, to the second-order, the Navier-Stokes equations with coefficient of viscosity $\tau/2$. Clearly, viscosity is proportional to the time step. The governing equations for the LBGK scheme are

$$f_i(\mathbf{x} + \mathbf{e}_i\tau, t + \tau) = (1 - \beta) f_i(\mathbf{x}, t) + \beta f_i^{mir}(\mathbf{x}, t),$$

where $f_i^{mir}(\mathbf{x}, t) = 2f_i^{eq}(\mathbf{x}, t) - f_i(\mathbf{x}, t)$, is the reflection of f_i in the quasiequilibrium manifold. The parameter $\beta = \beta(\tau) \in [0, 1]$ may be chosen to satisfy a physically relevant condition. This controls the viscosity in the model, with $\beta = 1$, the viscosity goes to zero. For $\beta = 0$, there is no change in f_i during collision. A choice of $\beta = 1/2$ corresponds to the Ehrenfests' step with viscosity proportional to the time step $\Delta t = \tau$. One variant of LBGK is the ELBM [12, 13, 14] in which instead of a linear mirror reflection $f \longmapsto f^{mir}$ an entropic involution $f \longmapsto \tilde{f}$ is used, where $\tilde{f} = (1 - \alpha)f + \alpha f^{eq}$. The number $\alpha = \alpha(f)$ is so chosen that the local constant entropy condition is satisfied: $S(f) = S(\tilde{f})$. The governing equations for ELBM become

$$f_i(\mathbf{x} + \mathbf{e}_i\tau, t + \tau) = (1 - \beta) f_i(\mathbf{x}, t) + \beta \tilde{f}_i(\mathbf{x}, t).$$

In [3], the authors have constructed the numerical method from the dynamics $\Theta_{-\tau/2}(f^{eq}_M) \to \Theta_{\tau/2}(f^{eq}_M)$, so that the first-order term in τ is canceled and an order τ^2 approximation to the Euler equation is obtained. The deviations of

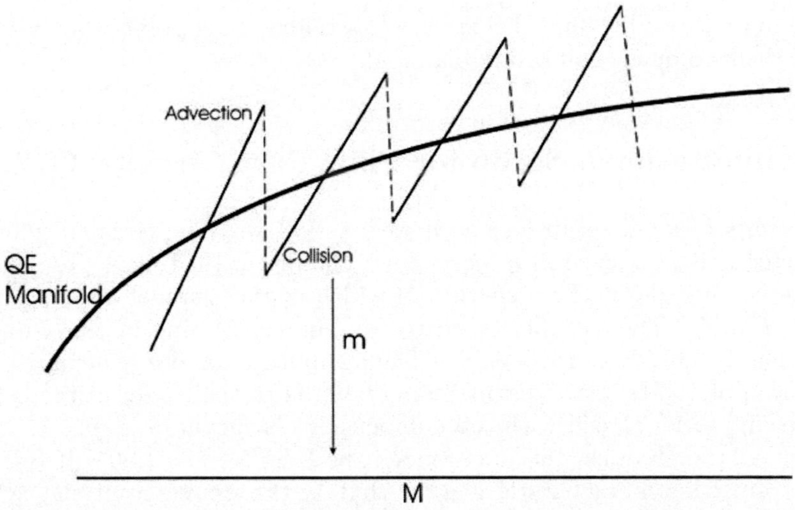

Fig. 2. Showing the alternating operations of free flight and collision chain in time near the quasi-equilibrium manifold and the linear map m from the microscopic populations to the macroscopic moments M.

the populations from the quasi-equilibrium manifold cause instabilities in both LBGK and ELBM simulations. By applying Ehrenfests' steps at a bounded number of sites, the populations are returned to the quasiequilibrium manifold and the simulation can be stabilized to order τ^2. This has been done by monitoring the local deviation of f from the corresponding quasiequilibrium, and the nonequilibrium entropy

$$\triangle S = S(f^{eq}) - S(f),$$

at every lattice site throughout the simulation. If a prespecified threshold value δ is exceeded, then an Ehrenfests' step is performed at the corresponding cite: $f \longmapsto f^{eq}$ at those points. For the discrete entropy

$$S(f) = -\sum_i f_i \log\left(\frac{f_i}{W_i}\right),$$

the non-equilibrium entropy is

$$\triangle S = \sum_i f_i \log\left(\frac{f_i}{f_i^{eq}}\right).$$

The governing equations can, now be written as

$$f_i(\mathbf{x} + e_i\tau, t+\tau) = \begin{cases} f_i(\mathbf{x},t) + 2\beta(f_i^{eq}(\mathbf{x},t) - f_i(\mathbf{x},t)), & \triangle S \leq \delta, \\ f_i(\mathbf{x},t), & \text{otherwise.} \end{cases}$$

In order that the Ehrenfests' steps are not allowed to degrade the accuracy of LBGK , it is pertinent to select the k sites with highest $\triangle S > \delta$ and return these to quasiequilibrium. If there are less than k such points then we return all of them to quasiequilibrium manifold.

4 Computational Setup for Flow Past Circular Cylinders

The computational setup for the flow is as follows: The circular cylinder of diameter D is immersed in a rectangular channel with its axis perpendicular to the flow direction. The length and width of the channel are respectively, 30D and 25D. The cylinder is placed on the center line in the y-direction resulting in a blockage ratio of 4%. The computational domain consists of an upstream of 10.5D and a downstream of 19.5D to the center of the cylinder. The computational grid with these dimensions is shown in Figure 3.

For all simulations, the inlet velocity is $(U_\infty, V_\infty) = (0.05, 0)$ (in lattice units) and the characteristic length, that is the diameter of the cylinder, is $D = 20$. The vortex shedding frequency f_ω is obtained from the discrete Fourier transform of the x-component of the instantaneous velocity at a monitoring point which is located at coordinates $(4D, -2D)$ with center of the cylinder being assumed at the origin. The simulations are recorded over $t_{max} = 1250D/U_\infty$ time steps. The parameter (k, δ) which controls the Ehrenfests' steps tolerances, are fixed at $(16, 10^{-3})$.

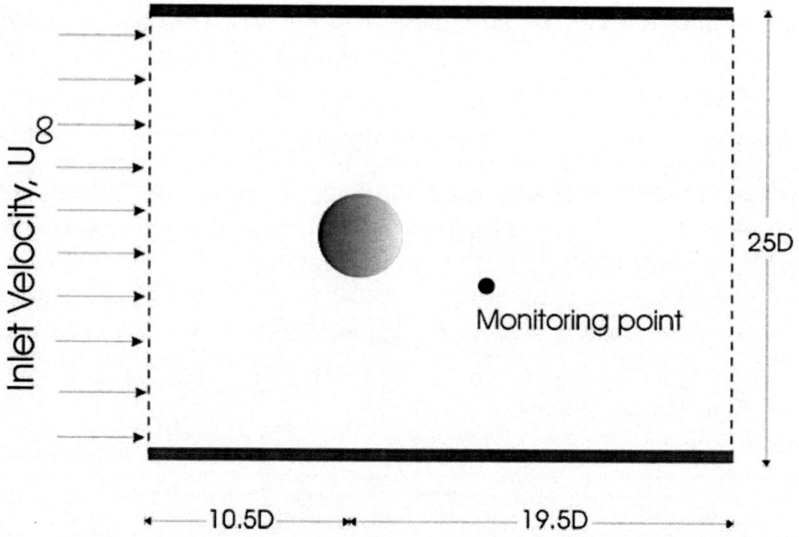

Fig. 3. Computational setup for flow past circular cylinder

5 Boundary Conditions

The free slip boundary condition [23] is imposed on the north and south channel walls. At the inlet, the populations are replaced with the quasi-equilibrium values that correspond to the free-streem velocity and density. As the simulation result is not very sensitive to the exact condition specified at the inlet boundary, this lower order approximation is sufficient there. The simulation is sensitive to the outlet boundary condition. The sensitivity for this problem has been known in [23]. We follow the prescription suggested in [2, 3]: at the outlet, the populations pointing towards the flow domain are replaced by the equilibrium values that correspond to the velocity and density of the penultimate row of the lattice.

On the cylinder walls, the interpolation based scheme of Filippova and Hänel (FH) model [7] with first-order and second-order improvements made by Renwei Mei [16] are applied. In Figure 4, a curved wall separates the solid region from the fluid region. The lattice nodes on the fluid and solid sides are denoted by $\mathbf{r}_f$ and $\mathbf{r}_s$ respectively. The filled small circles on the boundary $\mathbf{r}_w$, denote the intersections of the wall with different lattice links. The fraction of the intersected link in the fluid region is,

$$\triangle = \frac{|\mathbf{r}_f - \mathbf{r}_w|}{|\mathbf{r}_f - \mathbf{r}_s|}, \qquad where \qquad 0 < \triangle \leq 1.$$

The horizontal and vertical distance between $\mathbf{r}_f$ and $\mathbf{r}_w$ is $\triangle\delta x$ on the square lattice. After the collision step, $\tilde{f}_i(\mathbf{r}_f, t)$ on the fluid side is known and $\tilde{f}_{-i}(\mathbf{r}_s, t)$ on the solid side is to be determined. To find the unknown value $\tilde{f}_{-i}(\mathbf{r}_s, t) = f_{-i}(\mathbf{r}_f = \mathbf{r}_s + \mathbf{e}_{-i}\delta t, t + \delta t)$, based on information in the surrounding fluid nodes like $\tilde{f}_i(\mathbf{r}_f, t)$, $\tilde{f}_i(\mathbf{r}_{ff}, t)$ etc., FH [7] construct the following linear interpolation:

$$\tilde{f}_{-i}(\mathbf{r}_s, t) = (1 - \chi)\tilde{f}_i(\mathbf{r}_f, t) + \chi f_i^*(\mathbf{r}_s, t) - 2w_i\rho\frac{3}{c^2}(\mathbf{e}_{-i} \cdot \mathbf{u}_w),$$

where $\mathbf{u}_w = \mathbf{u}(\mathbf{r}_w, t)$ is the velocity at wall, χ is the weighting factor, and $f_i^*(\mathbf{r}_s, t)$ is a fictitious equilibrium distribution function defined as:

$$f_i^*(\mathbf{r}_s, t) = w_i\rho(\mathbf{r}_f, t)\left[1 + \frac{3}{c^2}(\mathbf{e}_i \cdot \mathbf{u}_{sf}) + \frac{9}{2c^2}(\mathbf{e}_i \cdot \mathbf{u}_f)^2 - \frac{3}{2c^2}(\mathbf{u}_f \cdot \mathbf{u}_f)\right],$$

where $\mathbf{u}_f = \mathbf{u}(\mathbf{r}_f, t)$ and $\mathbf{u}_{sf}$ is the fictitious velocity which is to be chosen. For FH model, the relevant equations for χ and $\mathbf{u}_{sf}$ are

$$\triangle < \frac{1}{2}: \quad \mathbf{u}_{sf} = \mathbf{u}_f, \qquad \chi = \frac{2\triangle - 1}{\tau - 1},$$

$$\triangle \geq \frac{1}{2}: \quad \mathbf{u}_{sf} = \frac{1}{\triangle}(\triangle - 1)\mathbf{u}_f + \frac{1}{\triangle}\mathbf{u}_w, \quad \chi = \frac{(2\triangle - 1)}{\tau}.$$

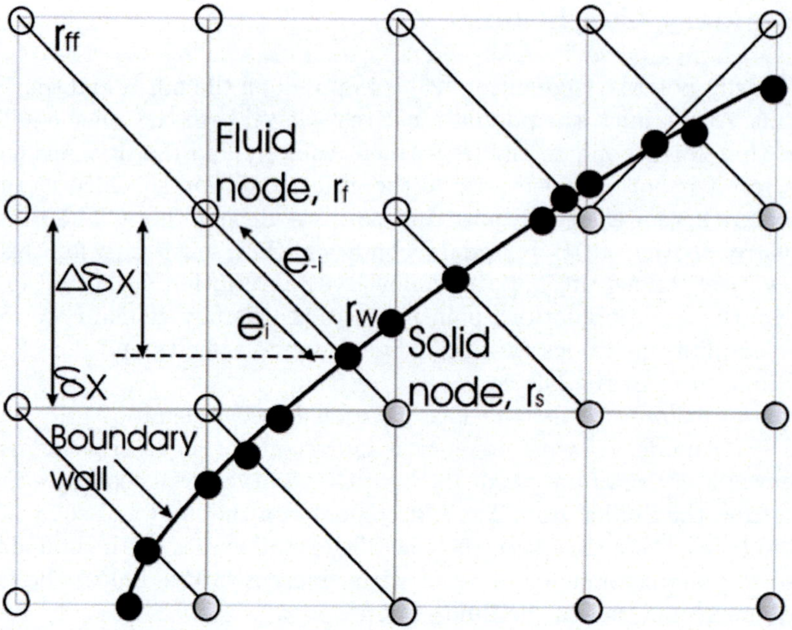

Fig. 4. Layout for the curved boundary on the lattice. The thick curve represents the boundary wall, the empty circles denote the fluid nodes, solid circles denote the wall nodes and the shaded solid circles denote the solid nodes, respectively

To improve the numerical stability, Renwei Mei et al. [16] suggested the following first-order modified equations for χ and $\mathbf{u}_{sf}$:

$$\Delta < \frac{1}{2}: \qquad \mathbf{u}_{sf} = \mathbf{u}_{ff}, \qquad \chi = \frac{2\Delta - 1}{\tau - 2},$$
$$\Delta \geq \frac{1}{2}: \qquad \mathbf{u}_{sf} = \frac{1}{\Delta}(\Delta - 1)\mathbf{u}_f + \frac{1}{\Delta}\mathbf{u}_w, \qquad \chi = \frac{(2\Delta - 1)}{\tau},$$

and the second-order improvements in the curve boundary give the following modified equations for χ and $\mathbf{u}_{sf}$:

$$\Delta < \frac{1}{2}: \qquad \mathbf{u}_{sf} = \mathbf{u}_{ff}, \qquad \chi = \frac{2\Delta - 1}{\tau - 2},$$
$$\Delta \geq \frac{1}{2}: \qquad \mathbf{u}_{sf} = \frac{1}{2\Delta}(2\Delta - 3)\mathbf{u}_f + \frac{3}{2\Delta}\mathbf{u}_w, \quad \chi = \frac{(2\Delta - 1)}{\tau + 1/2}.$$

In current simulations, Eqs. (33) and Eqs. (34) are used to obtain the values χ and $\mathbf{u}_{sf}$ and these values are then substituted into Eq. (31) to find $f_i^*(\mathbf{r}_s, t)$ and finally the unknown value $\tilde{f}_{-i}(\mathbf{r}_s, t)$ is found from Eq. (30).

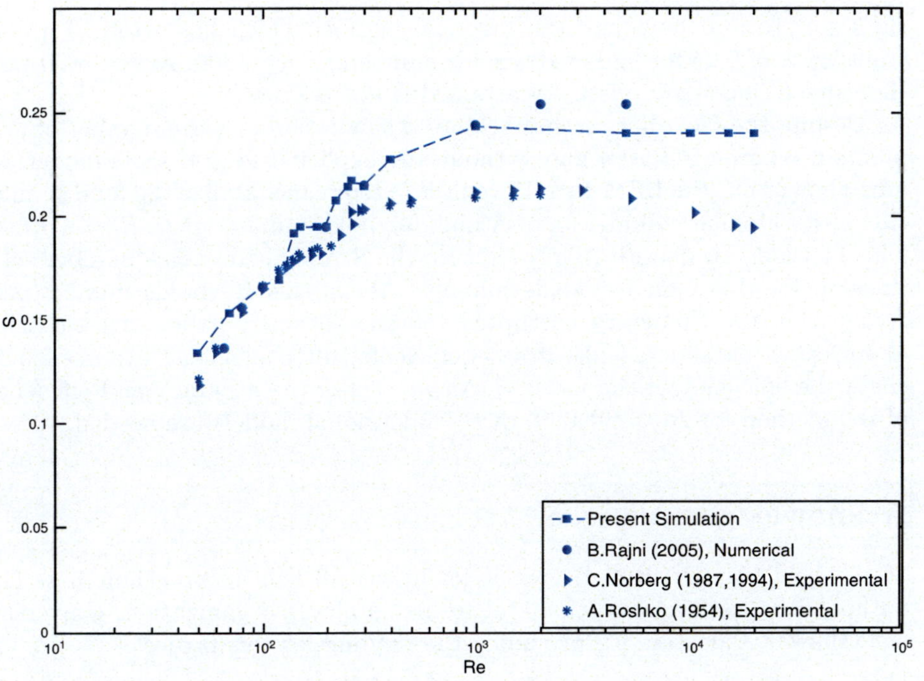

Fig. 5. Variation of Strouhal number with Reynolds number

6 Results and Discussion

In his experimental work [20], Roshko (1954) has found a laminar(periodic) vortex shedding regime for $Re = 40$ to 150, a transition regime for $Re = 150$ to 300, and an irregular regime for $Re = 300$ to 10,000+. Similar regimes have been confirmed by other experimental and numerical investigations. In the present work, the LBM with Ehrenfests' regularization described in the previous sections has been tested for the flow past a circular cylinder and the simulations are carried out from moderate to high Reynolds numbers up to Re=20,000.

In our simulations, we have observed the onset of vortex shedding at $Re = 49$ and found that the transition has started around $Re = 150$. A Strouhal-Reynolds number relationship is computed and compared with the experimental [17, 18, 20] and numerical [21, 22] simulations as shown in Figure 5. Numerical simulations often slightly overestimate the Strouhal number, since real experiments are three dimensional. We have been able to capture the asymptotic behavior of the Strouhal number which has attained a constant value of 0.24 for higher Reynolds numbers, $Re \geq 1000$. At $Re = 20,000$, the value of kinematic viscosity attained is $\nu = 5 \times 10^{-5}$.

Despite the fact that an underresolved simulation i.e., the number of grid points used are of $O(10^5)$ and without any explicit sub-grid scale model, we have shown that the LBM with Ehrenfests' steps can stabilize the fluid simulation past a circular cylinder for very high Reynolds number up to $Re = 20,000$. This method can quantitatively capture the Strouhal-Reynolds number relationship for this high Reynolds number. Above this Reynolds number, the errors from the boundary corrupted the simulation. Further investigations about other parameters like drag coefficients, lift coefficients are needed to check the efficiency of the method. Also to check the accuracy and efficiency of the method for $Re > 20,000$, three-dimensional models are needed.

Acknowledgement

We acknowledge Dr. Rob Brownlee for his useful help in programming. The author, T.S. Khan is also thankful to his employee organization, University of Peshawar, Pakistan, for providing the funding for this project.

References

1. S. Ansumali, I.V. Karlin, and H.C. Öttinger: Minimal entropic kinetic models for simulating hydrodynamics. Europhys. Lett. **63**, 2003, 798–804.
2. S. Ansumali, S.S. Chikatamarla, C.E. Frouzakis, and K. Boulouchos: Entropic lattice Boltzmann simulation of flow past square cylinder. International Journal of Modern Physics C **15**, 2004, 435–455.
3. R. Brownlee, A.N. Gorban, and J. Levesley: Stabilization of the lattice-Boltzmann method using the Ehrenfests' coarse-graining. Phs. Rev. E **74**, 2006, 037703.
4. R. Brownlee, A.N. Gorban, and J. Levesley: Stability and stabilization of the lattice Boltzmann method. Phys. Rev. E **75**, 2007, 036711.
5. R. Brownlee, A.N. Gorban, and J. Levesley: Stable simulation of fluid flow with high-Reynolds number using Ehrenfests' steps. Numerical Algorithms. **45**, 2007, 389–408.
6. R. Brownlee, A.N. Gorban, and J. Levesley: Nonequilibrium entropy limiters in lattice Boltzmann method. Physica A **387(2-3)**, 2008, 385–406.
7. O. Filippova and D. Hänel: Grid Refinement for lattice-BGK models. Journal of Computational Physics **147**, 1998, 219–228.
8. A.N. Gorban: Basic types of coarse-graining. In *Model Reduction and Coarse-graining Approaches for Multiscale Phenomena*, A.N. Gorban, N. Kazantzis, I.G. Kevrekidis, H.-C. Öttinger, and C. Theodoropoulos (eds.), Springer, Berlin, 2006, 117–176.
9. X. He and G. Doolen: Lattice Boltzmann method on a curvilinear coordinate system: vortex shedding behind a circular cylinder. Phys. Rev. E **56**, 1997, 434–440.

10. X. He and G. Doolen: Lattice Boltzmann method on a curvilinear coordinate system: flow around a circular cylinder. Journal of Computational Physics **134**, 1997, 306–315.
11. P.-H. Kao and R.-J. Yang: An investigation into curved and moving boundary treatments in the lattice Boltzmann method. Journal of Computational Physics **227**, 2008, 5671–5690.
12. I.V. Karlin, S. Ansumali, C.E. Frouzakis, and S.S. Chikatamarla: Elements of the lattice Boltzmann method I: linear advection equation. Communications in Computational Physics **1**(4), 2006, 616–655.
13. I.V. Karlin, S.S. Chikatamarla, and S. Ansumali: Elements of the lattice Boltzmann method II: kinetics and hydrodynamics in one dimension. Communications in Computational Physics **2**(2), 2007, 196–238.
14. I.V. Karlin, A.N. Gorban, S. Succi, and S. Boffi: Maximum entropy principle for lattice kinetic equations. Phys. Rev. Lett. **81**, 1998, 6–9.
15. J.H. Lienhard: Synopsis of lift, drag, and vortex frequency for rigid circular cylinder. College of Engg. Research Div. Bulletin 300, Technical. Extension Service, Washington State University, 1966.
16. R. Mei, L.S. Luo, and W. Shyy: An accurate curved boundary treatment in the lattice Boltzmann method. Journal of Computational Physics **155**, 1999, 307–330.
17. C. Norberg: Effects of Reynolds number and a low-intensity free-stream turbulence intensity on the flow around a circular cylinder. Publication No. 87/2 Dept. Applied Thermodynamics and Fluid Mechanics, Chalmers University of Technology, 1987.
18. C. Norberg: An experimental investigation of the flow around a circular cylinder: influence of aspect ratio. Journal of Fluid Mech. **258**, 1994, 287–316.
19. O. Posdziech and R. Grundmann: Numerical simulation of the flow around an infinitely long circular cylinder in the transition regime. Theoretical and Computational Fluid Dynamics **15**, 2001, 121–141.
20. A. Roshko: On the development of turbulent wakes from vortex streets. Rep. 1191, NACA, 1954.
21. B.N. Rajani, H.G. Lanka, and S. Majumdar: Laminar flow past a circular cylinder at Reynolds number varying from 50 to 5000. NAL PD CF 0501, 2005.
22. B.N. Rajani, A. Kandasamy, and S. Majumdar: Numerical simulation of laminar flow past a circular cylinder. Appl. Math. Modelling **33**, 2009, 1228–1247.
23. S. Succi: *The Lattice Boltzmann Equation for Fluid Dynamics and Beyond.* Oxford University Press, New York, 2001.
24. S. Succi, I.V. Karlin, and H. Chen: Colluquium: Role of the H theorem in lattice Boltzmann hydrodynamic simulations. Rev. Modern Phys. **74**, 2002, 1203–1220.
25. C.S. Sunder and V. Babu: Entropic lattice Boltzmann method, non-uniform grids. ICCS, LNCS 3516, 2005, 72–79.
26. C.H.K. Williamson: Vortex dynamics in the cylinder wake. Annual Rev. Fluid Mech. **28**, 1996, 477–539.
27. D.A. Wolf-Gladrow: *Lattice-Gas Cellular Automata and Lattice Boltzmann Models: An Introduction.* Lecture notes in Mathematics **1725**, Springer-Verlag, Berlin, 2000.
28. D. You, R. Mei, and W. Shyy: A unified boundary treatment in lattice Boltzmann method. AIAA 2003-0953, New York, 2003.
29. M.M. Zdravkovich: *Flow Around Circular Cylinders (Volume I: Fundamentals).* Oxford University Press, New York, 1997.

Fast and Stable Interpolation of Well Data Using the Norm Function

Brian Li and Jeremy Levesley

Department of Mathematics, University of Leicester, LE1 7RH, UK

Summary. We present an iterative algorithm for computing an approximation to data in wells using the norm function in three dimensions augmented with a tensor product of two dimensional Lagrange functions with one dimensional linear interpolants. This augmentation can be thought of as a trend. The algorithm avoids the stability problems associated with data which have very different separation in different directions. In this case, data are close in the vertical direction inside each well, but the wells are relatively far apart. We will give an estimate of the convergence rate of the algorithm in terms of the vertical point spacing and the horizontal well separation.

1 Introduction

In this paper we will describe a practical algorithm for interpolating data that are in m columns (wells), each of depth d, with $n+1$ pieces of data in each column. For simplicity we assume that the data are of the form

$$\mathbf{y}_{i,j} = (\mathbf{x}_i, z_j) \in \mathbb{R}^3,$$

where $\mathbf{x}_i \in \mathbb{R}^2$, $i = 1, 2, \ldots, m$ and $z_j = jh, j = 0, 1, \ldots, n$, with $h = d/n$. Thus we assume that the data are uniformly spaced within the wells and that the wells are all mutually parallel. We wish to interpolate data $f_{i,j}, i = 1, 2, \ldots, m, j = 0, 1, \ldots, n$.

In radial basis function interpolation one might form an approximation

$$s(\mathbf{y}) = \sum_{i=1}^{m} \sum_{j=0}^{n} \alpha_{i,j} \phi(\|\mathbf{y} - \mathbf{y}_{i,j}\|) + b, \qquad (1)$$

where $\phi : \mathbb{R}_+ \to \mathbb{R}$, $\|\cdot\|$ denotes the Euclidean norm, and b is a constant. In this paper we will use $\phi(r) = r$. The unknown coefficients $\alpha_{i,j}$ and b are determined by the $((n+1)m+1)$ conditions

$$s(\mathbf{y}_{i,j}) = f_{i,j}, \quad i = 1, 2, \ldots, m, \; j = 0, 1, \ldots, n,$$

E.H. Georgoulis, A. Iske, J. Levesley (eds.), *Approximation Algorithms for Complex Systems*, Springer Proceedings in Mathematics 3, DOI: 10.1007/978-3-642-16876-5_11,
© Springer-Verlag Berlin Heidelberg 2011

and

$$\sum_{i=1}^{m}\sum_{j=0}^{n}\alpha_{i,j}=0.$$

In the following table we consider the problem where $m = 2$, located at $(0, 0)$ and $(10, 10)$, and $d = 1$. Below we give the condition number of the interpolation matrix for this problem which is the ratio of the largest and smallest eigenvalue. If the condition number is large the problem cannot be solved in this way in most cases. Even when it could, the computed solution is not anywhere near the exact solution. We can see below that the condition number grows with order n^2. Given that the intended application of this mechanism involves data that will more than likely contain noise as they come from human measurements, these large condition numbers render the direct inversion of the interpolation matrix useless.

n	condition number
100	2.1 (5)
200	8.3 (5)
400	3.3 (6)
800	1.3 (7)

Table 1. Condition number of standard interpolation matrix.

In order to get around this conditioning problem we use the straightforward observation that the norm function is the mod function when restricted to one dimension. This makes it trivial to construct an interpolant to data in each individual column. As we will see, this interpolant behaves like a one dimensional polynomial at the other wells, in other words, in the far fields.

2 The Interpolant and Algorithm

In order to construct the linear interpolant in a column we introduce the functions

$$\psi_h(\mathbf{y}) = \frac{\phi(\mathbf{y}+h\mathbf{e}) - 2\phi(\mathbf{y}) + \phi(\mathbf{y}-h\mathbf{e})}{2h},$$

$$\chi_t(\mathbf{y}) = \frac{\phi(\mathbf{y}-h\mathbf{e}) - \phi(\mathbf{y}) + h}{2h},$$

$$\chi_b(\mathbf{y}) = \frac{\phi(\mathbf{y}+h\mathbf{e}) - \phi(\mathbf{y}) + h}{2h},$$

where $\mathbf{e} = (0, 0, 1)$ is a unit vector in the vertical direction.

It is straightforward to check the results of the following lemma.

Lemma 1. *For $h > 0$,*

1. $\psi_h(\lambda\mathbf{e}) = 0, \ |\lambda| \geqslant h,$
2. $\psi_h(\mathbf{0}) = 1,$
3. $\chi_t(\lambda\mathbf{e}) = 0, \ \lambda \geqslant h,$
4. $\chi_t(\mathbf{0}) = 1,$
5. $\chi_b(\lambda\mathbf{e}) = 0, \ \lambda \leqslant -h,$
6. $\chi_b(\mathbf{0}) = 1.$

Proof. For $h > 0$,

1.

$$\begin{aligned}
\psi_h(\lambda\mathbf{e}) &= \frac{\phi(\lambda\mathbf{e} + h\mathbf{e}) - 2\phi(\lambda\mathbf{e}) + \phi(\lambda\mathbf{e} - h\mathbf{e})}{2h} \\
&= \frac{\|\lambda\mathbf{e} + h\mathbf{e}\| - 2\|\lambda\mathbf{e}\| + \|\lambda\mathbf{e} - h\mathbf{e}\|}{2h} \\
&= \frac{|\lambda + h|\|\mathbf{e}\| - 2|\lambda|\|\mathbf{e}\| + |\lambda - h|\|\mathbf{e}\|}{2h} \\
&= \frac{2|\lambda| - 2|\lambda|}{2h} \qquad\qquad (\|\mathbf{e}\| = 1 \text{ and } |\lambda| \geqslant h) \\
&= 0.
\end{aligned}$$

2.

$$\begin{aligned}
\psi_h(\mathbf{0}) &= \frac{\phi(h\mathbf{e}) - 2\phi(\mathbf{0}) + \phi(-h\mathbf{e})}{2h} \\
&= \frac{\|h\mathbf{e}\| - 2\|\mathbf{0}\| + \|-h\mathbf{e}\|}{2h} \\
&= \frac{|h|\|\mathbf{e}\| + |-h|\|\mathbf{e}\|}{2h} \\
&= \frac{2h}{2h} \qquad\qquad (\|\mathbf{e}\| = 1) \\
&= 1.
\end{aligned}$$

3.

$$\begin{aligned}
\chi_t(\lambda\mathbf{e}) &= \frac{\phi(\lambda\mathbf{e} - h\mathbf{e}) - \phi(\lambda\mathbf{e}) + h}{2h} \\
&= \frac{\|\lambda\mathbf{e} - h\mathbf{e}\| - \|\lambda\mathbf{e}\| + h}{2h} \\
&= \frac{|\lambda - h|\|\mathbf{e}\| - |\lambda|\|\mathbf{e}\| + h}{2h} \\
&= \frac{|\lambda| - |h| - |\lambda| + h}{2h} \qquad\qquad (\|\mathbf{e}\| = 1 \text{ and } \lambda \leqslant -h) \\
&= 0.
\end{aligned}$$

The rest can be proved in the same way.

These functions form a set of Lagrange functions for interpolation in each column. In order to discuss our approximation algorithm we need to understand how these functions behave for large arguments.

Proposition 1. *Let* $\mathbf{y} = \mathbf{x} + \beta\mathbf{e}$ *for some* $\beta \in \mathbb{R}$, *with* $|\beta| \leqslant d \ll \|\mathbf{x}\|$. *Then*

1. $\psi_h(\mathbf{y}) = \dfrac{h}{2\|\mathbf{x}\|} + \mathcal{O}\left(\dfrac{h}{\|\mathbf{x}\|^3}\right)$,
2. $\chi_t(\mathbf{y}) = \dfrac{1}{2} - \dfrac{2\beta - h}{4\|\mathbf{x}\|} + \mathcal{O}\left(\dfrac{1}{\|\mathbf{x}\|^3}\right)$,
3. $\chi_b(\mathbf{y}) = \dfrac{1}{2} + \dfrac{2\beta + h}{4\|\mathbf{x}\|} + \mathcal{O}\left(\dfrac{1}{\|\mathbf{x}\|^3}\right)$.

Proof. Since $\mathbf{y} = \mathbf{x} + \beta\mathbf{e}$ and $\mathbf{x} \perp \mathbf{e}$,

$$\begin{aligned}
\phi(\mathbf{y}) &= \|\mathbf{x} + \beta\mathbf{e}\| \\
&= \sqrt{\|\mathbf{x}\|^2 + \beta^2} \\
&= \|\mathbf{x}\|\sqrt{1 + \frac{\beta^2}{\|\mathbf{x}\|^2}} \\
&= \|\mathbf{x}\|\left(1 + \frac{\beta^2}{\|\mathbf{x}\|^2}\right)^{\frac{1}{2}} \\
&= \|\mathbf{x}\|\left(\sum_{k=0}^{\infty}\binom{\frac{1}{2}}{k}\left(\frac{\beta^2}{\|\mathbf{x}\|^2}\right)^k\right)
\end{aligned}$$

using binomial expansion, in which the binomial coefficient

$$\binom{\frac{1}{2}}{k} = \binom{2k+1}{k}\frac{(-1)^{k+1}(k+1)}{2^{2k}(2k-1)(2k+1)}, \qquad k = 0, 1, 2, \ldots.$$

Given that $|(x+h)^k - 2x^k + (x-h)^k| \leqslant Ch^2x^{k-2}$ where C is a multiple of the combination coefficient of the form $2\binom{k}{2}$, for any $k \geqslant 2$,

$$\begin{aligned}
\psi_h(\mathbf{y}) &= \frac{\phi(\mathbf{y} + h\mathbf{e}) - 2\phi(\mathbf{y}) + \phi(\mathbf{y} - h\mathbf{e})}{2h} \\
&= \frac{\|\mathbf{x} + (\beta + h)\mathbf{e}\| - 2\|\mathbf{x} + \beta\mathbf{e}\| + \|\mathbf{x} + (\beta - h)\mathbf{e}\|}{2h} \\
&= \frac{\|\mathbf{x}\|}{2h}\sum_{k=0}^{\infty}\binom{\frac{1}{2}}{k}\left(\left(\frac{\beta + h}{\|\mathbf{x}\|}\right)^{2k} - 2\left(\frac{\beta}{\|\mathbf{x}\|}\right)^{2k} + \left(\frac{\beta - h}{\|\mathbf{x}\|}\right)^{2k}\right) \\
&= \frac{\|\mathbf{x}\|}{2h}\binom{\frac{1}{2}}{1}\left(\frac{(\beta + h)^2 - 2\beta^2 + (\beta - h)^2}{\|\mathbf{x}\|^2}\right) \\
&\quad + \frac{\|\mathbf{x}\|}{2h}\binom{\frac{1}{2}}{2}\left(\frac{(\beta + h)^4 - 2\beta^4 + (\beta - h)^4}{\|\mathbf{x}\|^4}\right) + \mathcal{O}\left(\frac{h}{\|\mathbf{x}\|^5}\right) \\
&= \frac{1}{2h\|\mathbf{x}\|}\left(\frac{1}{2}\right)(2h^2) + \frac{1}{2h\|\mathbf{x}\|^3}\left(-\frac{1}{8}\right)(12\beta^2h^2 + 2h^4) + \mathcal{O}\left(\frac{h}{\|\mathbf{x}\|^5}\right)
\end{aligned}$$

$$
\begin{aligned}
&= \frac{h}{2\|\mathbf{x}\|} - \frac{6\beta^2 h + h^3}{8\|\mathbf{x}\|^3} + \mathcal{O}\left(\frac{h}{\|\mathbf{x}\|^5}\right) \\
&= \frac{h}{2\|\mathbf{x}\|} + \mathcal{O}\left(\frac{h}{\|\mathbf{x}\|^3}\right).
\end{aligned}
$$

We also have, since $|(x-h)^k - x^k| \leqslant Chx^{k-1}$, for some C constant, for any $k \geqslant 1$,

$$
\begin{aligned}
\chi_t(\mathbf{y}) &= \frac{\phi(\mathbf{y} - h\mathbf{e}) - \phi(\mathbf{y}) + h}{2h} \\
&= \frac{\|\mathbf{x} + (\beta - h)\mathbf{e}\| - \|\mathbf{x} + \beta\mathbf{e}\| + h}{2h} \\
&= \frac{1}{2} + \frac{\|\mathbf{x}\|}{2h} \sum_{k=0}^{\infty} \binom{\frac{1}{2}}{k} \left(\left(\frac{\beta - h}{\|\mathbf{x}\|}\right)^{2k} - \left(\frac{\beta}{\|\mathbf{x}\|}\right)^{2k} \right) \\
&= \frac{1}{2} + \frac{\|\mathbf{x}\|}{2h} \binom{\frac{1}{2}}{1} \left(\frac{(\beta - h)^2 - \beta^2}{\|\mathbf{x}\|^2} \right) + \mathcal{O}\left(\frac{1}{\|\mathbf{x}\|^3}\right) \\
&= \frac{1}{2} - \frac{2\beta - h}{4\|\mathbf{x}\|} + \mathcal{O}\left(\frac{1}{\|\mathbf{x}\|^3}\right).
\end{aligned}
$$

Using a similar computation we can obtain the following for χ_b,

$$
\chi_b(\mathbf{y}) = \frac{1}{2} + \frac{2\beta + h}{4\|\mathbf{x}\|} + \mathcal{O}\left(\frac{1}{\|\mathbf{x}\|^3}\right).
$$

In order to compute an interpolant we first form the functions

$$
s_i^1(\mathbf{y}) = \sum_{j=1}^{n-1} f_{i,j}\psi_h(\mathbf{y} - \mathbf{y}_{i,j}) + f_{i,0}\chi_b(\mathbf{y} - \mathbf{y}_{i,0}) + f_{i,n}\chi_t(\mathbf{y} - \mathbf{y}_{i,n}) \quad i = 1, 2, \ldots, m.
$$

Proposition 2. *Let* $\mathbf{y} = \mathbf{x} + z\mathbf{e}$. *For each* $i = 1, 2, \ldots, m$,

$$
\begin{aligned}
s_i^1(\mathbf{y}) = &\left(1 + \frac{h}{2\|\mathbf{x} - \mathbf{x}_i\|}\right) \frac{f_{i,0} + f_{i,n}}{2} + \left(\frac{df_{i,n} + h\sum_{j=1}^{n-1} f_{i,j}}{2\|\mathbf{x} - \mathbf{x}_i\|} \right) \\
&+ \frac{(f_{i,0} - f_{i,n})z}{2\|\mathbf{x} - \mathbf{x}_i\|} + \mathcal{O}\left(\frac{1}{\|\mathbf{x} - \mathbf{x}_i\|^3}\right).
\end{aligned}
$$

Proof. For $i = 1, 2, \ldots, m$, we have

$$
\begin{aligned}
\sum_{j=1}^{n-1} f_{i,j}\psi_h(\mathbf{y} - \mathbf{y}_{i,j}) &= \sum_{j=1}^{n-1} f_{i,j}\psi_h((\mathbf{x} + z\mathbf{e}) - (\mathbf{x}_i + jh\mathbf{e})) \\
&= \sum_{j=1}^{n-1} f_{i,j}\psi_h((\mathbf{x} - \mathbf{x}_i) + (z - jh)\mathbf{e})
\end{aligned}
$$

$$
\begin{aligned}
&= \sum_{j=1}^{n-1} f_{i,j} \left(\frac{h}{2\|\mathbf{x}-\mathbf{x}_i\|} + \mathcal{O}\left(\frac{h}{\|\mathbf{x}-\mathbf{x}_i\|^3} \right) \right) \\
&= \frac{h}{2\|\mathbf{x}-\mathbf{x}_i\|} \sum_{j=1}^{n-1} f_{i,j} + \left(h \sum_{j=1}^{n-1} f_{i,j} \right) \cdot \mathcal{O}\left(\frac{1}{\|\mathbf{x}-\mathbf{x}_i\|^3} \right).
\end{aligned}
$$

Also,

$$
\begin{aligned}
f_{i,0}\chi_b(\mathbf{y}-\mathbf{y}_{i,0}) &= f_{i,0}\chi_b((\mathbf{x}+z\mathbf{e})-\mathbf{x}_i) \\
&= f_{i,0}\chi_b((\mathbf{x}-\mathbf{x}_i)+z\mathbf{e}) \\
&= f_{i,0} \left(\frac{1}{2} + \frac{2z+h}{4\|\mathbf{x}-\mathbf{x}_i\|} + \mathcal{O}\left(\frac{1}{\|\mathbf{x}-\mathbf{x}_i\|^3} \right) \right) \\
&= \frac{1}{2} f_{i,0} + \frac{2z+h}{4\|\mathbf{x}-\mathbf{x}_i\|} f_{i,0} + f_{i,0} \cdot \mathcal{O}\left(\frac{1}{\|\mathbf{x}-\mathbf{x}_i\|^3} \right).
\end{aligned}
$$

Finally, recalling that $d = nh$,

$$
\begin{aligned}
f_{i,n}\chi_t(\mathbf{y}-\mathbf{y}_{i,n}) &= f_{i,n}\chi_t((\mathbf{x}+z\mathbf{e})-(\mathbf{x}_i+d\mathbf{e})) \\
&= f_{i,n}\chi_t((\mathbf{x}-\mathbf{x}_i)+(z-d)\mathbf{e}) \\
&= f_{i,n} \left(\frac{1}{2} - \frac{2(z-d)-h}{4\|\mathbf{x}-\mathbf{x}_i\|} + \mathcal{O}\left(\frac{1}{\|\mathbf{x}-\mathbf{x}_i\|^3} \right) \right) \\
&= \frac{1}{2} f_{i,n} - \frac{2(z-d)-h}{4\|\mathbf{x}-\mathbf{x}_i\|} f_{i,n} + f_{i,n} \cdot \mathcal{O}\left(\frac{1}{\|\mathbf{x}-\mathbf{x}_i\|^3} \right).
\end{aligned}
$$

Aggregating the above three parts we have,

$$
\begin{aligned}
s_i^1(\mathbf{y}) &= \sum_{j=1}^{n-1} f_{i,j}\psi_h(\mathbf{y}-\mathbf{y}_{i,j}) + f_{i,0}\chi_b(\mathbf{y}-\mathbf{y}_{i,0}) + f_{i,n}\chi_t(\mathbf{y}-\mathbf{y}_{i,n}) \\
&= \left(\frac{h}{2\|\mathbf{x}-\mathbf{x}_i\|} \sum_{j=1}^{n-1} f_{i,j} + \underbrace{\left(h \sum_{j=1}^{n-1} f_{i,j} \right)}_{constant} \cdot \mathcal{O}\left(\frac{1}{\|\mathbf{x}-\mathbf{x}_i\|^3} \right) \right) \\
&\quad + \left(\frac{1}{2} f_{i,0} + \frac{2z+h}{4\|\mathbf{x}-\mathbf{x}_i\|} f_{i,0} + f_{i,0} \cdot \mathcal{O}\left(\frac{1}{\|\mathbf{x}-\mathbf{x}_i\|^3} \right) \right) \\
&\quad + \left(\frac{1}{2} f_{i,n} - \frac{2(z-d)-h}{4\|\mathbf{x}-\mathbf{x}_i\|} f_{i,n} + f_{i,n} \cdot \mathcal{O}\left(\frac{1}{\|\mathbf{x}-\mathbf{x}_i\|^3} \right) \right) \\
&= \frac{h}{2\|\mathbf{x}-\mathbf{x}_i\|} \sum_{j=1}^{n-1} f_{i,j} + \mathcal{O}\left(\frac{1}{\|\mathbf{x}-\mathbf{x}_i\|^3} \right) + \frac{f_{i,0}+f_{i,n}}{2} + \frac{2z+h}{4\|\mathbf{x}-\mathbf{x}_i\|} f_{i,0} \\
&\quad - \frac{2(z-d)-h}{4\|\mathbf{x}-\mathbf{x}_i\|} f_{i,n} + f_{i,0} \cdot \mathcal{O}\left(\frac{1}{\|\mathbf{x}-\mathbf{x}_i\|^3} \right) + f_{i,n} \cdot \mathcal{O}\left(\frac{1}{\|\mathbf{x}-\mathbf{x}_i\|^3} \right)
\end{aligned}
$$

$$= \frac{h}{2\|\mathbf{x}-\mathbf{x}_i\|}\sum_{j=1}^{n-1} f_{i,j} + \frac{f_{i,0}+f_{i,n}}{2} + \frac{z}{2\|\mathbf{x}-\mathbf{x}_i\|}f_{i,0} + \frac{h}{4\|\mathbf{x}-\mathbf{x}_i\|}f_{i,0}$$

$$-\frac{z}{2\|\mathbf{x}-\mathbf{x}_i\|}f_{i,n} + \frac{d}{2\|\mathbf{x}-\mathbf{x}_i\|}f_{i,n} + \frac{h}{4\|\mathbf{x}-\mathbf{x}_i\|}f_{i,n} + \mathcal{O}\left(\frac{1}{\|\mathbf{x}-\mathbf{x}_i\|^3}\right)$$

$$= \left(1+\frac{h}{2\|\mathbf{x}-\mathbf{x}_i\|}\right)\frac{f_{i,0}+f_{i,n}}{2} + \left(\frac{df_{i,n}+h\sum_{j=1}^{n-1} f_{i,j}}{2\|\mathbf{x}-\mathbf{x}_i\|}\right)$$

$$+\frac{(f_{i,0}-f_{i,n})z}{2\|\mathbf{x}-\mathbf{x}_i\|} + \mathcal{O}\left(\frac{1}{\|\mathbf{x}-\mathbf{x}_i\|^3}\right),$$

where $h\sum_{j=1}^{n-1} f_{i,j}$ is a constant whose value depends only on the dataset.

The above proof demonstrates the fact that if $\mathbf{y}$ is a long way from the ith column then $s_i^1(\mathbf{y})$ behaves linearly with variation in the $\mathbf{e}$ direction, i.e. changes in z. Let us construct

$$l_{i,k}^1(z) = \left(1+\frac{h}{2\|\mathbf{x}_k-\mathbf{x}_i\|}\right)\frac{f_{i,0}+f_{i,n}}{2} + \left(\frac{df_{i,n}+h\sum_{j=1}^{n-1} f_{i,j}}{2\|\mathbf{x}_k-\mathbf{x}_i\|}\right) + \frac{(f_{i,0}-f_{i,n})z}{2\|\mathbf{x}_k-\mathbf{x}_i\|},$$

and

$$p_i^1(z) = \sum_{\substack{k=1\\ i\neq k}}^{m} l_{i,k}^1(z).$$

Thus p_i^1 is the vertical variation at the ith well due to the sum of the vertical variations in the far fields of the interpolants along each well.

In order to construct our interpolant we need to build a two dimensional Lagrange basis for the points $\mathbf{x}_i$, $i=1,2,\ldots,m$. We can do this in any way we like, but for consistency we will use a radial basis function of the form (1)

$$\rho_i(\mathbf{x}) = \sum_{k=1}^{m} \alpha_{i,k}\|\mathbf{x}-\mathbf{x}_k\| + b_i,$$

where we compute the coefficients $\alpha_{i,k}$ and b_i by interpolation satisfying the conditions

$$\rho_i(\mathbf{x}_j) = \delta_{ij}, \quad i,j=1,2,\ldots,m,$$

with δ_{ij} being the Kronecker delta and

$$\sum_{k=1}^{m} \alpha_{i,k} = 0.$$

To compute our intermediate approximation we form

$$\sigma^1(\mathbf{y}) = \sum_{i=1}^{m} s_i^1(\mathbf{y}).$$

If we compute the residual

$$r_{i,j} := f_{i,j} - s^1(\mathbf{y}_{i,j}) \quad i = 1, 2, \ldots, m, \; j = 0, 1, \ldots, n,$$

then the previous proposition tells us that for each $i = 1, 2, \ldots, m$, $r_{i,j}$ varies approximately linearly with changes in j. Thus let us form the tensor product

$$q^1(\mathbf{y}) = \sum_{i=1}^{m} \rho_i(\mathbf{x}) p_i^1(z).$$

Our first approximate interpolant is then

$$s^1(\mathbf{y}) = \sigma^1(\mathbf{y}) - q^1(\mathbf{y}).$$

We produce better approximation to the interpolant by repeating the above procedure using successive residuals as the target function. Iterative algorithms for RBF approximation have been studied in detail by Faul and Powell [3, 5].

3 Convergence

In this section we shall consider the error of interpolation and see that it depends on the distance between the wells.

Theorem 1. *For $i = 1, 2, \ldots, m$ and $j = 0, 1, \ldots, n$,*

$$|s^1(\mathbf{y}_{i,j}) - f_{i,j}| \leqslant \frac{C}{\delta^3},$$

where

$$\delta = \min_{\substack{1 \leqslant i,k \leqslant m \\ i \neq k}} \|\mathbf{x}_i - \mathbf{x}_k\|.$$

Proof. For $i = 1, 2, \ldots, m$ and $j = 0, 1, \ldots, n$, using Lemma 1,

$$\begin{aligned}
|s^1(\mathbf{y}_{i,j}) - f_{i,j}| &= \sigma^1(\mathbf{y}_{i,j}) - q^1(\mathbf{y}_{i,j}) - f_{i,j} \\
&= f_{i,j} + \sum_{\substack{k=1 \\ i \neq k}}^{m} s_k^1(\mathbf{y}_{i,j}) - q^1(\mathbf{y}_{i,j}) - f_{i,j} \\
&= \sum_{\substack{k=1 \\ i \neq k}}^{m} \left(l_{i,k}^1(z_j) + \mathcal{O}\left(\frac{1}{\|\mathbf{x}_i - \mathbf{x}_k\|^3} \right) \right) - p_i^1(z_j),
\end{aligned}$$

since

$$
\begin{aligned}
q^1(\mathbf{y}_{i,j}) &= \sum_{k=1}^{m} \rho_k(\mathbf{x}_i) p_k^1(z_j) \\
&= \sum_{k=1}^{m} \delta_{i,k} p_k^1(z_j) \\
&= p_i^1(z_j).
\end{aligned}
$$

However, by definition,

$$
\sum_{\substack{k=1 \\ i \neq k}}^{m} l_{i,k}^1(z_j) = p_i^1(z_j),
$$

and the result is established.

Thus we see that, as opposed to the usual case where the larger the well-separation (in relative comparison to vertical separation amongst data points) the more difficult the interpolation (see Table 1), this algorithm converges faster the better separated the wells are.

4 Numerical Examples

In the first table below we list a set of numerical experiments. We generate m randomly spaced columns of data with bases in the square $[0 \ldots 10, 0 \ldots 10]$. Each column is of height 1. The target data is randomly generated in the interval $[0, 1]$ and the mean average of 10 repeated trials is analysed.

We can see that we get convergence to an interpolant in a small number of iterations (with the error threshold being 10^{-10}). The results confirm the analysis of the previous section that the number of iterations does not depend on the amount of data in each column, but on the number of columns, and more precisely the minimum separation between the columns.

We also observe that in two cases, $m = 4, n = 400$ and $m = 8, n = 25$, there are trials where convergence is slow. For the former, one of the trials required 17 iterations and the latter, 57 iterations. Upon close inspection (see Figure 1), we discover that both have wells that are positioned very close to each other. Further development of the algorithm will deal with wells which are close together.

More scrutiny of the well location plots reveals that the number of iterations depends on the minimal distance between the columns, as illustrated with the example below.

In this case, aside from Trial 1 which is shown above, both Trials 5 and 6 require a fairly large number of iterations in order to reach the error threshold (See Figure 2). Furthermore it can be seen that as the minimum distance between wells grew larger, the number of iterations gradually decreased (See Figure 3).

m	n	mean number of iterations
2	25	1.6
2	50	1.1
2	100	1.4
2	200	1.2
2	400	1.4
4	25	2.4
4	50	2.3
4	100	2.1
4	200	3.1
4	400	2.67
8	25	7.22
8	50	6.1
8	100	6.1
8	200	6.2
8	400	5.7
16	25	12.4
16	50	13.3
16	100	14.3

Table 2. Number of iterations for m wells in $[0\dots10, 0\dots10]$ and n data in each well.

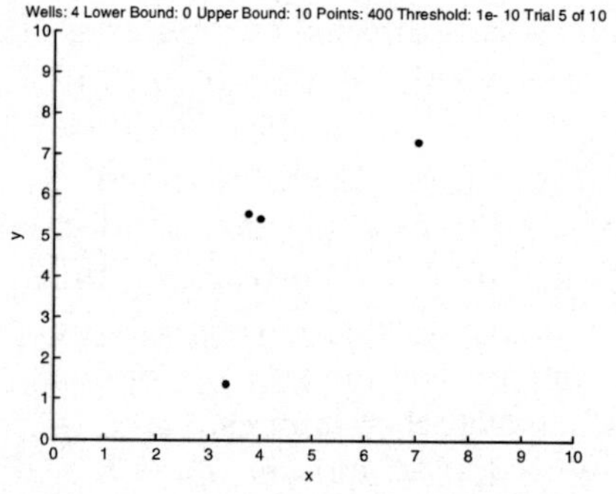

(a) Trial 5, $m = 4, n = 400$, 17 iterations

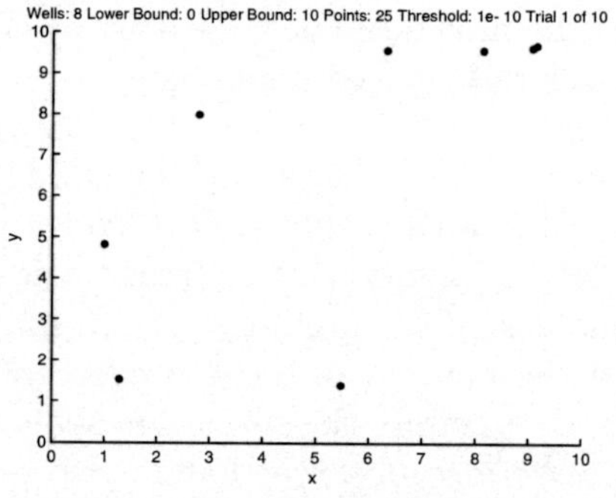

(b) Trial 1, $m = 8, n = 25$, 57 iterations

Fig. 1. Well location plot

In the next table we have fixed parameters $m = 8$ and $n = 50$. The side length of the square within which the wells are randomly positioned is M. We perform the algorithm for 10 sets of random data and compute the mean average number of iterations to achieve an error of less than 10^{-10}. We start with $M = 4$, as for smaller numbers we see instances of lack of convergence of the algorithm as there is a much greater chance of two wells being too near to each other. We see that the number of iterations decreases as the points become better spaced.

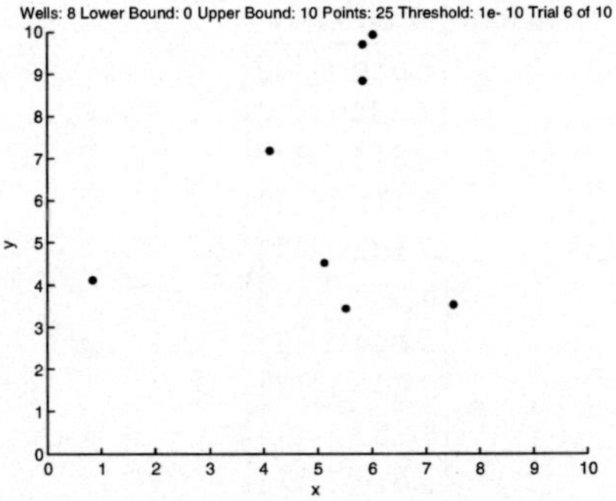

(a) Trial 6, $m = 8, n = 25$, 14 iterations

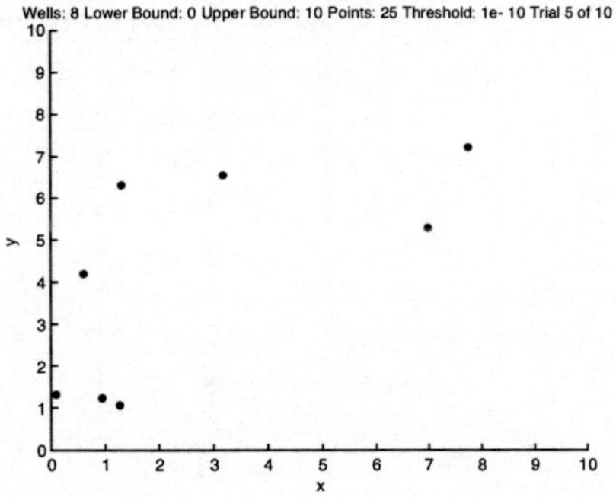

(b) Trial 5, $m = 8, n = 25$, 12 iterations

Fig. 2. Well location plot

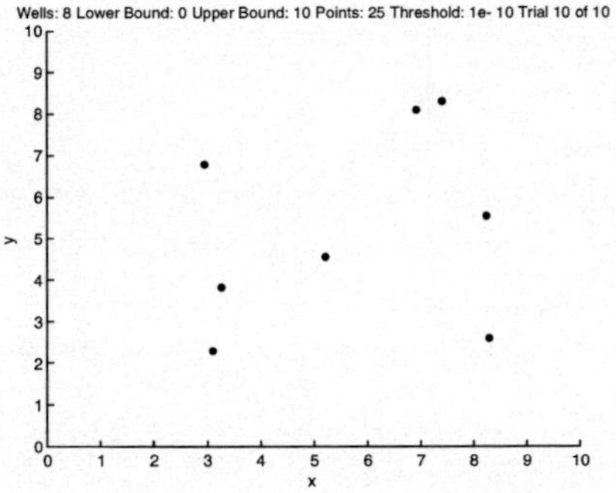

(a) Trial 10, $m = 8, n = 25$, 8 iterations

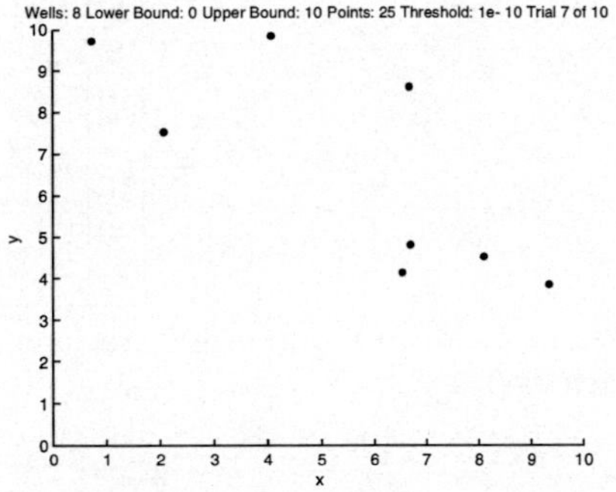

(b) Trial 7, $m = 8, n = 25$, 6 iterations

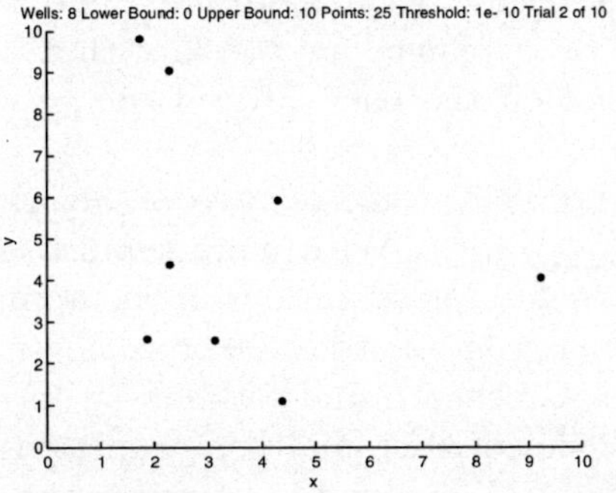

(c) Trial 2, $m = 8, n = 25$, 4 iterations

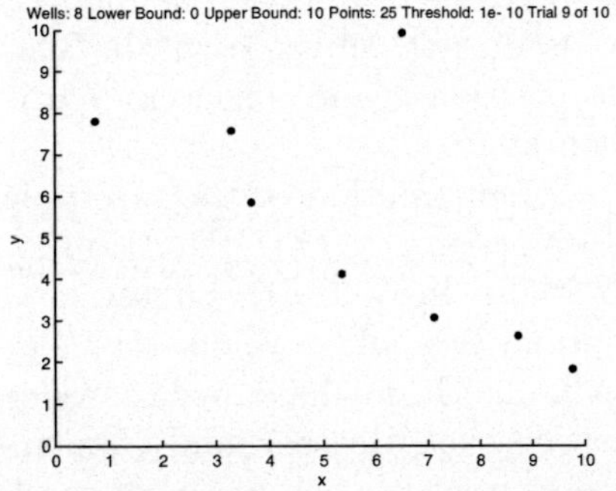

(d) Trial 9, $m = 8, n = 25$, 3 iterations

Fig. 3. Well location plot

trial number	number of iterations	final error
1	57	9.016 (-11)
2	4	3.634 (-11)
3	7	2.814 (-11)
4	8	5.118 (-11)
5	12	5.162 (-11)
6	14	5.675 (-11)
7	6	2.844 (-11)
8	3	8.636 (-11)
9	3	8.163 (-11)
10	8	2.578 (-11)

Table 3. Trial results where $m = 8, n = 25$.

M	number of iterations
4	9.67
6	8.89
8	8.2
10	6.1
12	4.44
16	2.7
20	2.8
24	2.4

Table 4. Mean number of iterations for 8 wells in $[0 \dots M, 0 \dots M]$ and 50 data in each well.

When two wells get too close to each other, the effects are not restricted to increases in iteration counts. Extreme results occur more often and some trials do not converge at all. This can be explained by the fact that both Proposition 2 and Theorem 1 are formed on the basis of wells being well-separated.

Direct solution of the interpolation problem would lead to an algorithm of order $(mn)^3$. This algorithm is of order $(mn)^2$. With more sophisticated programming and explicit use of the far-field representations of the approximations we could develop a fast algorithm, i.e. one of order mn or $mn \log(mn)$. Such algorithms have been developed by Greengard and Rokhlin [4] for potentials, and Beatson and Newsam [1, 2] for other radial basis functions. A good survey of such methods can be found in Wendland [6, Chapter 15].

5 Conclusions and Further Developments

The purpose of this paper is to provide a fast and stable algorithm for approximating data in columns. We have achieved this by blending univariate approximations, observing that in the far field, each of this behaves in a predictable way. We can easily improve the algorithm by increasing the length of the far field expansions used. In this way, a smaller number of iterations of the algorithm is required to get convergence to a specified tolerance. A implementation using the cubic terms in the far field expansion is almost complete.

An alternative approach to this method is to scale the points together and use standard approximation methods for more uniformly distributed data. This is a perfectly reasonable thing to do in Euclidean space, but fails to be appropriate on the surface of the sphere, and this is one of the directions we are interested in developing the algorithm. On the sphere there are often thin shell type approximations, in which the depth of the atmosphere may be small compared to the distances between columns where various quantities are being measured. Scaling on the sphere is not a continuous mapping, so the method proposed here of explicitly using the distance between the columns to stabilise the approximation process has some merit.

Acknowledgement

We thank the referee for useful remarks which have made this paper more coherent.

References

1. R.K. Beatson and G.N. Newsam: Fast evaluation of radial basis functions: I. Computers & Mathematics with Applications **24**, 1992, 7–19.
2. R.K. Beatson and G.N. Newsam: Fast evaluation of radial basis functions: moment-based methods. SIAM Journal on Scientific Computing **19**, 1998, 1428–1449.
3. A.C. Faul and M.J.D. Powell: Proof of convergence of an iterative technique for thin plate spline interpolation in two dimensions. Advances in Computational Mathematics **11**, 1999, 183–192.
4. L. Greengard and V. Rokhlin: A fast algorithm for particle simulations. Journal of Computational Physics **73**, 1987, 325–348.
5. M.J.D. Powell: A new iterative algorithm for thin plate spline interpolation in two dimensions. Annals of Numerical Mathematics **4**, 1997, 519–527.
6. H. Wendland: *Scattered Data Approximation.* Cambridge University Press, Cambridge, UK, 2005.

Algorithms and Literate Programs for Weighted Low-Rank Approximation with Missing Data

Ivan Markovsky

School of Electronics & Computer Science, Univ. of Southampton, SO17 1BJ, UK

Summary. Linear models identification from data with missing values is posed as a weighted low-rank approximation problem with weights related to the missing values equal to zero. Alternating projections and variable projections methods for solving the resulting problem are outlined and implemented in a literate programming style, using Matlab/Octave's scripting language. The methods are evaluated on synthetic data and real data from the MovieLens data sets.

1 Introduction

Low-Rank Approximation

We consider the following low-rank approximation problem: given a real matrix D of dimensions $q \times N$ and an integer $\mathtt{m}$, $0 < \mathtt{m} < \min(q, N)$, find a matrix $\widehat{D}$ of the same dimension as D, with rank at most $\mathtt{m}$, that is as "close" to D as possible, i.e.,

$$\text{minimize} \quad \text{over } \widehat{D} \quad \text{dist}(D, \widehat{D}) \quad \text{subject to} \quad \text{rank}(\widehat{D}) \leq \mathtt{m}. \tag{1}$$

The distance $\text{dist}(D, \widehat{D})$ between the given matrix D and its approximation $\widehat{D}$ can be measured by a norm of the approximation error $\Delta D := D - \widehat{D}$, i.e.,

$$\text{dist}(D, \widehat{D}) = \|D - \widehat{D}\|.$$

A typical choice of the norm $\|\cdot\|$ is the Frobenius norm

$$\|\Delta D\|_{\mathrm{F}} = \sqrt{\sum_{i=1}^{q} \sum_{j=1}^{N} \Delta d_{ij}^2},$$

i.e., the square root of the sum of squares of the elements. Assuming that a solution to 1 exists, the minimum value is the distance from D to the manifold

E.H. Georgoulis, A. Iske, J. Levesley (eds.), *Approximation Algorithms for Complex Systems*, Springer Proceedings in Mathematics 3, DOI: 10.1007/978-3-642-16876-5_12,
© Springer-Verlag Berlin Heidelberg 2011

of rank-$\mathtt{m}$ matrices and a minimum point $\widehat{D}^*$ is a "best" (in the sense specified by the distance measure "dist") rank-$\mathtt{m}$ approximation of D.

Apart from being an interesting mathematical problem, low-rank approximation has a large range of applications in diverse areas, e.g., in numerical analysis to find a rank estimate that is robust to "small" perturbations on the matrix. The intrinsic reasons for the widespread appearance of low-rank approximation in applications are 1) low-rank approximation has an interpretation as a data modeling tool and 2) any application area where mathematical methods are used is based on a model. Thus low-rank approximation can provide (approximate) models from data, to be used for analysis, filtering, prediction, control, etc., in the application areas.

In order to make a link between low-rank approximation and data modeling, next we define the notion of a linear static model. Let the observed variables be $d_1, \dots, d_q$ and let $d := \operatorname{col}(d_1, \dots, d_q)$ be the column vector of these variables. We say that the variables $d_1, \dots, d_q$ satisfy a *linear static model* if $d \in \mathcal{L}$, where $\mathcal{L}$, the model, is a subspace of the data space—the q-dimensional real vector space $\mathbb{R}^q$. The *complexity* of a linear model is measured by its dimension. Of interest is data fitting by low complexity models, in which case, generally, the model may only fit approximately the data. Consider a set of data points $\mathcal{D} = \{ d^{(1)}, \dots, d^{(N)} \} \subset \mathbb{R}^q$ and define the data matrix

$$D := \begin{bmatrix} d^{(1)} & \cdots & d^{(N)} \end{bmatrix} \in \mathbb{R}^{q \times N}.$$

Assuming that there are more measurements than data variables, i.e., $q < N$, it is easy to see that $\operatorname{rank}(D) \leq \mathtt{m}$ if and only if all data points satisfy a linear static model of complexity at most $\mathtt{m}$. This fact is the key link between low-rank approximation and data modeling.

The condition that the data satisfies exactly the model is too strong in practice. For example, if a "true" data $\bar{D}$ satisfies a linear model of low-complexity, i.e., $\operatorname{rank}(\bar{D}) < q$, but is measured subject to noise, i.e., $D = \bar{D} + \widetilde{D}$ ($\widetilde{D}$ being the measurement noise), the noisy measurements D generically do not satisfy a linear model of low-complexity, i.e., almost surely, $\operatorname{rank}(D) = q$. In this case, the modeling goal may be to *estimate* the true but unknown low-complexity model generating $\bar{D}$. Another example showing that the condition $\operatorname{rank}(D) < q$ is too strong is when the data is exact but is generated by a nonlinear phenomenon. In this case, the modeling goal may be to *approximate* the true nonlinear phenomenon by a linear model of bounded complexity. In both cases—estimation and approximation—the data modeling problem leads to low-rank approximation—the rank constraint ensures that the approximation $\widehat{D}$ satisfies exactly a low-complexity linear model. In the estimation example, this takes into account the prior knowledge about the true data generating phenomenon. The approximation criterion "$\min \operatorname{dist}(D, \widehat{D})$" ensures that the obtained model approximates "well" the data. In the estimation case, this corresponds to prior knowledge that the noise is zero mean and "small" in some sense.

Note 1 (Link to the principal component analysis). It can be shown that the well known principal component analysis method is equivalent to low-rank approximation in the Frobenius norm. The number of principal components in the principal component analysis corresponds to the rank constraint in the low-rank approximation problem and the span of the principal components corresponds to the column span of the approximation $\widehat{D}$, i.e., the model. Principal component analysis is typically presented and motivated in a stochastic context, however, the stochastic point of view is not essential and the method is also applicable as a deterministic approximation method.

Note 2 (Link to regression). The classical approach for data fitting involves, in addition to the rank constraint, a priori chosen input/output partition of the variables $\operatorname{col}(a, b) := \Pi d$, where Π is a permutation matrix. Then the low-rank approximation problem reduces to the problem of solving approximately an overdetermined system of equations $AX \approx B$ (from a stochastic point of view—regression), where $A\ B := (\Pi D)^\top$. By choosing specific fitting criteria, the classical approach leads to well know optimization problems, e.g., linear least squares, total least squares, robust least squares, and their numerous variations. The total least squares problem [6] is generically equivalent to low-rank approximation in the Frobenius norm.

Missing Data

A more general approximation criterion than $\|\Delta D\|_{\mathrm{F}}$ is the element-wise weighted norm of the error matrix

$$\|\Delta D\|_{\Sigma} := \|\Sigma \odot \Delta D\|_{\mathrm{F}}, \qquad \text{where } \odot \text{ denotes element-wise product,}$$

and $\Sigma \in \mathbb{R}^{q\times N}$ has positive elements. The low-rank approximation Problem 1 with $\operatorname{dist}(D, \widehat{D}) = \|D - \widehat{D}\|_{\Sigma}$, $\Sigma > 0$, is called (regular) weighted low-rank approximation [3, 21, 12, 14, 15]. The weights σ_{ij} allow us to emphasise or de-emphasise the importance of the individual elements of the data matrix D. If σ_{ij} is small, relative to the other weights, then the influence of d_{ij} on the approximation $\widehat{D}$ is small and vice verse.

In the extreme case of a zero weight, e.g., $\sigma_{ij} = 0$, the corresponding element d_{ij} of D is not taken into account in the approximation and therefore it may be missing. In this case, however, $\|\cdot\|_{\Sigma}$ is no longer a norm and the approximation problem is called singular. The above cited work on the weighted low-rank approximation problem treats the regular case and the methods fail in the singular case. The purpose of this paper is to extend the solution methods, derived for the regular weighted low-rank approximation problem to the singular case, so that these algorithms can treat missing data.

Note 3 (Missing rows and columns). The case of missing rows and/or columns of the data matrix is easy to take into account. It reduces the original singular problem to a smaller dimensional regular problem. The same reduction, however, is not possible when the missing elements have no simple pattern.

Low-rank approximation with missing data occurs in

- factor analysis of data from questioners due to questions left answered,
- computer vision due to occlusions,
- signal processing due to irregular measurements in time/space, and
- control due to malfunction of measurement devices.

An iterative solution method (called criss-cross multiple regression) for factor analysis with missing data was developed by Gabriel and Zamir [4]. Their method, however, does not necessarily converge to a minimum point (see the discussion in Section 6, page 491 of [4]). Grung and Manne proposed an alternating projections algorithm for the case of unweighted approximation with missing values, i.e., $\sigma_{ij} \in \{0, 1\}$. Their method was further generalized by Srebro [20] for arbitrary weights.

In this paper, apart from the alternating projections algorithm, we consider an algorithm for weighted low-rank approximation with missing data, based on the variable projections method [5]. The former has linear local convergence rate while the latter has super linear convergence rate which suggests that it may be faster. In addition, we present an implementation of the two algorithms in a literate programming style. A literate program is a combination of computer executable code and human readable description of this code [10]. From the source file, the user extracts both the computer code and its documentation. We use Matlab/Octave's scripting language for the computer code, LaTeX for its documentation, and `noweb` [18] for their combination, see Appendix A.

2 Low-Rank Approximation with Uniform Weights

Low-rank approximation in the Frobenius norm (equivalently weighted low-rank approximation uniform weights $\sigma_{ij} = \sigma$ for all i, j) can be solved analytically in terms of the singular value decomposition (SVD) of the data matrix D.

Lemma 1 (Matrix approximation lemma). *Let*

$$D = U\Sigma V^\top, \qquad \Sigma =: \operatorname{diag}(\sigma_1, \ldots, \sigma_q)$$

be the SVD of $D \in \mathbb{R}^{q\times N}$ and partition the matrices U, Σ, and V as follows:

$$U =: \begin{matrix} \mathtt{m} & \mathtt{p} \\ U_1 & U_2 \end{matrix}\; q, \quad \Sigma =: \begin{matrix} \mathtt{m} & \mathtt{p} & \\ \Sigma_1 & 0 & \mathtt{m} \\ 0 & \Sigma_2 & \mathtt{p} \end{matrix} \quad \text{and} \quad V =: \begin{matrix} \mathtt{m} & \mathtt{p} \\ V_1 & V_2 \end{matrix}\; N,$$

where $\mathtt{m} \in \mathbb{N}$, $0 \le \mathtt{m} \le \min(q, N)$, and $\mathtt{p} := q - \mathtt{m}$. Then

$$\widehat{D}^* = U_1 \Sigma_1 V_1^\top$$

is an, optimal in the Frobenius norm rank-$\mathtt{m}$*, approximation of* D*, i.e.,*

$$\|D - \widehat{D}^*\|_{\mathrm{F}} = \sqrt{\sigma_{\mathtt{m}+1}^2 + \cdots + \sigma_q^2} = \min_{\mathrm{rank}(\widehat{D}) \leq \mathtt{m}} \|D - \widehat{D}\|_{\mathrm{F}}.$$

The solution $\widehat{D}^*$ *is unique if and only if* $\sigma_{\mathtt{m}+1} \neq \sigma_{\mathtt{m}}$.

From a data modeling point of view of primary interest is the subspace

$$\operatorname{col\,span}(\widehat{D}^*) = \operatorname{col\,span}(U_1) \tag{2}$$

rather than the approximation $\widehat{D}$.

⟨*Matrix approximation (SVD)* 259a⟩≡ (259c)
```
[u,s,v] = svds(d,m);
p = u(:,1:m); % basis for the optimal model for D
```

Note that the subspace 2 depends only on the left singular vectors of D. Therefore, the model 2 is optimal for the data DQ, where Q is any orthogonal matrix. Let

$$D = R_1\ 0Q^\top \tag{3}$$

be the QR factorization of D (R_1 is lower triangular.) By the above argument, we can model R_1 instead of D. For $N \gg q$, computing the QR factorization 3 and the SVD of R_1 is a more efficient alternative for finding an image representation of the optimal subspace than computing the SVD of D.

⟨*Data compression (QR)* 259b⟩≡ (259c)
```
if nargout == 1
  d = triu(qr(d'))'; % = R,  where D = QR
  d = d(:,1:q);      % = R1, where R = [R1 0]
end
```

Putting the matrix approximation and data compression code together, we have the following function for low-rank approximation.

⟨lra 259c⟩≡ (263)
```
⟨lra header 271⟩
function [p,l] = lra(d,m)

d(isnan(d)) = 0; % Convert missing elements (NaNs) to 0s
[q,N] = size(d); % matrix dimension

⟨Data compression (QR) 259b⟩
⟨Matrix approximation (SVD) 259a⟩
if nargout == 2
  s = diag(s); % column vector
  l = s(1:m,ones(1,N)) .* v(:,1:m)'; % diag(S) * V'
end
```

3 Algorithms

In this section, we consider the weighted low-rank approximation problem:

$$\text{minimize} \quad \text{over } \widehat{D} \quad \|D - \widehat{D}\|_{\Sigma}^2 \quad \text{subject to} \quad \text{rank}(\widehat{D}) \leq \mathtt{m}, \tag{4}$$

where the weight matrix $\Sigma \in \mathbb{R}^{q \times N}$ has *nonnegative* elements. The rank constraint can be represented as follows:

$$\text{rank}(\widehat{D}) \leq \mathtt{m} \quad \iff \quad \text{there are } P \in \mathbb{R}^{q \times \mathtt{m}} \text{ and } L \in \mathbb{R}^{\mathtt{m} \times N}, \quad \text{such that } \widehat{D} = PL, \tag{5}$$

which turns problem 4 into the following parameter optimization problem:

$$\text{minimize} \quad \text{over } P \in \mathbb{R}^{q \times \mathtt{m}} \text{ and } L \in \mathbb{R}^{\mathtt{m} \times N} \quad \|D - PL\|_{\Sigma}^2. \tag{6}$$

Unfortunately the problem is nonconvex and there are no efficient methods to solve it. Next we present two local optimization approaches for finding locally optimal solutions, starting from a given initial approximation.

3.1 Alternating Projections

The first solution method is motivated by the fact that (6) is linear separately in either P or L. Indeed, by fixing either P or L in 6 the minimization over the free parameter is a (singular) weighted least squares problem, which can be solved globally and efficiently. This suggests an iterative solution method that alternates between the solution of the two weighted least squares problems. The solution of the weighted least squares problems can be interpreted as weighted projections, thus the name of the method—alternating projections.

The alternating projections method is started from an initial guess of one of the parameters P or L. An initial guess is a possibly suboptimal solution of 6, computed by a direct method. Such a solution can be obtained for example by solving the unweighted low-rank approximation problem where all missing elements are filled in by zeros. On each iteration step of the alternating projections algorithm, the cost function value is guaranteed to be non-increasing and is typically decreasing. It can be shown that the iteration converges [11, 9] and that the local convergence rate is linear.

A summary of the alternating projections method is given in Algorithm 1. We use the following Matlab-like notation for indexing a matrix. For a $q \times N$ matrix D and subsets $\mathcal{I}$ and $\mathcal{J}$ of the sets of, respectively, row and column indexes, $D_{\mathcal{I},\mathcal{J}}$ denotes the submatrix of D with elements whose indexes are in $\mathcal{I}$ and $\mathcal{J}$. Either of $\mathcal{I}$ and $\mathcal{J}$ can be replaced by ":" in which case all rows/columns are indexed.

The quantity $e^{(k)}$, computed on step 9 of the algorithm is the squared approximation error

$$e^{(k)} = \|D - D^{(k)}\|_{\Sigma}^{2}$$

on the kth iteration step. Convergence of the iteration is judged on the basis of the relative decrease of $e^{(k)}$ after an update step. This corresponds to choosing a tolerance on the relative decrease of the cost function value. More expensive alternatives are to check the convergence of the approximation $\widehat{D}^{(k)}$ or the size of the gradient of the cost function with respect to the model parameters.

3.2 Variable Projections

In the second solution methods, we view 6 as a double minimization problem:

$$\text{minimize} \quad \text{over } P \in \mathbb{R}^{q\times \mathtt{m}} \quad \underbrace{\min_{L\in\mathbb{R}^{\mathtt{m}\times N}} \|D - PL\|_{\Sigma}^{2}}_{f(P)}. \tag{7}$$

The inner minimization is a weighted least squares problem and therefore can be solved in closed form. Using Matlab's set indexing notation, the solution is

$$f(P) = \sum_{j=1}^{N} D_{\mathcal{J},j}^{\top}\operatorname{diag}(\Sigma_{\mathcal{J},j}^{2})P_{\mathcal{J},:}\left(P_{\mathcal{J},:}^{\top}\operatorname{diag}(\Sigma_{\mathcal{J},j}^{2})P_{\mathcal{J},:}\right)^{-1}P_{\mathcal{J},:}^{\top}\operatorname{diag}(\Sigma_{\mathcal{J},j}^{2})D_{\mathcal{J},j}, \tag{8}$$

where $\mathcal{J}$ is the set of indexes of the non-missing elements in the jth column of D.

The outer minimization is a nonlinear least squares problem and can be solved by general purpose local optimization methods. There are local optimization methods, e.g., the Levenberg–Marquardt method [16], that guarantee global convergence (to a locally optimal solution) with super-linear convergence rate. Thus, if implemented by such a method and started "close" to a locally optimal solution, the variable projections method requires fewer iterations than the alternating projections method.

The inner minimization can be viewed as a weighted projection on the subspace spanned by the columns of P. Consequently $f(P)$ has the geometric interpretation of the sum of squared distances from the data points to the subspace. Since the parameter P is modified by the outer minimization, the projections are on a varying subspace, hence the name of the method—variable projections.

Note 4 (Gradient and Hessian of f). In the implementation of the method in this version of the paper, we are using finite difference numerical computation of the gradient and Hessian of f. (These approximations are computed by the optimization method.) More efficient alternative, however, is to supply to the method analytical expressions for the gradient and the Hessian. This will be done in later versions of the paper. Please refer to [13] for the latest version.

Algorithm Alternating projections algorithm for weighted low-rank approximation with missing data.

Input: Data matrix $D \in \mathbb{R}^{q \times N}$, rank constraint $\mathtt{m}$, elementwise nonnegative weight matrix $\Sigma \in \mathbb{R}^{q \times N}$, and relative convergence tolerance ε.

Initial approximation: compute the Frobenius norm low-rank approximation of D with missing elements filled in with zeros

$$P^{(0)} := \mathtt{lra}(D, \mathtt{m}).$$

Let $k := 0$.

repeat

Let $e^{(k)} := 0$.

for $j = 1, \ldots, N$ **do**

Let $\mathcal{J}$ be the set of indexes of the non-missing elements in $D_{:,j}$.

Define

$$c := \operatorname{diag}(\Sigma_{\mathcal{J},j}) D_{\mathcal{J},j} = \Sigma_{\mathcal{J},j} \odot D_{\mathcal{J},j}$$

$$P := \operatorname{diag}(\Sigma_{\mathcal{J},j}) P^{(k)}_{\mathcal{J},:} = (\Sigma_{\mathcal{J},j} \mathbf{1}_{\mathtt{m}}^{\top}) \odot P^{(k)}_{\mathcal{J},:}.$$

Compute

$$l_j^{(k)} := \left(P^{\top} P\right)^{-1} P^{\top} c.$$

Let

$$e^{(k)} := e^{(k)} + \|c - P l_j^{(k)}\|^2.$$

end for

Define

$$L^{(k)} = l_1^{(k)} \cdots l_N^{(k)}.$$

Let $e^{(k+1)} := 0$.

for $i = 1, \ldots, q$ **do**

Let $\mathcal{I}$ be the set of indexes of the non-missing elements in the ith row $D_{i,:}$.

Define

$$r := D_{i,\mathcal{I}} \operatorname{diag}(\Sigma_{i,\mathcal{I}}) = D_{i,\mathcal{I}} \odot \Sigma_{i,\mathcal{I}}$$

$$L := L^{(k+1)}_{:,\mathcal{I}} \operatorname{diag}(\Sigma_{i,\mathcal{I}}) = L^{(k+1)}_{:,\mathcal{I}} \odot (\mathbf{1}_{\mathtt{m}} \Sigma_{i,\mathcal{I}}).$$

Compute

$$p_i^{(k+1)} := r L^{\top} \left(L L^{\top}\right)^{-1}.$$

Let

$$e^{(k+1)} := e^{(k+1)} + \|r - p_i^{(k+1)} L\|^2.$$

end for

Define

$$P^{(k+1)} = \begin{matrix} p_1^{(k+1)} \\ \vdots \\ p_q^{(k+1)} \end{matrix}.$$

$k = k + 1$.

until $|e^{(k)} - e^{(k-1)}| / e^{(k)} < \varepsilon$.

Output: Locally optimal solution $\widehat{D} = D^{(k)} := P^{(k)} L^{(k)}$ of 6.

4 Implementation

Both the alternating projections and the variable projections methods for solving weighted low-rank approximation problems with missing data are callable through the function `wlra`.

```
⟨wlra 263⟩≡
  ⟨wlra header 272⟩
  function [p,l,info] = wlra(d,m,s,opt)

  tic % measure the execution time
  ⟨Default parameters opt 270⟩
  switch lower(opt.Method)
  case {'altpro','ap'}
    ⟨Alternating projections method 264a⟩
  case {'varpro','vp'}
    ⟨Variable projections method 265c⟩
  otherwise
    error('Unknown method %s',opt.Method)
  end
  info.time = toc; % execution time

  ⟨lra 259c⟩ % needed for the initial approximation
  ⟨Cost function 265d⟩ % needed for the variable projections method
```

The output parameter `info` gives the approximation error $\|D - \widehat{D}\|_{\Sigma}^{2}$ (`info.err`), the number of iterations (`info.iter`), and the execution time (`info.time`) for computing the local approximation $\widehat{D}$. The optional parameter `opt` specifies which method and (in the case of the variable projections) which algorithm is to be used (`opt.Method` and `opt.Algorithm`), the initial approximation (`opt.P`), the convergence tolerance ε (`opt.TolFun`), an upper bound on the number of iterations (`opt.MaxIter`), and the level of printed information (`opt.Display`).

The initial approximation `opt.P` is a $q \times \mathtt{m}$ matrix, such that the columns of $P^{(0)}$ form a basis for the span of the columns of $D^{(0)}$, where $D^{(0)}$ is the initial approximation of D, see step 1 in Algorithm 1. If it is not provided via the parameter `opt`, the default initial approximation is chosen to be the unweighted low-rank approximation of the data matrix with all missing elements filled in with zeros.

Note 5 (Large scale, sparse data). In an application of (4) to building recommender systems [19], the data matrix D is large (q and N are several hundreds of thousands) but only a small fraction of the elements (e.g., one percent) are given. Such problems can be handled efficiently, encoding D and Σ as sparse matrices. The convention in this case is that missing elements are zeros. Of course, the S matrix indicates that they should be treated as missing. Thus the convention is a hack allowing us to use the powerful tool of sparse matrix representation and linear algebra available in Matlab/Octave.

4.1 Alternating Projections

The iteration loop for the alternating projections algorithm is:

⟨*Alternating projections method* 264a⟩≡ (263)

```
[q,N] = size(d); % define q and N
switch lower(opt.Display)
  case {'iter'}, sd = norm(s.*d,'fro')^2; % size of D
end

% Main iteration loop
k    = 0; % iteration counter
cont = 1;
while (cont)
  ⟨Compute L, given P 264b⟩
  ⟨Compute P, given L 264c⟩
  ⟨Check exit condition 265a⟩
  ⟨Print progress information 265b⟩
end
info.err  = el; % approximation error
info.iter = k;  % number of iterations
```

The main computational steps on each iteration of the algorithm are the two weighted least squares problems.

⟨*Compute L, given P* 264b⟩≡ (264 265)

```
dd = []; % vec(D - DH)
for j = 1:N
  J   = find(s(:,j));
  sJj = full(s(J,j));
  c   = sJj .* full(d(J,j));
  P   = sJj(:,ones(1,m)) .* p(J,:); % = diag(sJj) * p(J,:)
  l(:,j) = P \ c;
  dd  = [dd; c - P*l(:,j)];
end
ep = norm(dd)^2;
```

⟨*Compute P, given L* 264c⟩≡ (264a)

```
dd = []; % vec(D - DH)
for i = 1:q
  I   = find(s(i,:));
  sIi = full(s(i,I));
  r   = sIi .* full(d(i,I));
  L   = sIi(ones(m,1),:) .* l(:,I); % = l(:,I) * diag(sIi)
  p(i,:) = r / L;
  dd  = [dd, r - p(i,:)*L];
end
el = norm(dd)^2;
```

The convergence is checked by the size of the relative decrease in the approximation error $e^{(k)}$ after one update step.

⟨*Check exit condition* 265a⟩≡ (264a)

```
k    = k + 1;
re   = abs(el - ep) / el;
cont = (k < opt.MaxIter) & (re > opt.TolFun) & (el > eps);
```

If the optimal parameter `opt.Display` is set to `'iter'`, `wlra` prints on each iteration step the relative approximation error.

⟨*Print progress information* 265b⟩≡ (264a)

```
switch lower(opt.Display)
  case 'iter', fprintf('%2d : relative error = %18.8f\n', k, el/sd)
end
```

4.2 Variable Projections

We use Matlab's Optimization Toolbox for performing the outer minimization in (7), i.e., the nonlinear minimization over the P parameter. The parameter `opt.Algorithm` specifies the algorithm to be used. The available options are `fminunc` — a quasi-Newton type algorithm, and `lsqnonlin` — a nonlinear least squares algorithm. Both algorithm allow for numerical approximation of the gradient and Hessian/Jacobian through finite difference computations. In the current version of the code, we use the numerical approximation.

⟨*Variable projections method* 265c⟩≡ (263)

```
switch lower(opt.Algorithm)
  case {'fminunc'}
    [p,err,f,info] = fminunc(@(p)wlra_err(p,d,s),p,opt);
  case {'lsqnonlin'}
    [p,rn,r,f,info] = lsqnonlin(@(p)wlra_err_mat(p,d,s),p,[],[]);
  otherwise
    error('Unknown algorithm %s.',opt.Algorithm)
end
[info.err,l] = wlra_err(p,d,s); % in order to obtain the L parameter
```

The inner minimization in (7) has an analytical solution (8). The implementation of (8) is actually the chunk of code for computing the L parameter, given the P parameter, already used in the alternating projections algorithm.

⟨*Cost function* 265d⟩≡ (263) 265e ▷

```
function [ep,l] = wlra_err(p,d,s)
N = size(d,2); m = size(p,2);
⟨Compute L, given P 264b⟩
```

In the case of using a nonlinear least squares type algorithm, the cost function is not the sum of squares of the errors but the vector of the errors `dd`.

⟨*Cost function* 265d⟩+≡ (263) ◁ 265d

```
function dd = wlra_err_mat(p,d,s)
N = size(d,2); m = size(p,2);
⟨Compute L, given P 264b⟩
```

5 Test on Simulated Data

A "true" random rank-m matrix $\bar{D}$ is selected by generating randomly its factors $\bar{P}$ and $\bar{L}$ in a rank revealing factorization $\bar{D} = \bar{P}\bar{L}$, where $\bar{P} \in \mathbb{R}^{q \times \mathtt{m}}$ and $\bar{L} \in \mathbb{R}^{\mathtt{m} \times N}$.

⟨test 266a⟩≡ 266b ▷

```
randn('state',0); rand('state',0);
p0 = rand(q,m); l0 = rand(m,N);     % true data matrix
```

The location of the given elements is chosen randomly row by row. The number of given elements is such that the sparsity of the resulting matrix, defined as the ratio of the number of missing elements to the total number qN of elements, matches the specification r.

⟨test 266a⟩+≡ ◁ 266a 266c ▷

```
ne  = round((1-r)*q*N);  % number of given elements
ner = round(ne/q);       % number of given elements per row
I = []; J = [];          % row/column indices of the given elements
for i = 1:q
  I = [I i*ones(1,ner)]; % all selected elements are in the ith row
  rp = randperm(N);
  J = [J rp(1:ner)];     % and have random column indices
end
ne = length(I);
```

By construction there are ner given elements in each row of the data matrix, however, there may be columns with a few (or even zero) given elements. Columns with less than m given elements can not be recovered from the given observations, even when the data is noise-free. Therefore, we remove such columns from the data matrix.

⟨test 266a⟩+≡ ◁ 266b 266d ▷

```
% Find indexes of columns with less than M given elements
tmp = (1:N)';
J_del = find(sum(J(ones(N,1),:) == tmp(:,ones(1,ne)),2) < m);
% Remove them
l0(:,J_del) = [];
% Redefine I and J
tmp = sparse(I,J,ones(ne,1),q,N); tmp(:,J_del) = [];
[I,J] = find(tmp); N = size(l0,2);
```

Next, we construct a noisy data matrix with missing elements by adding to the true values of the given data elements independent, identically, distributed, zero mean, Gaussian noise, with a specified standard deviation s. The weight matrix Σ is binary: $\sigma_{ij} = 1$ if d_{ij} is given and $\sigma_{ij} = 1$ if d_{ij} is missing.

⟨test 266a⟩+≡ ◁ 266c 267a ▷

```
d0 = p0 * l0;          % full true data matrix
Ie = I + q * (J-1); % indexes of the given elements from d0(:)
d  = zeros(q*N,1);  d(Ie) = d0(Ie) + sigma*randn(size(d0(Ie)));
d  = reshape(d,q,N);
s  = zeros(q,N); s(Ie) = 1;
```

We apply the methods implemented in `lra` and `wlra` on the noisy data matrix D with missing elements and validate the results against the complete true data matrix $\bar{D}$.

⟨test 266a⟩+≡ ◁266d 267b▷

```
tic, [p0,l0] = lra(d,m); t0 = toc;
err0 = norm(s.*(d - p0*l0),'fro')^2; e0 = norm(d0 - p0*l0,'fro')^2;
[ph1,lh1,info1] = wlra(d,m,s);      e1 = norm(d0 - ph1*lh1,'fro')^2;
opt.Method = 'vp'; opt.Algorithm = 'fminunc';
[ph2,lh2,info2] = wlra(d,m,s,opt); e2 = norm(d0 - ph2*lh2,'fro')^2;
opt.Method = 'vp'; opt.Algorithm = 'lsqnonlin';
[ph3,lh3,info3] = wlra(d,m,s,opt); e3 = norm(d0 - ph3*lh3,'fro')^2;
```

For comparison, we use also a method for low-rank matrix completion, called singular value thresholding (SVT) [1]. Low-rank matrix completion is a low-rank approximation problem with missing data for exact data, i.e., data of a low-rank matrix. Although the SVT method is initially designed for the exact case, it is demonstrated to cope with noisy data as well, i.e., solve low-rank approximation problems with missing data. The method is based on convex relaxation of the rank constraint and does not require an initial approximation. A Matlab implementation of the SVT method is available at http://svt.caltech.edu/

⟨test 266a⟩+≡ ◁267a 267c▷

```
tau = 5*sqrt(q*N); delta = 1.2/(ne/q/N); % SVT calling parameters
try
    tic, [U,S,V] = SVT([q N],Ie,d(Ie),tau,delta); t4 = toc;
    dh4  = U(:,1:m)*S(1:m,1:m)*V(:,1:m)'; % approximation
catch
    dh4 = NaN; t4 = NaN; % SVT not installed
end
err4 = norm(s.*(d - dh4),'fro')^2; e4 = norm(d0 - dh4,'fro')^2;
```

The final result shows the relative approximation error $\|D-\widehat{D}\|_{\Sigma}^{2}/\|D\|_{\Sigma}^{2}$, the estimation error $\|\bar{D}-\widehat{D}\|_{\mathrm{F}}^{2}/\|\bar{D}\|_{\mathrm{F}}^{2}$, and the computation time for the five methods.

⟨test 266a⟩+≡ ◁267b

```
nd = norm(s.*d,'fro')^2; nd0 = norm(d0,'fro')^2;
format long
res = [err0/nd info1.err/nd  info2.err/nd  info3.err/nd err4/nd;
       e0/nd0  e1/nd0        e2/nd0        e3/nd0       e4/nd0;
       t0      info1.time    info2.time    info3.time   t4]
```

First, we call the test script with exact (noise-free) data.

⟨*Experiment 1: small sparsity, exact data* 267d⟩≡

```
q = 10; N = 100; m = 2; r = 0.1; sigma = 0; test
```

Table 1. Results for Experiment 1.

	`lra`	`ap`	`vp` + `fminunc`	`vp` + `lsqnonlin`	SVT
$\Vert D - \widehat{D}\Vert_\Sigma^2/\Vert D\Vert_\Sigma^2$	0.02	10^{-19}	10^{-12}	10^{-17}	10^{-8}
$\Vert \bar{D} - \widehat{D}\Vert_F^2/\Vert D\Vert_F^2$	0.03	10^{-20}	10^{-12}	10^{-17}	10^{-8}
Execution time (sec)	0.01	0.05	2	3	0.37

The experiment corresponds to a matrix completion problem [2]. The results, summarized in Table 1, show that all methods, except for `lra`, complete correctly (up to numerical errors) the missing elements. As proved by Candés in [2], exact matrix completion is indeed possible in the case of Experiment 1.

The second experiment is with noisy data.

⟨*Experiment 2: small sparsity, noisy data* 268a⟩≡

```
q = 10; N = 100; m = 2; r = 0.1; sigma = 0.1; test
```

The results, shown in Table 2, indicate that the methods implemented in `wlra` converge to the same (locally) optimal solution. The alternating projections method, however, is about 100 times faster than the variable projections methods, using the Optimization Toolbox functions `fminunc` and `lsqnonlin`, and about 10 times faster than the SVT method. The solution produces by the SVT method is suboptimal but close to being (locally) optimal.

Table 2. Results for Experiment 2.

	`lra`	`ap`	`vp` + `fminunc`	`vp` + `lsqnonlin`	SVT
$\Vert D - \widehat{D}\Vert_\Sigma^2/\Vert D\Vert_\Sigma^2$	0.037	0.0149	0.0149	0.0149	0.0151
$\Vert \bar{D} - \widehat{D}\Vert_F^2/\Vert D\Vert_F^2$	0.037	0.0054	0.0054	0.0055	0.0056
Execution time (sec)	0.01	0.03	2	3	0.39

In the third experiment we keep the noise standard deviation the same as in Experiment 2 but increase the sparsity.

⟨*Experiment 3: bigger sparsity, noisy data* 268b⟩≡

```
q = 10; N = 100; m = 2; r = 0.4; sigma = 0.1; test
```

The results, shown in Table 3, again indicate that the methods implemented in `wlra` converge to the same (locally) optimal solutions. In this case, the SVT method is further away from being (locally) optimal, but is still much better than the solution of `lra` — 1% vs 25% relative prediction error.

Table 3. Results for Experiment 3

	`lra`	`ap`	`vp` + `fminunc`	`vp` + `lsqnonlin`	SVT
$\Vert D - \widehat{D}\Vert_\Sigma^2/\Vert D\Vert_\Sigma^2$	0.16	0.0133	0.0133	0.0133	0.0157
$\Vert \bar{D} - \widehat{D}\Vert_F^2/\Vert D\Vert_F^2$	0.25	0.0095	0.0095	0.0095	0.0106
Execution time (sec)	0.01	0.04	3	5	0.56

6 Test on the MoviLens Data

The MoviLens date sets [7] were collected and published by the GroupLens Research Project at the University of Minnesota in 1998. Currently, they are recognized as a benchmark for predicting missing data in recommender systems. The "100K data set" consists of 100000 ratings of $q = 943$ users' on $N = 1682$ movies and demographic information for the users. (The ratings are encoded by integers in the range from 1 to 5.) In this paper, we use only the ratings, which constitute a $q \times N$ matrix with missing elements. The task of a recommender system is to fill in the missing elements.

Assuming that the true complete data matrix is rank deficient, building a recommender system is a problem of low-rank approximation with missing elements. The assumption that the true data matrix is low-rank is reasonable in practice because user ratings are influences by a few factors. Thus, we can identify typical users (related to different combinations of factors) and reconstruct the ratings of any user as a linear combination of the ratings of the typical users. As long as the typical users are fewer than the number of users, the data matrix is low-rank. In reality, the number of factors is not small but there are a few dominant ones, so that the true data matrix is approximately low-rank.

It turns out that two factors allow us to reconstruct the missing elements with 7.1% average error. The reconstruction results are validated by cross validation with 80% identification data and 20% validation data. Five such partitionings of the data are given on the MoviLens web site. The matrix $\Sigma_{\text{idt}}^{(k)} \in \{0,1\}^{q \times N}$ indicates the positions of the given elements in the kth partition ($\Sigma_{\text{idt},ij}^{(k)} = 1$ means that the element D_{ij} is used for identification and $\Sigma_{\text{idt},ij}^{(k)} = 0$ means that D_{ij} is missing). Similarly, $\Sigma_{\text{val}}^{(k)}$ indicates the validation elements in the kth partition.

Table 4 shows the mean relative identification and validation errors

$$e_{\text{idt}} := \frac{1}{5}\sum_{k=1}^{5} \frac{\|D - \widehat{D}^{(k)}\|_{\Sigma_{\text{idt}}^{(k)}}^2}{\|D\|_{\Sigma_{\text{idt}}^{(k)}}^2} \quad \text{and} \quad e_{\text{val}} := \frac{1}{5}\sum_{k=1}^{5} \frac{\|D - \widehat{D}^{(k)}\|_{\Sigma_{\text{val}}^{(k)}}^2}{\|D\|_{\Sigma_{\text{val}}^{(k)}}^2},$$

where $\widehat{D}^{(k)}$ is the reconstructed matrix in the kth partitioning of the data. The SVT method issues a message "Divergence!", which explains the poor results obtained by this method.

Table 4. Results on the MoviLens data.

	`lra`	`ap`	SVT
Mean identification error e_{idt}	0.100	0.060	0.298
Mean prediction error e_{val}	0.104	0.071	0.307
Mean execution time (sec)	1.4	156	651

7 Conclusions

Alternating projections and variable projections methods for element-wise weighted low-rank approximation were presented and implemented in a literate programming style. Some of the weights may be set to zero, which corresponds to missing or ignored elements in the data matrix. The problem is of interest for static linear modeling of data with missing elements. The simulation examples suggest that in the current version of the implementation overall most efficient is the alternating projections method, which is applicable to data with a few tens of thousands of rows and columns, provided the sparsity of the given elements is high. In the case of exact data with missing elements, the methods solve a matrix completion problem.

Acknowledgments

Research supported by PinView (Personal Information Navigator adapting through VIEWing), EU FP7 Project 216529. I would like to thank A. Prugel-Bennett and M. Ghazanfar for discussions on the topic of recommender systems and for pointing out reference [19] and the MovieLens data set.

Appendix A: Literate Programming with `noweb`

A literate program is composed of interleaved code segments, called chunks, and text. The program can be split into chunks in any way and the chunks can be presented in any order, deemed helpful for the understanding of the program. This allows us to focus on the logical structure of the program rather than the way a computer executes it. (The actual computer executable code is *weaved* from a *web* of the code chunks by skipping the text.) In addition, literate programming allow us to use a powerful typesetting system such as LaTeX (rather than ascii text) for the documentation of the code.

We use the `noweb` system for literate programming [17]. Its advantage over alternative systems is independence of the programming language being used. The usage of `noweb` is presented in [18]. Next, we explain the typographic conventions needed to follow the presentation.

The code is typeset in a true type font. A code chunk begins with a tag, consisting of a name and a number, identifying the chunk, e.g.,

⟨*Default parameters* `opt` 270⟩≡ (263)

```
try opt.MaxIter;   catch opt.MaxIter   = 100;  end
try opt.TolFun;    catch opt.TolFun    = 1e-5; end
try opt.Display;   catch opt.Display   = 'off'; end
try opt.Method;    catch opt.Method    = 'ap'; end
try opt.Algorithm; catch opt.Algorithm = 'lsqnonlin'; end
try p = opt.P;     catch
```

```
  switch lower(opt.Display)
    case 'iter', fprintf('Computing an initial approximation ...\n')
  end
    p = lra(d,m); % low-rank approximation
end
```

To the right of the identification tag in brackets is the page where the chunk is used, i.e., included in other chunks. To the left of the identification tag is a number identifying the part, called sub-chunks, of the current chunk. In case, like the one above, when the chunk is not split into sub-chunks, the sub-chunk identification number is the same as the chunk identification number. See page 265 for a chunk split into sub-chunks.

Appendix B: Function Headers

⟨lra *header* 271⟩≡ (259c)

```
% LRA - Low-Rank Approximation.
% [PH,LH] = LRA(D,M)
% Finads optimal solution to the problem:
% Minimize over DH norm(D - DH, 'fro') subject to rank(DH) <= M
%
% D  - data matrix of dimension qxN, q < N
% M  - rank constraint, M < q
% PH, LH - PH*LH is the rank-m approximation DH of D
```

⟨wlra *header* 272⟩≡ (263)

```
% WLRA - Weighted Low-Rank Approximation.
% [PH,LH,INFO] = WLRA(D,M,S,OPT)
% Finads locally optimal solution to the problem:
% Minimize over DH norm(S.*(D - DH),'fro') subject to rank(DH) <= M
%
% D - data matrix of dimension qxN, q < N
% M - rank constraint, M < q
% S - element-wise nonnegative weight matrix of dimension qxN
% OPT  - options for the optimization algorithm
%   OPT.Method has possible values
%     'ap' - alternating projections (default) and
%     'vp' - variable projections (requires Optimization Toolbox)
%   OPT.Algorithm - algorithm for the variable projections
%     'fminunc' - quasi-Newton type method
%     'lsqnonlin' - Levenberg-Marquardt method (default)
%   OPT.P        - initial approximation (default computed via svds)
%   OPT.TolFun  - convergence tolerance for the function value
%   OPT.MaxIter - maximum number of iterations
%   OPT.Display - level of printed information
%     'iter' - prints the cost function value per iteration
% PH, LH - PH*LH is the rank-m approximation DH of D
% INFO - exit information:
%   INFO.err   - approximation error
%   INFO.time  - execution time
%   INFO.iter  - number of iterations performed
%   Note: INFO.iter = OPT.MaxIter indicates lack of convergence
%
% Note: S(i,j) = 0 implies that D(i,j) is missing. Missing elements
% are ignored and in particular can be set to 0. This convention is
% convenient for large sparse data sets when SPARSE repr. is used.
```

References

1. J.-F. Cai, E. Candés, and Z. Shen: A singular value thresholding algorithm for matrix completion. http://www-stat.stanford.edu/~candes/papers/SVT.pdf, 2009.
2. E. Candés and B. Recht: Exact matrix completion via convex optimization. Found. of Comput. Math., 2009.
3. B. De Moor: Structured total least squares and L_2 approximation problems. Linear Algebra Appl., 188–189:163–207, 1993.
4. K. Gabriel and S. Zamir: Lower rank approximation of matrices by least squares with any choice of weights. Technometrics **21**, 1979, 489–498.
5. G. Golub and V. Pereyra: Separable nonlinear least squares: the variable projection method and its applications. Institute of Physics, Inverse Problems **19**, 2003, 1–26.
6. G. Golub and C. Van Loan: An analysis of the total least squares problem. SIAM J. Numer. Anal. **17**, 1980, 883–893.
7. GroupLens: Movielens data sets. http://www.grouplens.org/node/73.
8. N. Higham: Computing the nearest correlation matrix – a problem from finance. IMA Journal of Numerical Analysis **22**(3), 2002, 329–343.
9. H. Kiers: Setting up alternating least squares and iterative majorization algorithms for solving various matrix optimization problems. Comput. Statist. Data Anal. **41**, 2002, 157–170.
10. D. Knuth: *Literate Programming*. Cambridge University Press, 1992.
11. W. Krijnen: Convergence of the sequence of parameters generated by alternating least squares algorithms. Comput. Statist. Data Anal. **51**, 2006, 481–489.
12. J. Manton, R. Mahony, and Y. Hua: The geometry of weighted low-rank approximations. IEEE Trans. Signal Process. **51**(2), 2003, 500–514.
13. I. Markovsky: *Algorithms and Literate Programs for Weighted Low-Rank Approximation with Missing Data*. Technical Report 18296, ECS, Univ. of Southampton, http://eprints.ecs.soton.ac.uk/18296/, 2009.
14. I. Markovsky, M.-L. Rastello, A. Premoli, A. Kukush, and S. Van Huffel: The element-wise weighted total least squares problem. Comput. Statist. Data Anal. **50**(1), 2005, 181–209.
15. I. Markovsky and S. Van Huffel: Left vs right representations for solving weighted low rank approximation problems. Linear Algebra Appl. **422**, 2007, 540–552.
16. D. Marquardt: An algorithm for least-squares estimation of nonlinear parameters. SIAM J. Appl. Math. **11**, 1963, 431–441.
17. N. Ramsey: `noweb`. http://www.cs.tufts.edu/~nr/noweb/.
18. N. Ramsey: Literate programming simplified. IEEE Software **11**, 1994, 97–105.
19. T. Segaran: *Programming Collective Intelligence: Building Smart Web 2.0 Applications*. O'Reilly Media, 2007.
20. N. Srebro: *Learning with Matrix Factorizations*. PhD thesis, MIT, 2004.
21. P. Wentzell, D. Andrews, D. Hamilton, K. Faber, and B. Kowalski: Maximum likelihood principal component analysis. J. Chemometrics **11**, 1997, 339–366.

On Bivariate Interpolatory Mask Symbols, Subdivision and Refinable Functions

A. Fabien Rabarison[1] and Johan de Villiers[2]

[1] Dept of Mathematics and Statistics, University of Strathclyde, G1 1XQ, UK
[2] Dept of Mathematical Sciences, Stellenbosch University, Matieland 7602, SA

Summary. We present a full characterisation of interpolatory mask symbols where the dilation matrix is $M = 2I$. The characterization involves the analysis of polynomial identities in two variables by means of the Bezout theorem and the Euclidean algorithm. The convergence of the associated interpolatory subdivision scheme is closely related to the existence of a corresponding interpolatory refinable function. As a special case of our theory, we present the mask symbol corresponding to the Butterfly subdivision scheme.

1 Introduction

Subdivision methods, as initially introduced by de Rham (1956) and later by Chaikin (1974), play important roles in computer aided geometric design (CAGD) by generating curves and surfaces in computer graphics (see e.g. [4]). Several studies of refinement masks have been developed by using the associated mask symbols which often help to prove the convergence of the subdivision schemes with which they are associated (e.g. [6, pages 37-70], [1]). Moreover, a mask symbol A provides most of the essential features of the associated refinable function ϕ, that is a function that is expressible as a linear combination of the shifts of its own dilation.

In this paper, we center our attention to interpolatory mask symbols by presenting their full algebraic characterization for the case where the dilation matrix is given by $M = 2I$. The corresponding subdivision schemes are interpolatory, that is, the initial data points are preserved at all the steps of the recursive process. Such schemes are extremely relevant in CAGD where the initial data are required to be preserved while applying the subdivision process. Substantial progress in this area has been made, but computationally inefficient conditions are still often applied to refinement mask symbols in order to investigate the convergence of the associated subdivision schemes. For instance, the characterization by using the p-norm joint spectral radius of a collection of matrices associated with the refinement mask (see e.g. [8])

E.H. Georgoulis, A. Iske, J. Levesley (eds.), *Approximation Algorithms for Complex Systems*, Springer Proceedings in Mathematics 3, DOI: 10.1007/978-3-642-16876-5_13,
© Springer-Verlag Berlin Heidelberg 2011

can take impractically long to test computationally. The method based on contractivity criteria on the subdivision operator, as introduced in [6] (see also [5]), can also be a formidable computational task to perform.

We give sufficient conditions for the convergence of interpolatory subdivision schemes with respect to the interpolatory mask symbol and the associated interpolatory refinable functions, that is, refinable functions that take the value 1 at the origin and 0 at all other integers. Recall that refinable functions became popular in multiresolution analysis (MRA) to construct wavelets in order to study the local time-frequency behavior of a given signal, in contrast to methods based on Fourier transforms. Finally we introduce the cascade algorithm to recursively construct interpolatory refinable functions. Some numerical applications are given to illustrate the given theories.

Further studies on subdivision, refinable functions and their use in wavelet theory can be found in [2, 3] and [13]. For higher dimensions, we refer to [1, 10, 12] and to [9].

Notation

We shall denote the set of natural numbers by $\mathbb{N}$, the set of integers and non-negative integers respectively by $\mathbb{Z}$ and $\mathbb{Z}_+$, the set of real numbers by $\mathbb{R}$ and the set of complex numbers by $\mathbb{C}$. Similarly, the symbols $\mathbb{Z}^2$, $\mathbb{R}^2$ and $\mathbb{C}^2$ denote the set of ordered pairs with respectively integer, real number and complex number entries.

For the linear space $M(\mathbb{Z}^2)$ of all real-valued sequences $\boldsymbol{c} = \{c_{\mathbf{j}} \in \mathbb{R} : \mathbf{j} \in \mathbb{Z}^2\}$, the support of which is denoted by $\mathrm{supp}(c) := \{\mathbf{j} \in \mathbb{Z}^2 : c_{\mathbf{j}} \neq \mathbf{0}\}$, the subspace of finitely supported sequences constitute a linear subspace denoted by $M_0(\mathbb{Z}^2)$. In the same way, for the linear space $M(\mathbb{R}^2)$ of all real-valued bivariate functions f on $\mathbb{R}^2$, the support $\mathrm{supp}(f)$ of which is the smallest closed set containing $\{\mathbf{x} \in \mathbb{R}^2 : f(\mathbf{x}) \neq 0\}$, the set of finitely supported functions constitute a linear subspace denoted by $M_0(\mathbb{R}^2)$. Moreover, the subspaces of continuous functions respectively in $M(\mathbb{R}^2)$ and in $M_0(\mathbb{R}^2)$ are denoted by $C(\mathbb{R}^2)$ and $C_0(\mathbb{R}^2)$.

For a given 2×2 invertible matrix M with integer entries, a function $\phi \in M_0(\mathbb{R}^2)$ is termed M-refinable if there exists a sequence $\boldsymbol{a} = \{a_{\mathbf{j}} : \mathbf{j} \in \mathbb{Z}^2\} \in M_0(\mathbb{Z}^2)$ such that

$$\phi = \sum_{\mathbf{j}} a_{\mathbf{j}} \phi(M \cdot - \mathbf{j}). \tag{1}$$

We shall refer to M as the *dilation matrix*, whereas the sequence $\boldsymbol{a}$ is called the *refinement mask* (or simply the mask), and the equation (1) is referred to as the *refinement equation*. Note that an M-refinable function is therefore expressible as a linear combinations of the shifts of its own dilations with the factor of the dilation matrix M, as specified by the refinement mask $\boldsymbol{a}$. For convenience, we shall often simplify "M-refinable" to "refinable".

Considering a dilation matrix M and a finitely supported refinement mask a, the subdivision scheme S_a is defined as an operator which recursively produces denser and denser data points by means of linear combinations of the previous ones. We define the *subdivision operator* $S_a = S_{a,M} : c \mapsto S_a c$, by

$$(S_a c)_{\mathbf{j}} = \sum_{\mathbf{k}} a_{\mathbf{j}-M\mathbf{k}^T} c_{\mathbf{k}}, \quad \mathbf{j} \in \mathbb{Z}^2. \tag{2}$$

For any initial sequence c, S_a generates $\{c^{(r)} = S_a^r c : r \in \mathbb{Z}_+\}$ by means of

$$c^{(0)} = c, \qquad c^{(r+1)} = S_a(c^{(r)}), \quad r \in \mathbb{Z}_+. \tag{3}$$

Then, given a dilation matrix M and a refinement mask a such that

$$\begin{cases} a_{M\mathbf{j}^T} = \delta_{\mathbf{j}}, \quad \mathbf{j} \in \mathbb{Z}^2, \\ \sum_{\mathbf{j}} a_{\mathbf{j}} = |\det(M)|, \end{cases} \tag{4}$$

where $\delta_{\mathbf{0}} = 1$ and $\delta_{\mathbf{j}} = 0$, $\mathbf{j} \neq \mathbf{0}$, we have, for any sequence c,

$$(S_a c)_{M\mathbf{j}^T} = c_{\mathbf{j}}, \quad \mathbf{j} \in \mathbb{Z}^2.$$

By induction on $r \in \mathbb{Z}_+$, the sequence $\{c^{(r)} = S_a^r c : r \in \mathbb{Z}_+\}$ satisfies

$$c^{(r+1)}_{M\mathbf{j}^T} = c^{(r)}_{\mathbf{j}}, \quad \mathbf{j} \in \mathbb{Z}^2.$$

A mask a satisfying (4) is called an *interpolatory* mask.

We now introduce the concept of convergence for subdivision. Given a dilation matrix M and a mask a, we say that S_a is *convergent* on a subset $\mathcal{M} \subset M(\mathbb{Z}^2)$ if, for any $c \in \mathcal{M}$, there exists $f = f_c \in C(\mathbb{R}^2)$ such that

$$\lim_{r\to\infty} \|S_a^r c - f(M^{-r}\cdot)\|_\infty = 0, \tag{5}$$

where $f(M^{-r}\cdot) = \{f(M^{-r}\mathbf{j}^T) : \mathbf{j} \in \mathbb{Z}^2\}$, for $r \in \mathbb{Z}_+$, in which case we shall also use the notation $S_a^\infty c := f$. We shall simply say " convergent " for " convergent on $M(\mathbb{Z}^2)$".

As given before in (1), a function $\phi \in M_0(\mathbb{R}^2)$ is said to be *refinable* if it can be expressed as linear combination of its own dilations, that is, if it satisfies

$$\phi(\mathbf{x}) = \sum_{\mathbf{j}} a_{\mathbf{j}} \phi(M\mathbf{x} - \mathbf{j}), \quad \mathbf{x} \in \mathbb{R}^2, \tag{6}$$

where M is the associated dilation matrix and a the associated mask. We also say that $f \in M(\mathbb{R}^2)$ is *interpolatory* if

$$f(\mathbf{j}) = \delta_{\mathbf{j}}, \quad \mathbf{j} \in \mathbb{Z}^2.$$

2 Main Results

We proceed to establish a full a characterization of interpolatory refinement mask associated with an interpolatory refinable function. We have the following result.

Lemma 1. *Let ϕ be an interpolatory refinable function with associated dilation matrix M and associated mask $\mathbf{a}$, and suppose ϕ is integrable with non-zero integral over $\mathbb{R}^2$. Then*

$$\begin{cases} a_{M\mathbf{j}^T} = \delta_{\mathbf{j}}, \quad \mathbf{j} \in \mathbb{Z}^2, \\ \sum_{\mathbf{j}} a_{\mathbf{j}} = |\det(M)|, \end{cases} \tag{7}$$

according to which $\mathbf{a}$ is an interpolatory mask.

Proof. Since ϕ is refinable and interpolatory, the first line in (7) is obtained as follows:

$$\delta_{\mathbf{j}} = \phi(\mathbf{j}) = \sum_{\mathbf{k}} a_{\mathbf{k}} \phi(M\mathbf{j}^T - \mathbf{k}) = \sum_{\mathbf{k}} a_{\mathbf{k}} \delta_{M\mathbf{j}^T - \mathbf{k}} = a_{M\mathbf{j}^T}.$$

Writing $a_{i,j} = a_{\mathbf{j}}$, we can integrate the refinement equation (6) to obtain

$$\int\int_{\mathbb{R}^2} \phi(x,y) \mathrm{d}x \mathrm{d}y = \sum_{i,j} a_{i,j} \int\int_{\mathbb{R}^2} \phi(M(x,y)^T - (i,j)) \mathrm{d}x \mathrm{d}y,$$

and thus

$$\int\int_{\mathbb{R}^2} \phi(x,y) \mathrm{d}x \mathrm{d}y = \sum_{i,j} a_{i,j} \frac{1}{|\det(M)|} \int\int_{\mathbb{R}^2} \phi(x,y) \mathrm{d}x \mathrm{d}y,$$

from which the second line of (7) then follows. □

Considering the *dyadic* set $\mathcal{D}_M := \{M^{-r}\mathbf{j}^T : \mathbf{j} \in \mathbb{Z}^2, r \in \mathbb{Z}_+\}$, we give the following convergence result for interpolatory subdivision schemes.

Theorem 1. *Let ϕ be an interpolatory refinable function such that ϕ is integrable with non-zero integral over $\mathbb{R}^2$ and M a dilation matrix such that its diagonal elements satisfy $|M_{11}| \geq 2$ and $|M_{22}| \geq 2$. Then, for any sequence $c \in M(\mathbb{Z}^2)$, the function $\Phi \in M(\mathbb{R}^2)$ defined by*

$$\Phi(\mathbf{x}) = \sum_{\mathbf{j}} c_{\mathbf{j}} \phi(\mathbf{x} - \mathbf{j}), \quad \mathbf{x} \in \mathbb{R}^2, \tag{8}$$

satisfies

(i) $\Phi(\mathbf{m}) = c_{\mathbf{m}}, \quad \mathbf{m} \in \mathbb{Z}^2$;

(ii) $\Phi(M^{-r}\mathbf{m}) = (S_a^r c)_{\mathbf{m}}, \quad r \in \mathbb{Z}_+, \quad \mathbf{m} \in \mathbb{Z}^2,$

that is, S_a is trivially convergent, with $S_a^\infty c = \Phi$. In particular, $S_a^\infty \delta = \phi$.

Proof. Since ϕ is interpolatory, it follows from (8) that

$$\Phi(\mathbf{m}) = \sum_{\mathbf{j}} c_{\mathbf{j}} \phi(\mathbf{m} - \mathbf{j}) = c_{\mathbf{m}}, \quad \mathbf{m} \in \mathbb{Z}^2.$$

Since ϕ is refinable, it follows from (8), (2) and (3) that, for $r \in \mathbb{Z}_+$, $\mathbf{m} \in \mathbb{Z}^2$,

$$\begin{aligned}
\Phi\left(M^{-r}\mathbf{m}^T\right) =& \sum_{\mathbf{j}} c_{\mathbf{j}} \phi\left(M^{-r}\mathbf{m}^T - \mathbf{j}\right) \\
=& \sum_{\mathbf{j}} c_{\mathbf{j}} \sum_{\mathbf{k}} a_{\mathbf{k}} \phi\left(M^{-r+1}\mathbf{m}^T - M\mathbf{j}^T - \mathbf{k}\right) \\
=& \sum_{\mathbf{k}} (S_a c)_{\mathbf{k}} \phi\left(M^{-r+1}\mathbf{m}^T - \mathbf{k}\right) \\
&\vdots \\
=& \sum_{\mathbf{k}} (S_a^r c)_{\mathbf{k}} \phi(\mathbf{m} - \mathbf{k}) \\
=& (S_a^r c)_{\mathbf{m}},
\end{aligned} \tag{9}$$

by virtue of the interpolatory property of ϕ.

Since the dyadic set $\mathcal{D}_M$ is dense in $\mathbb{R}^2$, we deduce from (9) that $\|S_a^r c - \Phi(M^{-r}\cdot)\|_\infty = 0$, $r \in \mathbb{Z}_+$, and therefore (5) holds. Hence, for any sequence $c \in M(\mathbb{Z}^2)$, the subdivision scheme S_a converges to the function Φ given by (8), i.e. $S_a^\infty c = \Phi$. In particular, choosing $c = \delta$ in (8) yields $S_a^\infty \delta = \phi$. □

We now turn to the results for interpolatory mask symbols. Given a refinement mask a, the associated *mask symbol* A is the Laurent polynomial defined by

$$A(z_1, z_2) = \sum_{i,j} a_{i,j} z_1^i z_2^j, \quad z_1, z_2 \in \mathbb{C} \setminus \{0\}.$$

For $M = 2I$, we observe that a is an interpolatory mask if and only if the associated mask symbol A satisfies

$$\begin{cases} \text{The constant term in } A(z_1, z_2) \text{ is } 1; \\ A \text{ has no term in } z_1^{2i} z_2^{2j}, \quad (i,j) \neq (0,0); \\ A(1,1) = |\det(M)| = 4, \end{cases}$$

in which case A is called an *interpolatory mask symbol.* We proceed to establish the full characterisation of interpolatory mask symbols for $M = 2I$.

Theorem 2. *For integers $k_1, k_2 \in \mathbb{N}$ and a Laurent polynomial B such that $B(1,1) = 1$, consider the Laurent polynomial*

$$A(z_1, z_2) = 2^{2-k_1-k_2}(1+z_1)^{k_1}(1+z_2)^{k_2}B(z_1, z_2), \quad z_1, z_2 \in \mathbb{C} \setminus \{0\}. \tag{10}$$

Then A is an interpolatory mask symbol if and only if for any positive integers $\alpha_1 < 2k_1$ and $\alpha_2 < 2k_2$, with α_1, α_2 odd, B has the form

$$\begin{aligned} B(z_1, z_2) = 2^{k_1+k_2} z_1^{-2\alpha_1} z_2^{-2\alpha_2} \big[&T(z_1, z_2)(1-z_1)^{k_1}(1-z_2)^{k_2} \\ &+ \{S_1(z_1, z_2) + T_1(z_1, z_2)(1-z_1)^{k_1}\}\{S_2(z_1, z_2) + T_2(z_1, z_2)(1-z_2)^{k_2}\}\big], \end{aligned} \tag{11}$$

with S_1 odd in z_2 with degree less than k_1 in z_1, and S_2 odd in z_1 with degree less than k_2 in z_2, and such that

$$\begin{cases} (1+z_1)^{k_1}S_1(z_1, z_2) - (1-z_1)^{k_1}S_1(-z_1, z_2) = z_1^{\alpha_1} z_2^{\alpha_2}, \\ (1+z_2)^{k_2}S_2(z_1, z_2) - (1-z_2)^{k_2}S_2(z_1, -z_2) = z_1^{\alpha_1} z_2^{\alpha_2}, \end{cases} \tag{12}$$

whereas T_1 is even in z_1 but odd in z_2, T_2 is even in z_2 but odd in z_1, and T is odd in both z_1 and z_2.

The proof of Theorem 2 will rely on Lemmas 2 and 3 below.

Lemma 2. *If $\sum_{\mathbf{j}} a_{\mathbf{j}} = 4$, then $\boldsymbol{a}$ is interpolatory if and only if A satisfies, for $z_1, z_2 \in \mathbb{C} \setminus \{0\}$, the identity*

$$A(z_1, z_2) + A(-z_1, z_2) + A(z_1, -z_2) + A(-z_1, -z_2) = 4. \tag{13}$$

Also, if a Laurent polynomial A satisfies (13) and if there exist integers $k_1, k_2 \in \mathbb{N}$ and a Laurent polynomial B such that

$$A(z_1, z_2) = 2^{2-k_1-k_2}(1+z_1)^{k_1}(1+z_2)^{k_2}B(z_1, z_2),$$

with $B(1,1) = 1$, then A is an interpolatory mask symbol.

Proof. Suppose first that $\boldsymbol{a}$ is interpolatory. From (7), it holds that

$$A(z_1, z_2) + A(-z_1, z_2) + A(z_1, -z_2) + A(-z_1, -z_2) = 4\sum_{i,j} a_{2i,2j} z_1^{2i} z_2^{2j}, \tag{14}$$

which, since $a_{2i,2j} = \delta_{i,j}$, implies that (13) holds. Conversely, if (13) holds, we deduce from (14) that

$$\sum_{i,j} a_{2i,2j} z_1^{2i} z_2^{2j} = 1,$$

which proves that $a_{2i,2j} = \delta_{i,j}$.

The second statement of the lemma is an immediate consequence of the first statement. □

Lemma 3. *Let $k_1, k_2 \in \mathbb{N}$ and suppose α_1, α_2 are two odd integers in $\mathbb{N}$. Then:*

(a) if $\alpha_1 < 2k_1$, there exists a polynomial S_1 which is odd in z_2, with degree α_2 in z_2, and degree less than k_1 in z_1, such that the general Laurent polynomial solution K_1 of the identity

$$(1+z_1)^{k_1} K_1(z_1,z_2) - (1-z_1)^{k_1} K_1(-z_1,z_2) = z_1^{\alpha_1} z_2^{\alpha_2}, \qquad (15)$$

is the Laurent polynomial given by

$$K_1(z_1,z_2) = S_1(z_1,z_2) + T_1(z_1,z_2)(1-z_1)^{k_1},$$

with T_1 denoting an arbitrary Laurent polynomial which is even in z_1; also, K_1 is odd in z_2 if and only if T_1 is odd in z_2.

(b) if $\alpha_2 < 2k_2$, there exists a polynomial S_2 which is odd in z_1, with degree α_1 in z_1, and degree less than k_2 in z_2, such that the general Laurent polynomial solution K_2 of the identity

$$(1+z_2)^{k_2} K_2(z_1,z_2) - (1-z_2)^{k_2} K_2(z_1,-z_2) = z_1^{\alpha_1} z_2^{\alpha_2},$$

is the Laurent polynomial given by

$$K_2(z_1,z_2) = S_2(z_1,z_2) + T_2(z_1,z_2)(1-z_2)^{k_2},$$

with T_2 denoting an arbitrary Laurent polynomial which is even in z_2; also, K_2 is odd in z_1 if and only if T_2 is odd in z_1.

Proof. (a) Since the two univariate polynomials $(1+z_1)^{k_1}$ and $(1-z_1)^{k_1}$ have no common factor, there exist by the Bezout theorem two univariate polynomials U_1 and V_1 such that

$$(1+z_1)^{k_1} U_1(z_1) + (1-z_1)^{k_1} V_1(z_1) = 1, \quad z_1 \in \mathbb{C}. \qquad (16)$$

Multiplying both sides of (16) by $z_1^{\alpha_1} z_2^{\alpha_2}$ yields, for $z_1, z_2 \in \mathbb{C}$,

$$(1+z_1)^{k_1} \left[z_1^{\alpha_1} z_2^{\alpha_2} U_1(z_1)\right] + (1-z_1)^{k_1} \left[z_1^{\alpha_1} z_2^{\alpha_2} V_1(z_1)\right] = z_1^{\alpha_1} z_2^{\alpha_2}, \quad z_1, z_2 \in \mathbb{C}. \qquad (17)$$

Using the polynomial division theorem, we deduce the existence of two polynomials Q_1 and R_1 satisfying

$$z_1^{\alpha_1} V_1(z_1) = Q_1(z_1)(1+z_1)^{k_1} + R_1(z_1), \quad z_1 \in \mathbb{C},$$

such that the degree of R_1 is less than k_1, and where Q_1 and R_1 are uniquely determined by α_1 and V_1. It then follows from (17) that

$$(1+z_1)^{k_1} S_1(z_1,z_2) + (1-z_1)^{k_1} \tilde{R}_1(z_1,z_2) = z_1^{\alpha_1} z_2^{\alpha_2}, \quad z_1, z_2 \in \mathbb{C}, \qquad (18)$$

where S_1 is the polynomial defined by $S_1(z_1,z_2) = z_1^{\alpha_1} z_2^{\alpha_2} U_1(z_1) + (1-z_1)^{k_1} z_2^{\alpha_2} Q_1(z_1)$, and $\tilde{R}_1$ is the polynomial given by $\tilde{R}_1(z_1,z_2) = z_2^{\alpha_2} R_1(z_1)$, for all $z_1, z_2 \in \mathbb{C}$. We claim that the degree in z_1 of S_1 is less than k_1. To prove this, we first note from (18) that

$$(1+z_1)^{k_1} S_1(z_1,z_2) = z_1^{\alpha_1} z_2^{\alpha_2} - (1-z_1)^{k_1} \tilde{R}_1(z_1,z_2), \quad z_1, z_2 \in \mathbb{C},$$

according to which, since the degree of $\tilde{R}_1$ in z_1 is less than k_1, and since $\alpha_1 < 2k_1$, we necessarily have that the degree in z_1 of S_1 is less than k_1.

Replacing z_1 by $-z_1$ in (18), and using the fact that α_1 is odd, we deduce that

$$(1-z_1)^{k_1} \left[-S_1(-z_1,z_2)\right] + (1+z_1)^{k_1} \left[-\tilde{R}_1(-z_1,z_2)\right] = z_1^{\alpha_1} z_2^{\alpha_2}. \tag{19}$$

Subtracting the identities (18) and (19) now yields

$$(1+z_1)^{k_1} [S_1(z_1,z_2) + \tilde{R}_1(-z_1,z_2)] = -(1-z_1)^{k_1} [S_1(-z_1,z_2) + \tilde{R}_1(z_1,z_2)],$$

and thus

$$S_1(z_1,z_2) + \tilde{R}_1(-z_1,z_2) = M_1(z_1,z_2)(1-z_1)^{k_1}, \tag{20}$$

for some polynomial M_1. Since the degree in z_1 of the polynomial in the left-hand-side of (20) is less than k_1, we necessarily have $M_1 = 0$ in (20), or, equivalently,

$$S_1(z_1,z_2) = -\tilde{R}_1(-z_1,z_2), \quad z_1, z_2 \in \mathbb{C}, \tag{21}$$
$$\tilde{R}_1(z_1,z_2) = -S_1(-z_1,z_2), \quad z_1, z_2 \in \mathbb{C}. \tag{22}$$

Using (18), (21) and (22), we find that the polynomial S_1 satisfies

$$(1+z_1)^{k_1} S_1(z_1,z_2) - (1-z_1)^{k_1} S_1(-z_1,z_2) = z_1^{\alpha_1} z_2^{\alpha_2}, \quad z_1, z_2 \in \mathbb{C}, \tag{23}$$

which means that S_1 is a particular polynomial solution of the identity (15) with degree less than k_1 in z_1. Moreover, from (21), we see that $S_1(z_1,z_2) = -z_2^{\alpha_2} R_1(-z_1)$. Since α_2 is odd, we conclude that S_1 is odd in z_2, and that its degree in z_2 is α_2.

Now, let K_1 denote the general Laurent polynomial solution of (15). Subtracting (15) from (23), we obtain, for $z_1, z_2 \in \mathbb{C} \setminus \{0\}$,

$$(1+z_1)^{k_1} \left[K_1(z_1,z_2) - S_1(z_1,z_2)\right] = (1-z_1)^{k_1} \left[K_1(-z_1,z_2) - S_1(-z_1,z_2)\right]. \tag{24}$$

Since $(1+z_1)^{k_1}$ and $(1-z_1)^{k_1}$ have no common factor, it follows from (24) that there exists a Laurent polynomial T_1 satisfying

$$K_1(z_1,z_2) - S_1(z_1,z_2) = T_1(z_1,z_2)(1-z_1)^{k_1}. \tag{25}$$

Substituting (25) into (24) yields that $T_1(z_1,z_2) = T_1(-z_1,z_2)$ for $z_1, z_2 \in \mathbb{C} \setminus \{0\}$, i.e T_1 is even in z_1. Thus, we deduce from (25) that K_1 is given by

$$K_1(z_1,z_2) = S_1(z_1,z_2) + T_1(z_1,z_2)(1-z_1)^{k_1}, \tag{26}$$

where T_1 is an arbitrary even Laurent polynomial in z_1.

Also, since S_1 is odd in z_2, we get from (26) that, for $z_1, z_2 \in \mathbb{C} \setminus \{0\}$,

$$\begin{aligned} K_1(z_1, -z_2) &= S_1(z_1, -z_2) + T_1(z_1, -z_2)(1 - z_1)^{k_1} \\ &= -S_1(z_1, z_2) + T_1(z_1, -z_2)(1 - z_1)^{k_1}, \end{aligned} \tag{27}$$

whereas also

$$-K_1(z_1, z_2) = -S_1(z_1, z_2) - T_1(z_1, z_2)(1 - z_1)^{k_1}. \tag{28}$$

Subtracting the identities (27) and (28) gives

$$K_1(z_1, -z_2) + K_1(z_1, z_2) = (1 - z_1)^{k_1} [T_1(z_1, -z_2) + T_1(z_1, z_2)],$$

from which it then immediately follows that K_1 is odd in z_2 if and only if T_1 is odd in z_2.

(b) The proof is similar to (a), and is omitted here. □

Proof of Theorem 2.

(a) By defining the Laurent polynomial H as

$$H(z_1, z_2) = A(z_1, z_2) + A(z_1, -z_2), \quad z_1, z_2 \in \mathbb{C} \setminus \{0\}, \tag{29}$$

we observe that the identity (13) is equivalent to

$$H(z_1, z_2) + H(-z_1, z_2) = 4. \tag{30}$$

Also, by using (10) and (29), we have that

$$H(z_1, z_2) = 2^{2-k_1-k_2}(1 + z_1)^{k_1} G(z_1, z_2), \tag{31}$$

where the Laurent polynomial G is defined by

$$G(z_1, z_2) = (1 + z_2)^{k_2} B(z_1, z_2) + (1 - z_2)^{k_2} B(z_1, -z_2), \tag{32}$$

with B denoting the Laurent polynomial for which (10) is satisfied.

It then follows from (30) and (31) that G satisfies the identity

$$2^{-k_1-k_2}(1 + z_1)^{k_1} G(z_1, z_2) + 2^{-k_1-k_2}(1 - z_1)^{k_1} G(-z_1, z_2) = 1. \tag{33}$$

Now, choose any pair of odd integers $\alpha_1, \alpha_2 \in \mathbb{N}$ such that $\alpha_1 < 2k_1$ and $\alpha_2 < 2k_2$. Then, for the Laurent polynomial G given by (32), we define the Laurent polynomial K_1 by

$$G(z_1, z_2) = 2^{k_1+k_2} z_1^{-\alpha_1} z_2^{-\alpha_2} K_1(z_1, z_2), \quad z_1, z_2 \in \mathbb{C} \setminus \{0\}. \tag{34}$$

It follows from (34) and (33) that K_1 satisfies the identity

$$(1+z_1)^{k_1}K_1(z_1,z_2)-(1-z_1)^{k_1}K_1(-z_1,z_2)=z_1^{\alpha_1}z_2^{\alpha_2}. \tag{35}$$

Hence, according to Lemma 3 (a), there exist a polynomial S_1 and a Laurent polynomial T_1 such that

$$K_1(z_1,z_2)=S_1(z_1,z_2)+(1-z_1)^{k_1}T_1(z_1,z_2),$$

with the polynomial S_1 and the Laurent polynomial T_1 satisfying the properties as stated in Lemma 3 (a).
Besides, (31) and (34) yield

$$H(z_1,z_2)=4(1+z_1)^{k_1}z_1^{-\alpha_1}z_2^{-\alpha_2}K_1(z_1,z_2), \tag{36}$$

according to which, since the Laurent polynomial H defined by (29) is even in z_2, we deduce that K_1 is odd in z_2, and hence also, from Lemma 3 (a), T_1 is also odd in z_2.

Similarly, by defining the Laurent polynomial J as

$$J(z_1,z_2)=A(z_1,z_2)+A(-z_1,z_2), \tag{37}$$

the identity (13) is equivalent to

$$J(z_1,z_2)+J(z_1,-z_2)=4. \tag{38}$$

Also, by using (10) and (37), we have that

$$J(z_1,z_2)=2^{2-k_1-k_2}(1+z_1)^{k_1}L(z_1,z_2), \tag{39}$$

where the Laurent polynomial L is defined by

$$L(z_1,z_2)=(1+z_1)^{k_1}B(z_1,z_2)+(1-z_1)^{k_1}B(-z_1,z_2), \tag{40}$$

with B denoting the Laurent polynomial for which (10) is satisfied.

It then follows from (38) and (39) that L satisfies the identity

$$2^{-k_1-k_2}(1+z_2)^{k_2}L(z_1,z_2)+2^{-k_1-k_2}(1-z_2)^{k_2}L(z_1,-z_2)=1. \tag{41}$$

Define the Laurent polynomial K_2 by

$$L(z_1,z_2)=2^{k_1+k_2}z_1^{-\alpha_1}z_2^{-\alpha_2}K_2(z_1,z_2). \tag{42}$$

It follows from (42) and (41) that K_2 satisfies the identity

$$(1+z_2)^{k_2}K_2(z_1,z_2)-(1-z_2)^{k_2}K_2(z_1,-z_2)=z_1^{\alpha_1}z_2^{\alpha_2}.$$

Hence, according to Lemma 3 (a), there exist a polynomial S_2 and a Laurent polynomial T_2 such that

$$K_2(z_1,z_2)=S_2(z_1,z_2)+(1-z_2)^{k_2}T_2(z_1,z_2),$$

with the polynomial S_2 and the Laurent polynomial T_2 satisfying the properties as stated in Lemma 3 (a).
Besides, (39) and (42) yield

$$J(z_1, z_2) = 4(1 + z_2)^{k_2} z_1^{-\alpha_1} z_2^{-\alpha_2} K_2(z_1, z_2),$$

according to which, since the Laurent polynomial J defined by (37) is even in z_1, we deduce that K_2 is odd in z_1, and hence also, from Lemma 3 (a), T_2 is also odd in z_1.

Next, we deduce from (40) and (42) that

$$(1+z_1)^{k_1} B(z_1, z_2) + (1-z_1)^{k_1} B(-z_1, z_2) = 2^{k_1+k_2} z_1^{-\alpha_1} z_2^{-\alpha_2} K_2(z_1, z_2). \quad (43)$$

Consider the Laurent polynomial $\tilde{B}$ such that

$$B(z_1, z_2) = 2^{k_1+k_2} z_1^{-2\alpha_1} z_2^{-2\alpha_2} \tilde{B}(z_1, z_2). \quad (44)$$

Then equation (43) becomes

$$(1 + z_1)^{k_1} \tilde{B}(z_1, z_2) + (1 - z_1)^{k_1} \tilde{B}(-z_1, z_2) = z_1^{\alpha_1} z_2^{\alpha_2} K_2(z_1, z_2). \quad (45)$$

Since $K_2(z_1, z_2)$ is odd in z_1, by virtue of (35), the Laurent polynomial $B_1(z1, z2) := K_1(z1, z2)K_2(z1, z2)$ is a particular solution of (43), that is, for $z_1, z_2 \in \mathbb{C} \setminus \{0\}$,

$$(1 + z_1)^{k_1} B_1(z_1, z_2) + (1 - z_1)^{k_1} B_1(-z_1, z_2) = 2^{k_1+k_2} z_1^{\alpha_1} z_2^{\alpha_2} K_2(z_1, z_2). \quad (46)$$

Subtracting (45) from (46) yields

$$(1+z_1)^{k_1} [\tilde{B}(z_1, z_2) - B_1(z_1, z_2)] = -(1-z_1)^{k_1} [\tilde{B}(-z_1, z_2) - B_1(-z_1, z_2)], \quad (47)$$

which shows that there exists a Laurent polynomial $T_3(z_1, z_2)$ such that

$$\tilde{B}(z_1, z_2) - B_1(z_1, z_2) = T_3(z_1, z_2)(1 - z_1)^{k_1}. \quad (48)$$

Substituting this expression into (47) gives

$$(1 + z_1)^{k_1} (1 - z_1)^{k_1} T_3(z_1, z_2) = -(1 - z_1)^{k_1} (1 + z_1)^{k_1} T_3(-z_1, z_2),$$

showing that T_3 is odd in z_1. It follows from (10) and (48) that

$$A(z_1, z_2) = 4(1 + z_1)^{k_1} (1 + z_2)^{k_2} z_1^{-\alpha_1} z_2^{-\alpha_2} [B_1(z_1, z_2) + T_3(z_1, z_2)(1 - z_1)^{k_1}]. \quad (49)$$

Since K_1 is odd in z_2, we deduce from (49) and (35) that

$$A(z_1, z_2) + A(z_1, -z_2) = 4(1 + z_1)^{k_1} z_1^{-2\alpha_1} z_2^{-2\alpha_2} [z_1^{\alpha_1} z_2^{\alpha_2} K_1(z_1, z_2) \\ +(1 - z_1)^{k_1} \{(1 + z_2)^{k_2} T_3(z_1, z_2) + (1 - z_2)^{k_2} T_3(z_1, -z_2)\}]. \quad (50)$$

Using (29) and (36), we deduce from (50) that

$$(1+z_2)^{k_2}T_3(z_1,z_2)+(1-z_2)^{k_2}T_3(z_1,-z_2)=0,$$

showing that there exists a Laurent polynomial T, such that $T_3(z_1, z_2) = T(z_1, z_2)(1 - z_2)^{k_2}$. Note that T odd in z_2 and since T_3 is odd in z_1, T is also odd in z_1. The result (11) follows from (44) and (48).

(b) Conversely, suppose that for any pair of odd (positive) integers α_1 and α_2 such that $\alpha_1 < 2k_1$ and $\alpha_2 < 2k_2$, the Laurent polynomial B has the form given by (11). To show that the Laurent polynomial A is an interpolatory mask symbol, it will suffice to prove that A satisfies the identity (13) in Lemma 2.

To this end, since by assumption S_2, T_2 and T are odd in z_1, observe from (10) and (11) that, for $z_1, z_2 \in C \setminus \{0\}$,

$$\begin{aligned}
&A(z_1,z_2)+A(-z_1,z_2)\\
=&4z_1^{-2\alpha_1}z_2^{-2\alpha_2}(1+z_1)^{k_1}(1+z_2)^{k_2}\left[T(z_1,z_2)(1-z_1)^{k_1}(1-z_2)^{k_2}\right.\\
&+\left\{S_1(z_1,z_2)+T_1(z_1,z_2)(1-z_1)^{k_1}\right\}\left.\left\{S_2(z_1,z_2)+T_2(z_1,z_2)(1-z_2)^{k_2}\right\}\right]\\
&+4z_1^{-2\alpha_1}z_2^{-2\alpha_2}(1-z_1)^{k_1}(1+z_2)^{k_2}\left[-T(z_1,z_2)(1+z_1)^{k_1}(1-z_2)^{k_2}\right.\\
&+\left\{S_1(-z_1,z_2)+T_1(z_1,z_2)(1+z_1)^{k_1}\right\}\left.\left\{-S_2(z_1,z_2)-T_2(z_1,z_2)(1-z_2)^{k_2}\right\}\right],
\end{aligned}$$

which, together with (12), yields

$$\begin{aligned}
&A(z_1,z_2)+A(-z_1,z_2)\\
=&4z_1^{-2\alpha_1}z_2^{-2\alpha_2}(1+z_2)^{k_2}\left[z_1^{\alpha_1}z_2^{\alpha_2}\left\{S_2(z_1,z_2)+T_2(z_1,z_2)(1-z_2)^{k_2}\right\}\right].
\end{aligned}\tag{51}$$

Replacing z_2 by $-z_2$ in (51), and using the fact that T_2 is even in z_2, we obtain

$$\begin{aligned}
&A(z_1,-z_2)+A(-z_1,-z_2)\\
=&4z_1^{-2\alpha_1}z_2^{-2\alpha_2}(1-z_2)^{k_2}\left[-z_1^{\alpha_1}z_2^{\alpha_2}\left(S_2(z_1,-z_2)+T_2(z_1,z_2)(1+z_2)^{k_2}\right)\right].
\end{aligned}\tag{52}$$

Since S_2 satisfies (12), adding (51) with (52) yields

$$A(z_1,z_2)+A(-z_1,z_2)+A(z_1,-z_2)+A(-z_1,-z_2)=4,$$

thereby showing that the Laurent polynomial A satisfies the identity (13), which concludes our proof. □

3 Existence Results for Refinable Functions

We now present methods to construct interpolatory refinable functions for $M = 2I$. Consider the torus T and its subset $\tilde{T}$ defined respectively by

$$T=\{(e^{ix_1},e^{ix_2}):x_1,x_2\in\mathbb{R}\},\ \text{and}\ \tilde{T}=\{(e^{ix_1},e^{ix_2})\in T:\ \ |x_1|,|x_2|\le\pi/2\}.$$

We say that a refinement mask a is *non-negative* if its associated mask symbol A satisfies

$$A(e^{ix_1}, e^{ix_2}) \geq 0, \quad x_1, x_2 \in \mathbb{R}.$$

Then we have the following result from [11].

Theorem 3. *Consider $M = 2I$, and a non-negative interpolatory mask a. If $k_1, k_2 \in \mathbb{N}$ and B are such that*

$$A(z_1, z_2) = 2^{2-k_1-k_2}(1+z_1)^{k_1}(1+z_2)^{k_2}B(z_1, z_2),$$

with $B(1,1) = 1$ and $B(z_1, z_2) \neq 0$ for $(z_1, z_2) \in \tilde{T}$, then there exists a corresponding interpolatory refinable function $\phi_a \in C_0(\mathbb{R}^2)$.

Given two univariate functions $\tilde{\phi} \in C^{\alpha_1}(\mathbb{R})$ and $\tilde{\psi} \in C^{\alpha_2}(\mathbb{R})$, the *tensor product* $\phi = \tilde{\phi} \cdot \tilde{\psi}$ is defined by

$$\phi(x, y) = \tilde{\phi}(x)\tilde{\psi}(y), \quad (x, y) \in \mathbb{R}^2,$$

so that $\phi \in C^{\alpha}(\mathbb{R}^2)$ where $\alpha = \min\{\alpha_1, \alpha_2\}$.

The following result is straightforward.

Theorem 4. *Let $\tilde{\phi} \in C_0^{\alpha_1}(\mathbb{R})$ and $\tilde{\psi} \in C_0^{\alpha_2}(\mathbb{R})$ be interpolatory and refinable with masks $\tilde{a}$ and $\tilde{b}$.*

Then the function $\phi = \tilde{\phi} \cdot \tilde{\psi}$, that is,

$$\phi(x, y) = \tilde{\phi}(x)\tilde{\psi}(y), \quad (x, y) \in \mathbb{R}^2,$$

is interpolatory and refinable, with $M = 2I$ and with mask a given by

$$a_{j,k} = \tilde{a}_j \tilde{b}_k, \quad (j, k) \in \mathbb{Z}^2.$$

Another method to construct refinable functions is by using the cascade algorithm. Given a dilation matrix M and a refinement mask a, we define the *cascade operator* $T_a = T_{a,M} : f \in M(\mathbb{R}^2) \mapsto T_a f \in M(\mathbb{R}^2)$ by

$$(T_a f)(\mathbf{x}) = \sum_{\mathbf{j}} a_{\mathbf{j}} f(M\mathbf{x} - \mathbf{j}), \quad \mathbf{x} \in \mathbb{R}^2. \tag{53}$$

For a given initial function $g \in M(\mathbb{R}^2)$, T_a generates the sequence $\{f_r = T_a^r g : r \in \mathbb{Z}_+\}$ by means of the *cascade algorithm*:

$$f_0 = g, \qquad f_{r+1} = T_a f_r, \quad r \in \mathbb{Z}_+. \tag{54}$$

We say that T_a is *convergent* on a subset $\mathcal{M} \subset C_0(\mathbb{R}^2)$ if, for $g \in \mathcal{M}$, there exists $f = f_g \in C(\mathbb{R}^2)$ such that

$$\lim_{r \to \infty} \|T_a^r g - f\|_\infty = 0,$$

in which case we write $T_a^\infty g := f$. We shall simply say " convergent " for " convergent on $C_0(\mathbb{R}^2)$ ". We proceed to establish the following result with respect to the cascade algorithm.

Theorem 5. *Suppose M is a dilation matrix and $\boldsymbol{a}$ an interpolatory mask such that*

$$\sum_{\mathbf{j}} a_{\mathbf{k}} - M\mathbf{j}^T = 1, \quad \mathbf{k} \in \mathbb{Z}^2, \quad \text{and} \quad [2\alpha, 2\beta]^2 \subseteq M[\alpha, \beta]^2, \quad \alpha, \beta \in \mathbb{Z}. \tag{55}$$

Also, for an initial function $g \in C_0(\mathbb{R}^2)$, let the sequence $\{\phi_r : r \in \mathbb{Z}_+\}$ be generated by the cascade algorithm as in (54). Then:

(a) $\phi_r \in C_0(\mathbb{R}^2)$, $r \in \mathbb{Z}_+$;
(b) If g is interpolatory, then ϕ_r is interpolatory for each $r \in \mathbb{Z}_+$;
(c) If g is interpolatory and $T_{\boldsymbol{a}}$ is convergent with $\phi = T_{\boldsymbol{a}}^{\infty} g$, then:
  (i) $\phi \in C_0(\mathbb{R}^2)$;
  (ii) ϕ is an interpolatory refinable function with mask $\boldsymbol{a}$ and dilation matrix M, and $\phi = T_{\boldsymbol{a}}\phi$.

Proof. (a), (b) We proceed by induction on r. Recall from the recursive formula (54), together with (53), that

$$\phi_{r+1} = T_{\boldsymbol{a}}\phi_r = \sum_{\mathbf{j}} a_{\mathbf{j}} \phi_r(M \cdot - \mathbf{j}), \quad r \in \mathbb{Z}_+. \tag{56}$$

For $r = 0$, $\phi_0 = g \in C_0(\mathbb{R}^2)$. Let us fix $r \in \mathbb{Z}_+$. The following holds:
If $\operatorname{supp}(\phi_r) \subseteq [N_1, N_2]^2$, it holds that, for $\mathbf{x} \in \mathbb{R}^2$ and $\mathbf{j} \in [N_1, N_2]^2$,

$$\begin{aligned} M\mathbf{x}^T - \mathbf{j} \in [N_1, N_2]^2 &\Longrightarrow M\mathbf{x}^T \in \mathbf{j} + [N_1, N_2]^2 \subseteq [2N_1, 2N_2]^2 \\ &\Longrightarrow \mathbf{x} \in M^{-1}\left(\mathbf{j} + [N_1, N_2]^2\right) \subseteq M^{-1}\left[2N_1, 2N_2\right]^2. \end{aligned} \tag{57}$$

Since $\boldsymbol{a}$ is supported on $[N_1, N_2]^2$, and since there is only a finite number of integers $\mathbf{j}$ in $[N_1, N_2]^2$, we deduce from (57) and (56) that the support of ϕ_{r+1} satisfies

$$\begin{aligned} \operatorname{supp}(\phi_{r+1}) \subseteq \bigcup_{\mathbf{j} \in [N_1, N_2]^2} M^{-1}\left(\mathbf{j} + [N_1, N_2]^2\right) &\subseteq \bigcup_{\mathbf{j} \in [N_1, N_2]^2} M^{-1}\left[2N_1, 2N_2\right]^2 \\ &\subseteq [N_1, N_2]^2, \end{aligned}$$

by virtue of (55). Hence ϕ_r is compactly supported.

If ϕ_r is continuous, then the shifts with respects to $\mathbb{Z}^2$ of its dilations are continuous, so that, from (56), we deduce that ϕ_{r+1} is also continuous.

If ϕ_r is interpolatory, we deduce from (56) and (7) that, for $\mathbf{j} \in \mathbb{Z}^2$,

$$\phi_{r+1}(\mathbf{j}) = \sum_{\mathbf{k}} a_{\mathbf{k}} \phi_r(M\mathbf{j}^T - \mathbf{k}) = a_{M\mathbf{j}^T} = \delta_{\mathbf{j}}.$$

(c) If, moreover, $T_{\boldsymbol{a}}$ is convergent, it follows from the uniform convergence result $\|\phi - \phi_r\|_\infty \to 0$, $r \to \infty$, that $\phi \in C_0(\mathbb{R}^2)$. Also, since ϕ_r is interpolatory for every $r \in \mathbb{Z}_+$, we have, for $\mathbf{j} \in \mathbb{Z}^2$,

$$|\phi(\mathbf{j}) - \delta_{\mathbf{j}}| = |\phi(\mathbf{j}) - \phi_r(\mathbf{j})| \leq \|\phi - \phi_r\|_\infty \to 0, \quad r \to \infty,$$

and it follows that ϕ is interpolatory.

To prove that ϕ satisfies the refinement equation (6), we use (54) and (53) to obtain

$$\begin{aligned}\|\phi - T_a\phi\|_\infty \leq& \|\phi - \phi_{r+1}\|_\infty + \|T_a(\phi_r - \phi)\|_\infty \\ \leq& \|\phi - \phi_{r+1}\|_\infty + \left[\sum_{\mathbf{j}} |a_{\mathbf{j}}|\right] \|\phi_r - \phi\|_\infty \to 0, \quad r \to \infty,\end{aligned}$$

i.e. $\phi = T_a\phi$, which is equivalent to (6).

4 Numerical Application

The previous results can be summarized as follows. Given a dilation matrix M and an interpolatory mask $\boldsymbol{a}$ such that the diagonal elements of M satisfy $|M_{11}| \geq 2$, $|M_{22}| \geq 2$,

$$\sum_{\mathbf{j}} a_{\mathbf{k}-M\mathbf{j}^T} = 1, \quad \mathbf{k} \in \mathbb{Z}^2, \quad \text{and} \quad [2\alpha, 2\beta]^2 \subseteq M[\alpha, \beta]^2, \quad \alpha, \beta \in \mathbb{Z},$$

we have

Convergence of $T_a \Longrightarrow$ Existence of interpolatory refinable function $\phi \Longrightarrow$ Convergence of S_a.

We proceed to apply the results of this paper to numerically investigate the existence of the interpolatory refinable function corresponding to the Butterfly subdivision scheme ([5, 4, 7]). Given a parameter $w \in \mathbb{R}$, the interpolatory mask symbol $\mathcal{B}_w$ of the Butterfly scheme is defined by

$$\mathcal{B}_w(z_1, z_2) = \frac{1}{2}(1 + z_1)(1 + z_2)(1 + z_1^{-1}z_2^{-1})(1 - wC(z_1, z_2)),$$

where the Laurent polynomial C is given by

$$\begin{aligned}C(z_1, z_2) = & 2z_1^{-2}z_2^{-1} + 2z_1^{-1}z_2^{-2} - 4z_1^{-1}z_2^{-1} - 4z_1^{-1} - 4z_2^{-1} \\ & + 2z_1^{-1}z_2 + 2z_1z_2^{-1} + 12 - 4z_1 - 4z_2 - 4z_1z_2 + 2z_1^2z_2 + 2z_1z_2^2.\end{aligned}$$

Note that $\mathcal{B}_w$ is supported on the square $[-3, 3]^2$ for any w, which is then also the support of the associated refinable function $\phi_{\mathcal{B}_w}$. One can verify that the Butterfly mask symbol is a special case of Theorem 2 where the polynomials S_1 and S_2 are given by

$$S_1(z_1, z_2) = \frac{1}{2}z_2, \quad S_2(z_1, z_2) = \frac{1}{2}z_1,$$

the Laurent polynomials $T_1 = T_{w,1}$ and $T_2 = T_{w,2}$ are given by

$$T_1(z_1,z_2) = \frac{1}{4} z_1^{-2} z_2^{-1} (2wz_1^4 z_2^4 - 2wz_1^2 z_2^4 - z_1^2 z_2^2 + 2wz_2^2 - 2w), \tag{58}$$

$$T_2(z_1,z_2) = \frac{1}{4} z_1^{-1} z_2^{-2} (2wz_1^4 z_2^4 - 2wz_1^4 z_2^2 - z_1^2 z_2^2 + 2wz_2^2 - 2w), \tag{59}$$

and the Laurent polynomial $T = T_w$ is given by

$$\begin{aligned} T(z_1,z_2) = -\frac{1}{16} z_1^{-3} z_2^{-3} \big(& 4w^2 z_1^8 z_2^8 - 4w^2 z_1^6 z_2^8 - 4w^2 z_1^8 z_2^6 + 4w^2 z_1^6 z_2^6 \\ & - 4wz_1^6 z_2^6 + 4w^2 z_1^4 z_2^6 + 2wz_1^4 z_2^6 - 4w^2 z_1^2 z_2^6 + 4w^2 z_1^6 z_2^4 \\ & + 2wz_1^6 z_2^4 - 8w^2 z_1^4 z_2^4 + 16wz_1^4 z_2^4 - z_1^4 z_2^4 + 4w^2 z_1^2 z_2^4 \\ & + 2wz_1^2 z_2^4 - 4w^2 z_1^6 z_2^2 + 4w^2 z_1^4 z_2^2 + 2wz_1^4 z_2^2 + 4w^2 z_1^2 z_2^2 \\ & - 4wz_1^2 z_2^2 - 4w^2 z_2^2 - 4w^2 z_1^2 + 4w^2 \big) . \end{aligned} \tag{60}$$

We apply the cascade algorithm to the Butterfly mask symbol $\mathcal{B}_w$ to generate Figure 1$(a)-(b)$ and Figure 2(a), thereby obtaining the graph of the refinable function $\phi_{\mathcal{B}_w}$. We also applied the Butterfly subdivision scheme in Figure 2(b) for an initial control point sequence as shown. Based on Figure 2(a), we conjecture that $\phi_{\mathcal{B}_w}$, $w = \frac{1}{16}$ exists and is of class C^1, i.e. $\phi_{\mathcal{B}_w} \in C_0^1(\mathbb{R}^2)$.

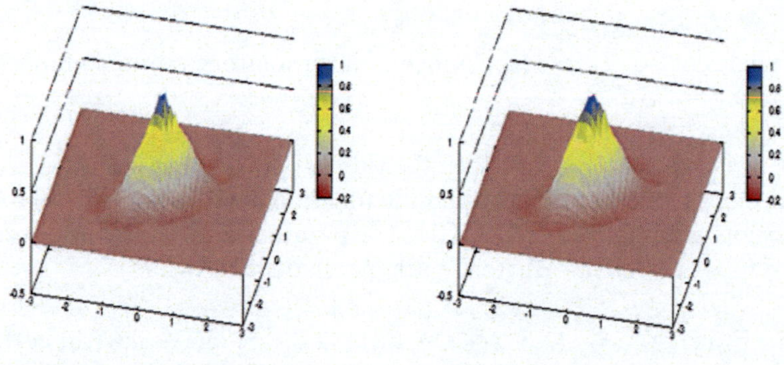

(a) Graph of $T_{\mathcal{B}_w} g$, $w = \frac{1}{16}$ (b) Graph of $T^2_{\mathcal{B}_w} g$, $w = \frac{1}{16}$

Fig. 1. Cascade algorithm associated with $\mathcal{B}_w$, $w = \frac{1}{16}$

5 Conclusion

We give a characterization result for interpolatory mask symbols for the case where the dilation matrix is given by $M = 2I$, the proof of which requires the

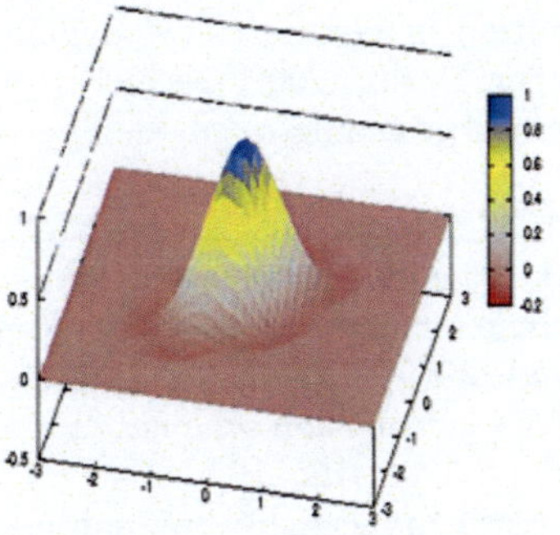

(a) Graph of $\phi_{\mathcal{B}_w}$, $w = \frac{1}{16}$

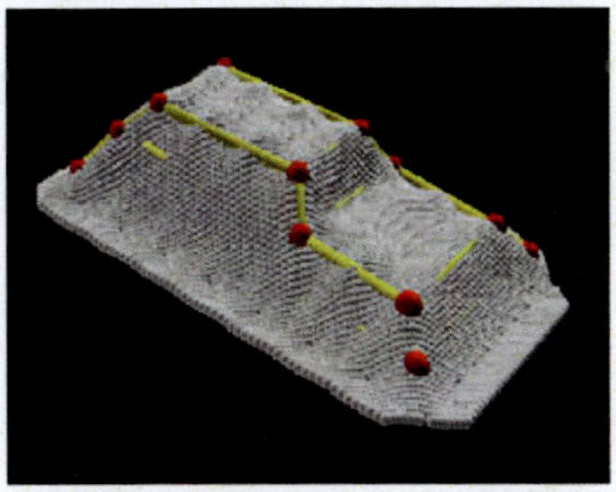

(b) Graph of $\Phi_{\mathcal{B}_w}$, $w = \frac{1}{16}$

Fig. 2. Interpolatory refinable function $\phi_{\mathcal{B}_w}$ and subdivision limit function $\Phi_{\mathcal{B}_w}$

Bezout theorem and and the Euclidean algorithm to solve for the polynomials S_1 and S_2 in Theorem 2. The convergence of the associated interpolatory subdivision scheme is related to the existence of the corresponding interpolatory refinable function. Methods to find such refinable functions are given for non-negative mask symbols and for masks obtained from tensor products. Interpolatory refinable functions can also be obtained by using the cascade algorithm which is a method consisting of a recursively built sequence of functions which eventually converge to an interpolatory refinable function.

A further continuation of this work would consist in studying the properties of the Laurent polynomials T_1, T_2 and T in Theorem 2 and their effects on the smoothness properties of the associated subdivision schemes and the interpolatory refinable functions. A possible difficulty is that one may encounter complicated formulas such as for the Butterfly mask symbol in (58), (59) and (60), in which case analysis becomes formidable.

References

1. A.S. Cavaretta, W. Dahmen, and C.A. Micchelli: Stationary subdivision. Memoirs of Amer. Math. Soc. **93**(453), 1991.
2. J.M. de Villiers and K.M. Hunter: Interpolatory subdivision based on local interpolation. East Journal on Approximations **12**(3), 2006, 303–330.
3. J.M. de Villiers and K.M. Hunter: On the construction and convergence of a class of symmetric interpolatory subdivision schemes. East Journal of Approximations **12**(2), 2006, 151–188.
4. N. Dyn: Subdivision schemes in computer aided geometric design. In *Advances in Numerical Analysis II: Wavelets, Subdivision Algorithms and Radial Basis Functions*, W.A. Light (ed.), Oxford University Press, Oxford, UK, 1992, 36–104.

5. N. Dyn: Analysis of convergence and smoothness by the formalism of Laurent polynomials. In *Tutorials on Multiresolution in Geometric Modelling*, A. Iske, E. Quak, and M.S. Floater (eds.), Springer, Berlin, 2002, 51–68.
6. N. Dyn and D. Levin: Subdivision schemes in geometric modelling. Acta Numerica, 2002, 1–72.
7. N. Dyn, D. Levin, and J.A. Gregory: A butterfly subdivision scheme for surface interpolation with tension control. ACM Trans. Graphics **9**(2), 1990, 160–169.
8. B. Han and R.Q. Jia: Multivariate refinement equations and convergence of subdivision schemes. SIAM J. Math. Anal. **29**(5), 1998, 1177–1199.
9. B. Han and R.Q. Jia: Optimal interpolatory subdivision schemes in multidimensional spaces. SIAM J. Numer. Anal. **36**(1), 1998, 105–124.
10. R.Q. Jia: Interpolatory subdivision schemes induced by box splines. Applied and Computational Harmonic Analysis **8**, 2000, 286–292.
11. C.A. Micchelli: Interpolatory subdivision schemes and wavelets. J. Approx. Theory **86**, 1996, 41–71.
12. S.D. Riemenschneider and Z.W. Shen: Multidimensional interpolatory subdivision schemes. SIAM J. Numer. Anal. **34**, 1997, 2357–2381.
13. J.M. De Villiers, K.M. Goosen, and B.M. Herbst: Dubuc-Deslauriers subdivision for finite sequences and interpolation wavelets on an interval. SIAM J. Math. Anal. **35**(2), 2003, 423–452.

Model and Feature Selection in Metrology Data Approximation

Xin-She Yang and Alistair B. Forbes

National Physical Laboratory, Teddington TW11 0LW, UK

Summary. For many data approximation problems in metrology, there are a number of competing models which can potentially fit the observed data. A crucial task is to quantify the extent to which one model performs better than others, taking into account the influence of random effects associated with the data. For example, for a given data set, we can use a series of polynomials of various degrees to fit the data using a least squares criterion. The residual sum of squares is a measure of how well the model fits the data. However, it is generally required to balance goodness of fit with minimising the model complexity. We consider a number of criteria that aim to do this: the Akaike information criterion (AIC), the Bayesian/Schwarz information criterion, and the AIC with a correction for small sample size (AICc). In this paper, we compare the performance of these criteria for polynomial regression and show that for the examples tested the AICc criterion performs best. A second element of model selection is to determine from a set of feature vectors, the subset that defines a model space most suitable for describing the observed response. Since there are 2^N possible model spaces defined by N feature vectors, for even a modest number of feature vectors it is necessary to reduce or prioritise the number of candidate models. Partial least squares and the least angle regression algorithms can be used as model reduction tools. We describe these algorithms in the context of feature selection and how they can be used with a model selection criterion such as AICc and illustrate their performance using simulations and on an application from human sensory perception.

1 Introduction

In many data fitting and modelling problems, we often have to find the best model to fit a given data set. Typically the model space is parametrized by parameters $\boldsymbol{\alpha}$, e.g., polynomial coefficients, and the best-fit model is found by minimising some goodness of fit measure to determine optimal parameter values $\mathbf{a}$. In this paper, we consider the case where there are a number of competing model spaces. In such cases, the model fitting has two interacting elements: the selection of an appropriate model space, and the determination

E.H. Georgoulis, A. Iske, J. Levesley (eds.), *Approximation Algorithms for Complex Systems*, Springer Proceedings in Mathematics 3, DOI: 10.1007/978-3-642-16876-5_14,
© Springer-Verlag Berlin Heidelberg 2011

of the best fit from within the selected model space. In the model selection process, a balance has to be made between fitting the data and minimising the model complexity, particularly with regard to gaining insight into the underlying physical mechanisms. If there are more than one model which fit the data equally well, we usually choose the simplest model according to the Occam's razor principle or some form of parsimony. The idea is that we should use as few assumptions as possible to explain any given data [8, 12, 20]. Because of their practical importance, there has been much interest in developing effective model selection techniques in recent years [11, 12, 14, 15, 17, 20].

In this paper, we focus on the selection of the best model space using well known criteria: root mean square residual (RMS), the Akaike information criterion (AIC), the Bayesian/Schwarz information criterion (BIC), and the AIC with a correction for small sample size (AICc). We compare the performance of these criteria.

Another important element of model selection is to determine the best subset of features from a set of feature vectors, that is to obtain a reduced model space most suitable for describing the observed response. Partial least squares (PLS) and least angle regression (LARS) algorithms can be used as model reduction/feature selection tools.

The paper is organized as follows. In Section 2, we briefly introduce the problem of model fitting and selection and in Section 3 we describe the selection criteria considered in this paper and their performance on polynomial regression. In Section 4, we describe algorithms that attempt to build up a nested sequence of model spaces from features to which selection criteria can be applied. Our concluding remarks are given in Section 5.

2 Model Fitting and Selection

We assume an observational model of the form

$$y = \phi(\mathbf{x}, \boldsymbol{\alpha}) + \epsilon, \quad \epsilon \in \mathrm{N}(0, \sigma^2),$$

where ϕ is a known function specified by parameters $\boldsymbol{\alpha} = (\alpha_1, \ldots, \alpha_n)^T$ that models the response of a system that depends on the covariates $\mathbf{x}$ (that can be measured accurately), y is the measured response and ϵ represents a random effect drawn from a Gaussian distribution with known standard deviation σ. Given a data set $\{(\mathbf{x}_i, y_i)\}_{i=1}^{m}$ with $m \geq n$ data points, the least squares estimate $\mathbf{a}$ of the parameters $\boldsymbol{\alpha}$ minimises

$$F(\boldsymbol{\alpha}) = \sum_{i=1}^{m} (y_i - \phi(\mathbf{x}_i, \boldsymbol{\alpha}))^2.$$

The statistical model for the random effects defines the probability density $p(\mathbf{y}|\boldsymbol{\alpha})$ of observing a response vector $\mathbf{y} = (y_1, \ldots, y_m)^T$ for a system specified by parameters $\boldsymbol{\alpha}$. Regarded as a function of $\boldsymbol{\alpha}$, this density is known as the likelihood function $\mathcal{L}(\boldsymbol{\alpha}; \mathbf{y})$ and, for Gaussian noise,

$$\mathcal{L}(\boldsymbol{\alpha}; \mathbf{y}) = \exp\left\{-\frac{F(\boldsymbol{\alpha})}{2\sigma^2}\right\},$$

up to an additive constant. Thus, for Gaussian noise, the least squares estimate is also the maximum likelihood estimate.

If the model is linear in the parameters $\boldsymbol{\alpha}$ so that the observational model takes the form

$$\mathbf{y}|\boldsymbol{\alpha} \in \mathrm{N}(C\boldsymbol{\alpha}, \sigma^2 I),$$

where C is the $m \times n$ observation matrix, then the least squares estimate is given by

$$\mathbf{a} = \left(C^T C\right)^{-1} C^T \mathbf{y}.$$

If C has QR factorisation [6],

$$C = QR = \begin{bmatrix} Q_1 \; Q_2 \end{bmatrix} \begin{bmatrix} R_1 \\ \mathbf{0} \end{bmatrix} = Q_1 R_1,$$

then

$$\mathbf{a} = R_1^{-1} Q_1^T \mathbf{y}, \quad \hat{\mathbf{y}} = C\mathbf{a} = Q_1 Q_1^T \mathbf{y}, \quad \mathbf{r} = \mathbf{y} - C\mathbf{a} = Q_2 Q_2^T \mathbf{y}. \tag{1}$$

We let $\mathrm{RSS} = F(\mathbf{a}) = \|Q_2^T \mathbf{y}\|_2^2$, the sum of squares function evaluated at the least squares solution.

We now consider the case where there are K competing models defined by functions $\phi_k(\mathbf{x}, \boldsymbol{\alpha}_k)$ involving parameter vectors $\boldsymbol{\alpha}_k$ of length n_k. For a given data set $\{(\mathbf{x}_i, y_i)\}_{i=1}^m$, for each model space we can calculate the least squares best estimate $\mathbf{a}_k$ of the model parameters and $\mathrm{RSS}_k = F(\mathbf{a}_k)$, the residual sum of squares at the least squares estimate. For the linear case,

$$\mathbf{a}_k = (\mathbf{C}_k^T \mathbf{C}_k)^{-1} \mathbf{C}_k^T \mathbf{y},$$

where C_k is the observation matrix associated with the kth model.

3 Selection Criteria

In this section we consider selection criteria that aim to balance goodness of fit with model complexity: root mean square, Akaike information criterion (AIC) and variants, and the Bayesian/Schwarz information criterion (BIC) [16]. All the criteria are simple to implement as they can be defined in terms RSS_k, representing goodness of fit, and n_k, representing model complexity. We recall that for Gaussian noise RSS is related to the likelihood and the probability of observing the response vector $\mathbf{y}$.

3.1 Root Mean Square

The root mean square residual (RMS) is commonly used for model selection and is defined by

$$\mathrm{RMS}_k = \sqrt{\frac{\mathrm{RSS}_k}{m - n_k}},$$

where n_k is the number of parameters in the kth model. The model with the smallest RMS value is selected. For a fixed data set, the number of data points m is fixed and the minimising k is the same as that which minimises

$$m \ln(\mathrm{RSS}_k/m) + m \ln\left(\frac{m}{m - n_k}\right). \tag{2}$$

The first term represents the penalty associated with the goodness of fit and is defined by the likelihood while the second term specifies penalty associated with the model complexity.

3.2 Akaike Information Criterion

See [1]. For the Gaussian noise model considered here,

$$\mathrm{AIC} = m \ln(\mathrm{RSS}_k/m) + 2n_k. \tag{3}$$

Compared to (2), the penalty for model complexity is stronger, reducing the tendency for overfitting. AIC is derived in terms of the information loss that would be incurred if the true model is replaced by the approximating model. The model selected by AIC minimises a measure of information loss. For small sample sizes, a second order effect becomes more important which leads to AICc [9]

$$\mathrm{AICc} = m \ln(\mathrm{RSS}_k/m) + 2n_k \frac{m}{(m - n_k - 1)}, \tag{4}$$

giving more weight to the model complexity term for smaller m/n_k. Since AICc approaches AIC for large m/n_k, Burnham and Anderson suggest that AICc should always be used, regardless of the sample size [3].

3.3 Bayes Information Criterion

Schwarz used a Bayesian analysis to develop the BIC [16]. For the Gaussian case,

$$\mathrm{BIC} = m \ln(\mathrm{RSS}_k/m) + n_k \ln m. \tag{5}$$

Comparing with the standard AIC, BIC imposes a stronger penalty on model complexity.

Comparing (2)–(5), it is seen that the goodness of fit penalty is the same for all criteria but that the criteria put different weight on model complexity as specified by the number of model parameters n_k.

3.4 Example: Polynomial Regression

We have performed a range of simulations involving a polynomial model. The simulation engine has, as input, the number of Monte Carlo simulations M, the number of data points m, the order k_0 of the generating polynomial and the standard deviation σ associated with the noise. If C_{k_0} is the observation matrix associated with m uniformly spaced data points $x_i \in [0,\ 1]$, for $q = 1, \ldots, M$, the script generates, at random, coefficients $\mathbf{a}_q$ for an order k_0 polynomial, response vector $\mathbf{y}_q \in \mathrm{N}(C_{k_0}\mathbf{a}_q, \sigma^2 I)$, and then for $k = 1, \ldots, m-1$, calculates the criteria values and notes the order $k_{\min,q}$ that generates the minimum values for the criteria. The required values of RSS_k can be calculated efficiently from the QR factorisation of C_{k_0} using the relationships in (1).

Figure 1 graphs the mean criteria values for $k = 1, \ldots, 20$, over $M = 5000$ simulations for the case $m = 30$, $k_0 = 4$ and $\sigma = 0.1$. All four criteria have a distinct behaviour at $k = 4$ with AICc and BIC having recognisable minima there. As k approaches m only AICc remains increasing. Figure 2 shows the results for the case $k_0 = 8$ and $\sigma = 1.0$. The increased noise means that the behaviour at $k = 8$ is less pronounced. Figures 3 and 4 show the fraction of the simulations for which k was identified as the minimum criterion value for the two cases. Only for AICc and BIC is k_0 the most probable choice of order. On the basis of these simulations, AICc performs best.

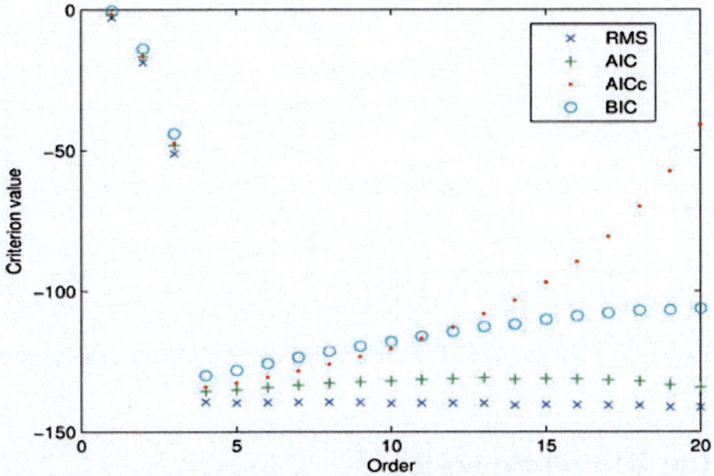

Fig. 1. Information criteria versus order of polynomial for the case $m = 30, k_0 = 4$ and $\sigma = 0.1$.

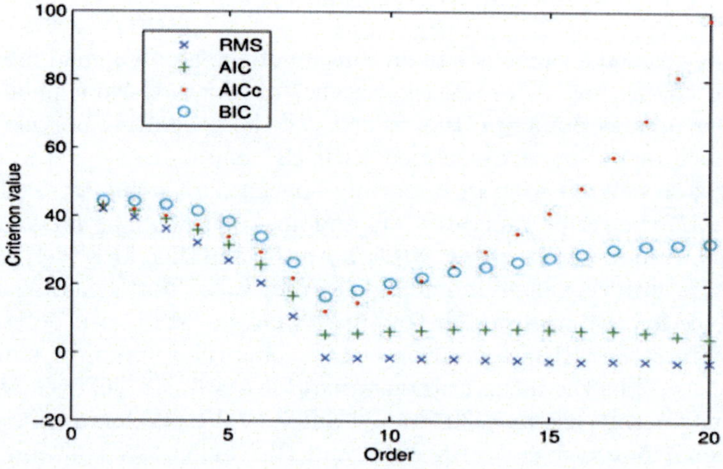

Fig. 2. As Fig. 1 for the case $k_0 = 8$ and $\sigma = 1.0$.

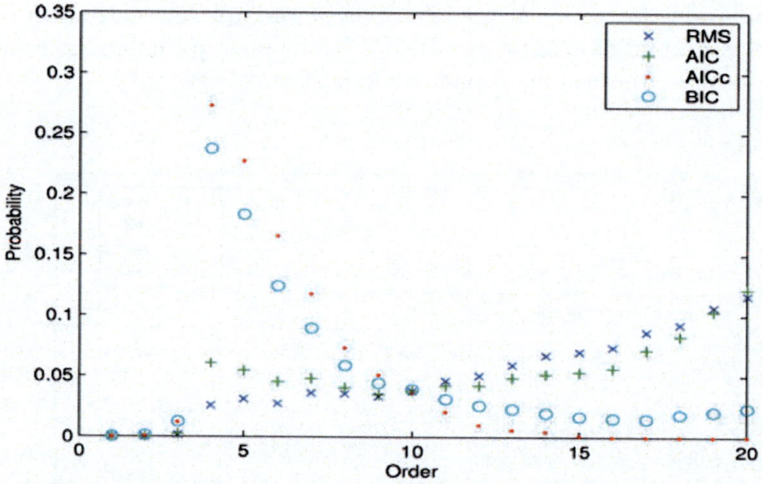

Fig. 3. Probability of the information criteria choosing k for the case $m = 30$, $k_0 = 4$ and $\sigma = 0.1$.

4 Directed Model Selection

The selection criteria above choose amongst all models the one with the optimal criterion. In order to implement this approach, all models have to be tested. For many model selection problems, such as feature selection to be discussed below, the number of models grows exponentially with the number of features, so that a complete scan of all the models becomes infeasible. For practical implementations, we need to prioritize the models so that those

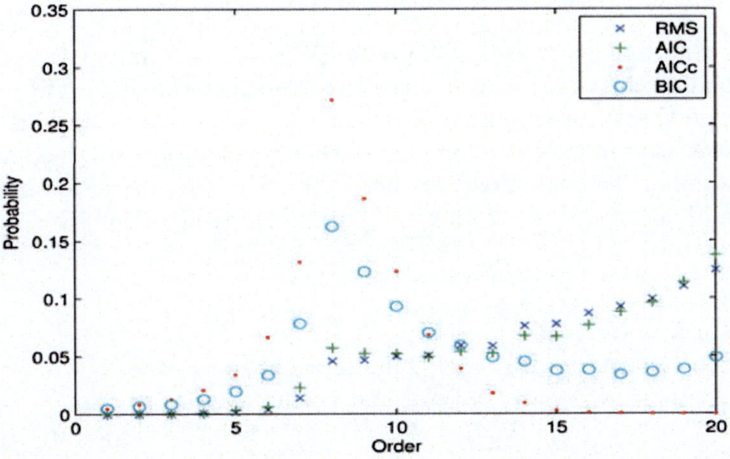

Fig. 4. As Fig. 3 for the case $k_0 = 8$ and $\sigma = 1.0$.

models that are likely to be supported by the data are examined first, that is, some prior ordering of the models is required. For model spaces such as polynomials, there is a natural nesting of the spaces. For general sets of features, there is no *a priori* ordering available.

4.1 Partial Least Squares

Partial least squares (PLS) is a well-known method for approximating a response vector from a low dimensional subspace. For a given $m \times n$ array of feature vectors $X = [\mathbf{x}_1 \ \ldots \ \mathbf{x}_n]$, and a response vector $\mathbf{y}$, the main idea is to generate a series of $n \times k$ matrices V_k satisfying

$$V_k = [V_{k-1}\mathbf{v}_k],$$

where $\mathbf{v}_k$ is a new column, and to set $\mathbf{a}_k$ to be the least squares solution of

$$\min_{\boldsymbol{\alpha}_k} \|\mathbf{y} - X_k\boldsymbol{\alpha}_k\|^2, \quad X_k = XV_k.$$

Ideally, the models should explain $\mathbf{y}$ using a small number of linear combinations of the feature vectors $\mathbf{x}_j$. Mathematically, the solutions depend only on the choice of the matrices V_k. One choice for the sequence V_k is motivated by the conjugate gradient method for solving symmetric systems of equations [6]

$$Y\mathbf{b} = \mathbf{z},$$

where Y is an $n \times n$ symmetric matrix ($Y^T = Y$). In this method the matrices V_k have a column space (Krylov space) generated by the vectors

$$\{\mathbf{z}, Y\mathbf{z}, Y^2\mathbf{z}, \ldots, Y^{k-1}\mathbf{z}\}.$$

Since the normal equations $X^T X\boldsymbol{\alpha} = X^T\mathbf{y}$ are a symmetric system, the corresponding choice for V_k is derived from the above with $\mathbf{z} = X^T\mathbf{y}$ and $Y = X^T X$. The theory of conjugate gradients [6] shows that with this sequence of matrices (and in exact arithmetic), either the solution obtained at the kth step is to minimise $\|\mathbf{y} - X\boldsymbol{\alpha}\|_2^2$ or there will be a strict reduction in the norm of the residuals at the $(k+1)$th step. The following algorithm is based on this approach [13]:

1. Set $X_1 = X$, $\mathbf{y}_1 = \mathbf{y}$.
2. For $k = 1, \ldots, K$,
   Set $\mathbf{v}_k = X_k^T\mathbf{y}_k$, $\gamma_k = \|\mathbf{v}_k\|$ and normalize $\mathbf{v}_k = \mathbf{v}_k/\gamma_k$.
   Set $\mathbf{u}_k = X_k^T\mathbf{v}_k$, $\alpha_k = \|\mathbf{u}_k\|$ and normalize $\mathbf{u}_k = \mathbf{u}_k/\alpha_k$.
   Set $\mathbf{w}_k = X_k^T\mathbf{u}_k$, $X_{k+1} = X_k - \mathbf{u}_k\mathbf{w}_k^T$ and $\mathbf{y}_{k+1} = \mathbf{y}_k - \mathbf{u}_k(\mathbf{u}_k^T\mathbf{y}_k)$.
3. Next k.

The matrices $U_k = [\mathbf{u}_1, \ldots, \mathbf{u}_k]$ and $V_k = [\mathbf{v}_1, \ldots, \mathbf{v}_k]$ are orthogonal, $B_k = U_k^T X V_k$ is upper-bidiagonal and this factorisation can be used to determine the solution $\mathbf{a}_k$ at the kth step efficiently.

A PLS approach can be combined with the AICc, for example, so that the iterative scheme can be stopped when the AICc value starts to increase.

4.2 Least Angle Regression (LARS)

A more recent approach to subset selection is given by least angle regression (LARS), developed by Efron *et al* in 2004 [5]. The LARS algorithm works as follows. The matrix of feature vectors X is scaled and centred so that all columns have zero mean and norm unity and the response vector is centred to have zero mean. The algorithm starts by finding the feature vector, say $\mathbf{x}_{j_1}$, which is most correlated with the response $\mathbf{y}$. Then the largest step in the direction of this predictor is used until another predictor, say x_{j_2}, has as much correlation with the current residual. From this point, the LARS algorithm moves in a direction equiangular between the two predictors until a third feature vector becomes equally correlated. Then, LARS proceeds equiangularly between three directions along the 'least angle direction', until a fourth variable earns its way into the set, and so on. The algorithm can be implemented efficiently and involves computing the QR factorisation of the observation matrix a column at a time as a feature vector is added to the active set.

The LARS algorithm generates a sequence of submatrices of X with

$$X_{k+1} = \left[X_k \; \mathbf{x}_{j_{k+1}} \right]$$

along with coefficients $\tilde{\mathbf{a}}_k$ such that $\tilde{\mathbf{y}}_k = X_k\tilde{\mathbf{a}}_k$ approximates $\mathbf{y}$. In general $\tilde{\mathbf{a}}_k$ will not be the least squares solution $\mathbf{a}_k$ associated with $\mathbf{y}$ and X_k but the least squares solution can equally easily be calculated along the way. In order to use a criterion such as AICc as a stopping criterion, it is more appropriate to calculate the residual sum of squares associated with the least squares solution as it is related to the likelihood on which the selection criterion is based.

The LARS algorithm is related to the LASSO (least absolute shrinkage and selection operator) method developed by Tibshirani [18]. For a linear model with feature vectors $X = [\mathbf{x}_1, \ldots, \mathbf{x}_n]$ and response vector $\mathbf{y}$, the LASSO algorithm solves

$$\min_{\boldsymbol{\alpha}} \|\mathbf{y} - X\boldsymbol{\alpha}\|_2^2 \quad \text{subject to} \quad \sum_{j=1}^{n} |\alpha_j| \leq \Gamma, \tag{6}$$

where $\Gamma \geq 0$ is a tuning parameter [10, 18]. The effect of the constraint is to limit the number of nonzero coefficients so that the LASSO solution provides an element of feature selection. The LASSO algorithm can be implemented using the machinery of mathematical programming, but the LARS algorithm can be adapted to perform LASSO-type calculations in which feature vectors can be dropped as well as added, providing simpler implementation of LASSO. More details are given in [5].

4.3 Simulations to Assess the Performance of PLS and LARS

We have tested the PLS and LARS algorithms for randomly generated $m \times n$ observation matrices X_q. Given a fixed number of features k_0, we generate randomly a k_0 index set $I_q \subset \{j\}_{j=1}^n$ and associated random coefficients $\mathbf{a}_q$ having k_0 non-zero elements in the indices specified by I_q. The response vector $\mathbf{y}_q$ is generated according to $\mathbf{y}_q \in \mathrm{N}(C\mathbf{a}_q, \sigma^2 I)$, as before. Figure 5 shows the average AICc criteria for the PLS and LARS algorithms for a fit of 20 feature vectors to data generated with 4 and 8 feature vectors. For a response vector $\mathbf{y}$ generated from 4 feature vectors, the optimal fit for the LARS algorithm, according to the AICc criterion, is usually determined in 4 to 6 steps while that for the PLS algorithm is determined in 8 to 10 steps. The LARS algorithm picks out the 4 feature vectors used to generate the data 66% of the time in 4 steps, 82% in 5 steps. For data generated with 8 feature vectors the optimal fit for LARS algorithm usually requires 10 to 13 steps while for the PLS algorithm it requires between 8 and 12 steps. The LARS algorithm can only be expected to produce a good fit after it has a chance to include all 8 feature vectors used to generate the data. After 8 steps, the LARS algorithm picks out the 8 generating only 20% of the time, but after 10 steps, this percentage feature rises to 61%.

Figure 6 shows a typical evolution of the parameter estimates $\mathbf{a}_k$ at each iteration for the case of 8 generating features. Initially, all the coefficients are zero and at each iteration, a new coefficient becomes nonzero. After the 8th iteration, the values of the coefficients change little from iteration to iteration. The actual coefficients used to generate the data are denoted by '+'. Figure 7 similarly shows the evolution of the approximant $\mathbf{y}_k = X_k \mathbf{a}_k$ to the response vector $\mathbf{y}$ at each stage. After iteration 8, the approximant changes little.

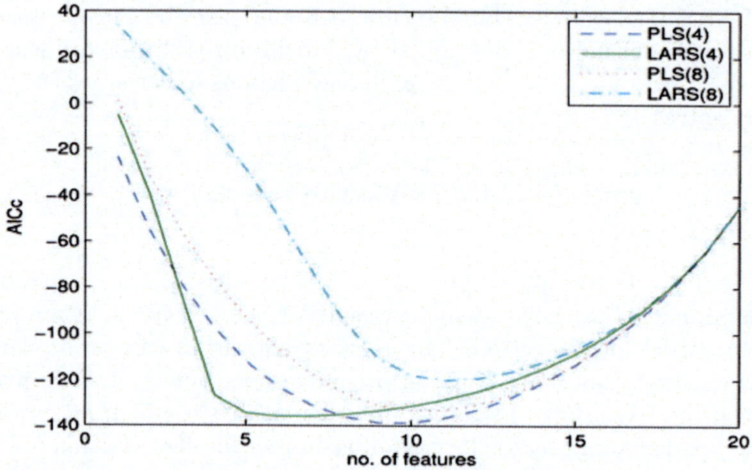

Fig. 5. Average AICc criteria for the PLS and LARS algorithms for a fit of 20 feature vectors to data generated with 4 and 8 feature vectors.

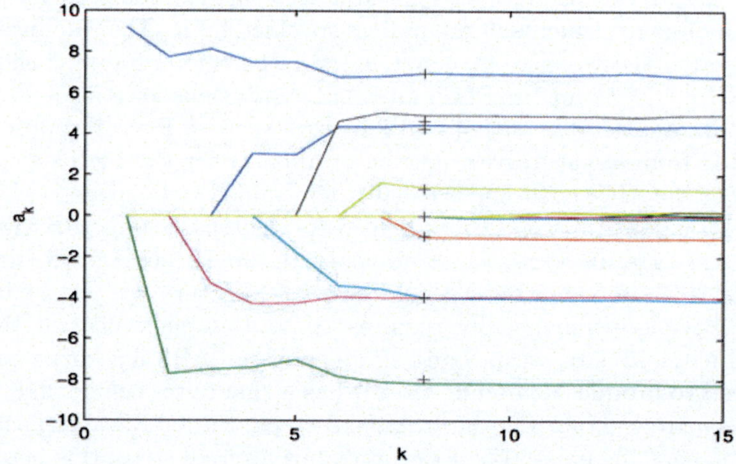

Fig. 6. Evolution of the LARS algorithm parameter estimates for the first 15 steps for data generated with 8 feature vectors with corresponding coefficients indicated by '+'.

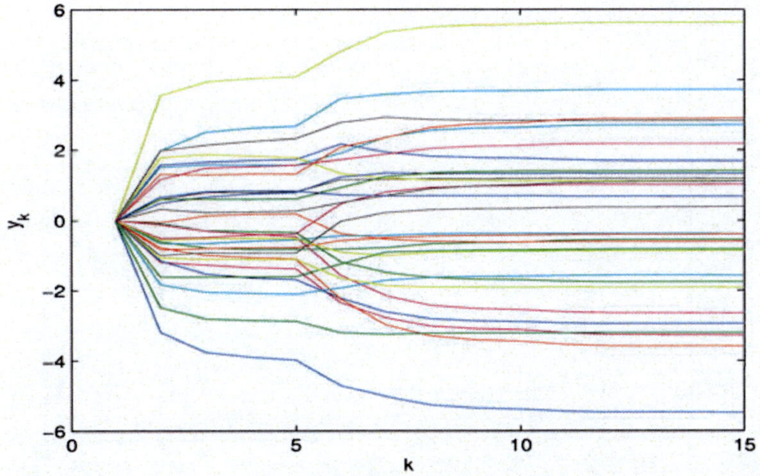

Fig. 7. Evolution of the LARS algorithm approximations $\mathbf{y}_k$ to the response vector $\mathbf{y}$ for the first 15 steps for data generated with 8 feature vectors.

We have performed simular calculations using polynomial model spaces. Figure 8 shows the average AICc criteria for the PLS and LARS algorithms for a fit of polynomials of order up to 20 to 30 data points generated with 4 and 8 feature vectors, i.e., subsets of 4 and 8 basis vectors chosen at random from the 20 polynomial basis vectors. The results are similar to those of Figure 5 except that the form of the feature vectors allows the PLS algorithm to determine better approximants in few iterations for the case of 8 feature vectors.

4.4 Application: Measurement of Naturalness

The LARS algorithm has been applied to data associated with a European project on the measurement of naturalness [2, 7]. An aim of the project was to predict how a person would classify a material such as wood, stone or fabric as real or synthetic on the basis of vision, touch or both senses. One data set involved an experiment in which a number of subjects were asked to report on the naturalness of 30 samples of wood or wood-like surfaces on the basis of tactile senses, i.e., through touching, stroking, etc. Averaging over the subjects produced a response between 0 and 1 for each of the samples, yielding a 30×1 response vector $\mathbf{y}$. A large number of physical measurements were made on the wood samples, including hardness, friction and many measures of surface roughness and texture making a total of 78 feature vectors $X = [\mathbf{x}_1, \ldots, \mathbf{x}_{78}]$ with which to explain the psycho-physical responses $\mathbf{y}$ of the subjects.

Since the matrix X of features is 30×78, it is required to reduce the number of feature vectors to at most 30 (29 if the feature and response vectors are centred). If we were not interested in feature selection, we would simply

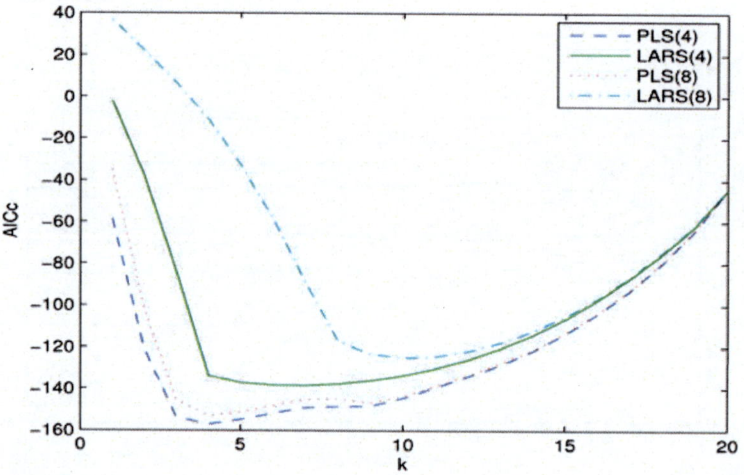

Fig. 8. Average AICc criteria for the PLS and LARS algorithms for a fit of 20 polynomial basis vectors to data generated with 4 and 8 basis vectors vectors.

perform a QR factorisation of X and use the first 30 (29) columns of the orthogonal factor as basis vectors. In order to be able to explain the response in terms of specific features we instead try to select the 30 (29) feature vectors that give the best conditioned submatrix. One approach uses the QR factorisation with column pivoting and the singular value decomposition (SVD) [6, Section 12.2] as follows. Calculate the SVD of $X = USV^T$ and set $k_{\max}$ to be less than or equal to the rank of X. Let V_1 be the first $k_{\max}$ columns of V and calculate the QR decomposition, with pivoting, of V_1^T so that $Q^T V_1^T P$ is upper triangular where Q is orthogonal and P is a permutation matrix. The selected features are the first $k_{\max}$ columns of XP. This algorithm can in fact be used for feature selection but makes its selection on the basis of maximising the degree of orthogonality of the matrix of selected features (i.e., minimising the condition of the selected submatrix) without reference to the observed response vector $\mathbf{y}$. By contrast, both the PLS and LARS algorithms build the nested model spaces in order to approximate the response vector $\mathbf{y}$ optimally.

With the feature vectors reduced to 30 (29) in number, the LARS algorithm can be applied directly. Figure 9 shows the four criteria for model selection for up to 20 features to response data. On the basis of AICc (and other criteria) there is evidence that a model involving 7 features is optimal. Further cross-validation tests also suggested that a subset of 6 or 7 feature vectors was optimal. Figure 10 shows an example response vector along with uncertainty bars representing two standard deviations and the model fit determined from 7 feature vectors. The model fit is generally good but the differences between model prediction and data cannot all be accounted for by uncertainty associated with the response data, however, adding more features does not appreciably improve matters (as suggested by the AICc scores).

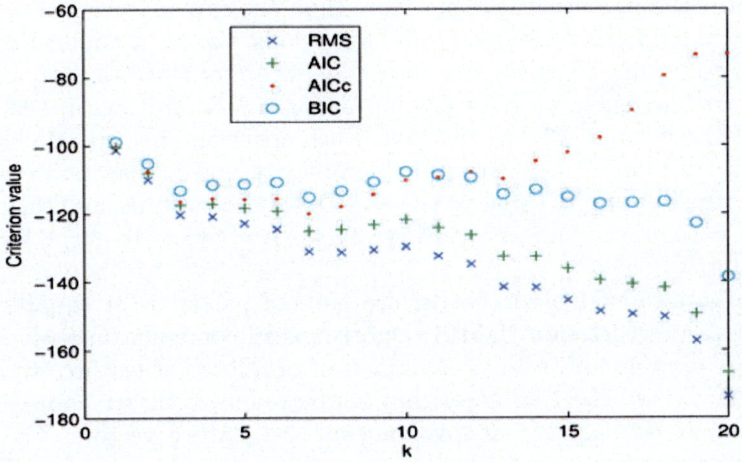

Fig. 9. Criteria scores for LARS feature selection to fit psycho-physical data representing degree of naturalness for tactile data from wood samples.

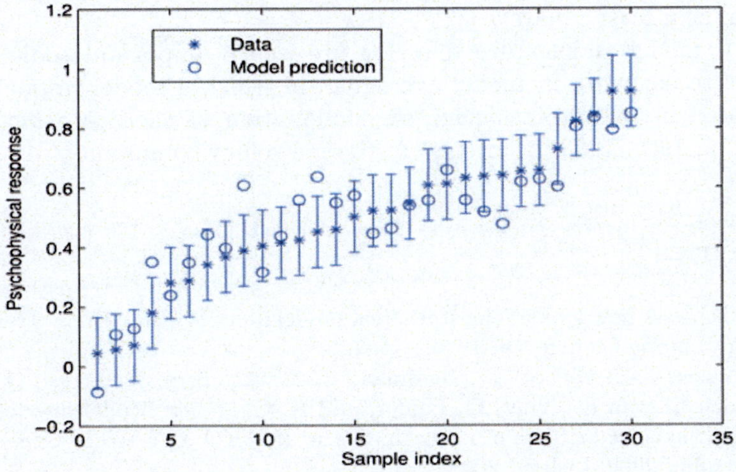

Fig. 10. Model fit to psycho-physical data based on 7 selected features.

5 Concluding Remarks

As metrology moves into more challenging application areas such as biotechnology, human sensory perception, environmental monitoring and climate change, it becomes more likely that more than one explanatory model may be put forward to characterise a set of observations. For such applications, it is necessary to be able to assess how one model or set of features compares with another. In this paper we have examined four selection criteria, denoted by RMS, AIC, AICc and BIC in the text. Each criterion aims to balance goodness of fit (likelihood) against model complexity and all can be determined from the residual sum of squares, the number of observations and the number of model parameters. Our study shows that AICc generally outperforms the other criteria.

We have also evaluated the performance of partial least squares (PLS) and least angle regression (LARS) algorithms in combination with AICc in order to determine an optimal combination or subset of features to explain a response vector. The PLS algorithm tends to deliver better approximation sooner but at the expense of involving all the feature vectors. The LARS algorithm seems effective in picking out those features which are most likely to have given rise to the observed response and seems an attractive alternative to the more involved LASSO algorithm (6). The LARS algorithm, in combination with AICc (and other criteria), has been used extensively at NPL to analyse psycho-physical data arising in a study of the perception of 'naturalness'. It is worth pointing that the methodology and conclusions obtained in this paper based on the polynomials are still valid for the cases of Chebyshev polynomials and nonlinear least squares.

While model and feature selection are clearly important tools for data analysis, recent work in model averaging, in which a model response is approximated in terms of a weighted combination of candidate models (see e.g., [4, 15, 19]), is also of interest to the metrology community.

References

1. H. Akaike: A new look at the statistical model identification. IEEE Transactions on Automatic Control **19**, 1974, 716–723.
2. A. Bialek, A.B. Forbes, T. Goodman, R. Montgomery, M. Rides, G. van der Heijden, H. van der Voet, G. Polder, and K. Overvliet: Model development to predict perceived degree of naturalness. In: *IMEKO XIX World Congress*, 6-11 September 2009, Lisbon, 2009.
3. K.P. Burnham and D.R. Anderson: *Model Selection and Multimodel Inferences: A Practical Information-Theoretic Approach*. 2nd edition, Springer, New York, 2002.
4. H. Chipman, E.I. George, and R.E. McCulloch: The practical implementation of Bayesian model selection. In: *IMS Lecture Notes – Monograph Series, Vol. 38*, P. Lahiri (ed.), Institute of Mathematical Statistics, Beachwood, Ohio, 2001.

5. B. Efron, T. Hastie, I. Johnstone, and R. Tibshirani: Least angle regression. Annals of Statistics **32**(2), 2004, 407–499.
6. G.H. Golub and C.F. Van Loan: *Matrix Computations*. John Hopkins University Press, Baltimore, third edition, 1996.
7. T.G. Goodman, R. Montgomery, A. Bialek, A. Forbes, M. Rides, A. Whitaker, K. Overvliet, F. McGlone, and G. van der Heijden: The measurement of naturalness (MONAT). In: *Man, Science & Measurement*. Proceedings of the 12th IMEKO TC1–TC7 joint symposium, Annecy, France, September 3–5, 2008.
8. R. Hoffman, V.I. Minkin, and B.K. Carpenter: Ockham's razor and chemistry. International Journal for Philosophy of Chemistry **3**, 1997, 3–28.
9. C.M. Hurvich and C. Tsai: Regression and time series model selection in sample samples. Biometrika **76**, 1989, 297–307.
10. K. Knight and W. Fu: Asymptotics for LASSO-type estimators. Annals of Statistics **28**, 2000, 1356–1378.
11. H. Linhart and W. Zucchini: *Model Selection*. Wiley, New York, 1986.
12. D. Madigan and A.E. Raftery: Model selection and accounting for model uncertainty in graphical models using Occam's window. Journal of the American Statistical Association **89**, 1535–1546.
13. R. Manne: Analysis of two partial-least-squares algorithms for multivariate calibration. Chemometrics and Intelligent Laboratory Systems **2**(1-3), August 1987, 187–198.
14. A.J. Miller: *Subset Selection in Regression*. Chapman-Hall, New York, 1990.
15. A.E. Raftery, D. Madigan, and J.A. Hoeting: Bayesian model averaging for linear regression. Journal of the American Statistical Association **92**, 1997, 179–191.
16. G. Schwarz: Estimating the dimension of a model. Annals of Statistics **6**, 1978, 461–464.
17. M. Sewell: Statistical inference (and what is wrong with classical statistics). In: *The Social Construction of Statistics*, S.M. Harding, T. and R. Thomas (eds.), Pluto Press, London, 2008.
18. R. Tibshirani: Regression shrinkage and selection via LASSO. Journal of Royal Statistical Society, Series B **58**, 1996, 267–288.
19. L. Wasserman: Bayesian model selection and model averaging. Journal of Mathematical Psychology **44**, 2000, 92–107.
20. A. Zellner, H.A. Keuzenkamp, and M. McAleer: *Simplicity, Inference and Modelling: Keeping it Sophisticated Simple*. Cambrige University Press, 2001.